Nondestructive Biomarkers in Vertebrates

Edited by
M. Cristina Fossi
Claudio Leonzio

Department of Environmental Biology
Siena University, Siena, Italy

T0166246

LEWIS PUBLISHERS
Boca Raton   Ann Arbor   London   Tokyo

Library of Congress Cataloging-in-Publication Data

Nondestructive biomarkers in vertebrates / edited by M. Cristina Fossi
and Claudio Leonzio.
    p. cm.
    Includes bibliographical references and index.
    ISBN 0-87371-648-5
    1. Vertebrates--Effect of pollution on--Congresses.
2. Biochemical markers--Congresses. 3. Environmental monitoring-
-Congresses. I. Fossi, M. Cristina. II. Leonzio, Claudio.
QL605.N66   1993
596'.05'072--dc20
                    93-10836
                    CIP

© 1994 by CRC Press, Inc.
Lewis Publishers is an imprint of CRC Press

No claim to original U.S. Government works
International Standard Book Number 0-87371-648-5
Library of Congress Card Number 93-10836
Printed in the United States of America 1 2 3 4 5 6 7 8 9 0
Printed on acid-free paper

Preface

In the last twenty years, ecotoxicology has been increasingly concerned with the use of biomarkers to evaluate the biological hazard of toxic chemicals and, as an integrated approach, in the assessment of environmental health. The concept of biomarkers in the evaluation of environmental risk has captured the attention of regulatory agencies and is currently being assessed by several research commissions. This interest is confirmed by the increasing number of specialist manuals (see other publications by McCarthy and Shugart,[1] Huggett et al.,[2] Peakall,[3] and Shugart and Peakall[4]).

The central feature of this methodological approach is to "quantify exposure and its potential impact by monitoring biological end-points (biomarkers) in feral animals and plants as indicators of exposure to and the effects of environmental contaminants".[4] Sometimes, however, in environmental contamination problems, the terms of investigation may shift from evaluation of environmental health, using sentinel species as bioindicators, to a more specific investigation of the "health" of a population or an endangered species in a situation of already-ascertained environmental pollution. This inversion of terms inevitably leads to a demand for analytical and sampling methods that are compatible with the protection and conservation of the organism to be studied. In light of this increasingly important requirement, this book focuses on the use of nondestructive biomarkers (NDB) in the hazard assessment of vertebrate populations.

The choice of nondestructive biomarkers over destructive biomarkers is not only an ethical one. The editors do not wholly agree with the ideology of certain radical environmental movements in which the animal organism, as an individual, must be saved at all costs. From the ecological point of view, the value of a population or a community is greater than that of an individual. With this in mind, the loss of a few individuals for research purposes is permissible if the data obtained contribute to the conservation of the population or community studied. On the other side of the scale, there is the problem of the "ethic of the researcher". One may often ask whether the researcher is more harmful to the population than the contaminants studied. Several examples exist of "case studies" in which populations of protected species, already heavily stressed by anthropogenic disturbance and contaminants, have been further reduced in number by "wildcat" sampling on the part of shortsighted ecotoxicologists. Apart from ethical considerations, destructive testing in vertebrates may be unacceptable under many conditions, for example, in the hazard assessment of protected or threatened species, or when the number of animals available at a site is limited, or when sequential samples from the same individual are required for time-course studies.

The use of noninvasive methods of monitoring the health of species and populations at risk has rarely been the subject of investigation by the "biomarker scientific community".[5-9] In this book we present an alternative approach for hazard assessment in high vertebrates based on nondestructive, rather than destructive, methods. World experts in the biomarker field have been co-opted in this "editorial adventure" in which we attempt to review the state of the art and to define the development and validation procedures of this new strategy.

In November 1991, after a stimulating discussion with John McCarthy and Lee Shugart at Oak Ridge National Laboratory (U.S.), we conceived the idea of organizing an international workshop to discuss the current state of the nondestructive biomarker approach with the main experts in the sector, many of whom are authors of chapters in this book. In the winter of 1992, an application was made to the U.S. Department of Energy (USDOE) to hold an International Workshop on "Nondestructive Biomarkers in Vertebrates". The Organizing Committee consisted of M. Cristina Fossi and Claudio Leonzio, as directors, together with Lee Shugart, John McCarthy, David Peakall, Colin Walker, Silvano Focardi, and Aristeo Renzoni. The application was approved in the spring of 1992 and additional financial support to supplement the USDOE award was obtained from the Consiglio Nazionale delle Ricerche (Italy) and the Italian bank, Monte dei Paschi di Siena (MPS). On May 25 through 27, 19 scientists from six countries — United States, 6; United Kingdom, 6; Italy, 4; Spain, 2; Denmark, 1 — met in a medieval monastery, the Certosa di Pontignano, now owned by the University of Siena, for high-level scientific discussions on a new strategy for hazard assessment in vertebrates based on nondestructive biomarkers. This workshop provided a forum for the comprehensive review of the state of the art and for establishing an international consensus on the most useful and sensitive nondestructive biomarkers. Research priorities for the development and validation of this promising new method were also defined. This book makes the results of the workshop available to the international scientific community.

The chapters in this volume describe different types of nondestructive biomarkers for hazard assessment in vertebrate species. The biomarkers are classified according to the nature of the toxic endpoint being probed. Particular attention is paid to the study of endangered species such as marine mammals. Each chapter contains an introduction in which the scientific basis and rationale for the endpoint being used as biomarker is explained, followed by a brief history of its application to environmental problems, together with available analytical techniques and possible destructive and nondestructive uses. The book is organized in eight sections: Overview (Chapter 1), Enzymatic Biomarkers (Chapters 2 and 3), Metabolic Products as Biomarkers (Chapter 4), Genotoxic Responses (Chapters 5, 6, and 7), Cellular Biomarkers (Chapter 8), Biomarkers in Eggs (Chapter 9), Biomarkers in Studies of Endangered Species (Chapters 10 and 11), and Remarks on Nondestructive Biomarker Strategy (Chapters 12, 13, and 14). A summary of the chapters follows.

Chapter 1 — The Use of Nondestructive Biomarkers in the Hazard Assessment of Vertebrate Populations

The aim of this chapter is to present some of most important aspects of the nondestructive biomarker strategy. The following topics are discussed:

- The advantages of the use of nondestructive strategies in biomonitoring programs
- The fields in which nondestructive biomarkers can or should be applied
- The biological materials suitable for nondestructive biomarkers and residue analysis
- Which biomarkers lend themselves to noninvasive sampling techniques
- The relationship between residue analysis and biomarkers
- The development and validation of the nondestructive biomarker approach

Chapter 2 — Blood Esterases as Indicators of Exposure to Organophosphorus and Carbamate Insecticides

A description of general aspects of the use of esterases, particularly blood esterases (butyrylcholinesterase and carboxylesterases), as indicators of exposure to organophosphorus and carbamate insecticides in vertebrates is reported. A number of technical and interpretative problems are considered, for example, the possible reactivation of inhibited enzyme, the difficulty of obtaining reliable control values for blood esterases, and interindividual and interspecific variability and temporal variations. Examples are given of monitoring exposure to organophosphorus insecticides in the field using blood esterases as biomarkers.

Chapter 3 — Clinical Biochemistry

The concept of the potential application of clinical biochemistry panels to free-ranging vertebrates is proposed. Clinical biochemistry can be divided into several subdisciplines: clinical enzymology, products of metabolism, hematology, clinical endocrinology, and diagnostic immunology. Many of the approaches described are noninvasive (samples of blood, biopsies, urine, feces, cerebral spinal fluid, synovial fluid, ocular fluids, and salivary secretions) and all are conducted on live animals. Clinical biochemistry provides an extremely useful and relatively inexpensive approach for monitoring and assessing the health of free-living species ranging from soil invertebrates to fish, birds, and mammals.

Chapter 4 — Porphyrins as Nondestructive Indicators of Exposure to Environmental Pollutants

Many foreign chemicals cause inhibition of the activity of one or more of the enzymes of the pathway of haem biosynthesis, inducing alterations in the profile of the porphyrins which accumulate or are excreted. Varied and often diagnostic patterns of porphyrin accumulation (protoporphyrin, uroporphyrin, and coproporphyrin) are observed in certain types of intoxication (herbicides, polyhalogenated aromatic compounds, arsenic, mercury, and lead) so that porphyrins can be used as sensitive indicators of exposure to environmental contaminants in vertebrates. The use of HPLC techniques in blood and excreta samples is proposed to the authors as a nondestructive method for monitoring exposure to several environmental pollutants in free-ranging animals.

Chapter 5 — Genotoxic Responses in Blood

Genetic toxicology is the science of the interaction between DNA-damaging agents and genetic material in relation to subsequent effects on the health of the organism. Environmental genotoxicology applies the principles and techniques of genetic toxicology to assess the potential effects of genotoxic agents on the health of biota. In this chapter the *in situ* approach is described with its methods for investigating indigenous species living in an environment contaminated by genotoxins. Three common types of structural changes that genotoxic agents cause to DNA (adducts, strand breakage, and modification to bases) will be discussed.

Chapter 6 — Genotoxic Responses in Blood Detected by Cytogenetic and Cytometric Assays

The use of cytological and cytometric techniques as nondestructive biomarkers of environmental pollution is proposed in this chapter. Two cytological assays have traditionally been used: chromosome and micronucleus assays. These are sensitive procedures that detect the effects of mutagenic chemicals and can be adapted to nondestructive field studies. However, advances in cytometric technology have led to more rapid and sensitive tests involving the use of flow cytometers that can measure slight changes in certain cellular components in large numbers of cells.

Chapter 7 — Hemoglobin Adducts

Hemoglobin is proposed in this chapter as a surrogate for DNA in the estimation of the *in vivo* dose of chemicals subsequent to exposure. Hemoglobin has reactive nucleophilic sites and the products of reaction with electrophilic agents are stable. It has a well-established life-span and is readily available in large quantities from humans and animals. Modification of hemoglobin has been shown to give an indirect measure of the dose to DNA in target cells of genotoxic agents. Protein adducts, especially hemoglobin, have only recently received attention as nondestructive biomarkers for the assessment of environmental contamination.

Chapter 8 — An Integrated Approach to Cellular Biomarkers in Fish

This chapter considers the development and application of molecular and cellular biomarkers for the assessment of exposure to toxic chemical pollutants and their impact on the health of fish populations. The authors describe the use of functionally and hierarchically linked multiparameter biomarkers in liver samples from *Limanda limanda* along a contamination gradient in the North Sea. The study focuses on the pathobiology of cell membranes such as endoplasmic reticulum and the endocytic-lysosomal system.

Chapter 9 — Biomarkers in Egg Samples

In this chapter the author considers the advantages of using eggs in biomarker studies. Besides the relatively nondestructive nature of this material (it is far less destructive to the species than using adult organisms), there are several advantages with respect to adult organisms. For example, the large clutch size of many species means that sufficient samples can be readily obtained. In many species, the outer shell is transparent and abnormalities can be observed. The early stages are frequently the most sensitive to the effects of chemicals. Bird eggs have been widely used to screen for teratogens and to measure esterase and mixed function oxidase activities. Studies of biomarkers in eggs are most advanced in the class *Aves*. These studies could be profitably extended to other classes such as fish, amphibians, and reptiles.

Chapter 10 — An Improved Whale Biopsy System Designed for Multidisciplinary Research

Cetaceans have been subjected to heavy predation pressure by humans in the last century. Now the need to study pollutants and their effects on marine mammals is often in contrast with the need to protect these animals. In this chapter, a new whale biopsy system is proposed. This sampling method is designed to obtain a larger quantity of tissue for genetic, toxicological, and pharmacokinetic research, and was field tested in the southern Gulf of Maine on humpback whales *(Megaptera novaeangliae)*.

Chapter 11 — Assessment of Organochlorine Pollutants in Cetaceans by Means of Skin and Hypodermic Biopsies

Cetaceans are good integrators of mid- and long-term changes in marine pollution. Interest in the study of organochlorine contamination in these mammals has greatly increased in recent years. Monitoring of organochlorine residues in cetaceans, however, presents some difficulties. Most past studies have relied on stranded or commercially caught specimens. This chapter reviews the potentials and difficulties involved in monitoring organochlorine pollutants in cetaceans through biopsy techniques and presents data on the reliability and utility of these techniques.

Chapter 12 — The Rational Basis for the Use of Biomarkers as Ecotoxicological Tools

This chapter discusses the concepts and strategies involved in the use of biomarkers, underlining important aspects of the interpretation of biomarker data. It opens with a discussion of biomarkers in the context of dose-response relationships, particularly the sequential biochemical, physiological, and behavioral responses shown by organisms undergoing increasing exposure to pollutants. These "multi-responses" can be used to evaluate the severity of exposure. Also discussed in this chapter is the difficulty of predicting lethal exposure levels from sublethal exposures and the consequences for populations and communities of pollutant exposure. A new classification of biomarkers is proposed.

Chapter 13 — Nondestructive Biomarker Strategy: Perspectives and Applications

In this chapter the editors make a critical review of the information supplied by the individual contributors in an attempt to lay the practical foundations for a nondestructive biomarker strategy. In the first part, essential features are recalled. For example, criteria for nondestructive sampling and the main features (quality, sensitivity, levels of response, and ecotoxicological meaning of response) of the possible techniques are reviewed. The development of procedures for a nondestructive approach is extensively discussed. In the second part, details are given of the sequence in which nondestructive biomarkers may be applied in field studies: identification of ecosystems (phase 1), species (phase 2), and populations (phase 3) at risk.

Chapter 14 — The Future of Nondestructive Biomarkers

This chapter summarizes and discusses, in critical terms, the information coming from the discussion section of the "Nondestructive Biomarkers in Vertebrates" workshop. The following three topics are discussed:

• Scientific and regulatory applications
• Nondestructive biomarkers in studies of protected or threatened species
• Advanced methodology and innovative biomarker techniques

REFERENCES

1. McCarthy, J. F. and Shugart, L. R., *Biomarkers of Environmental Contamination*, Lewis Publishers, Boca Raton, FL, 1990.
2. Huggett, R. J., Kimerle, R. A., Mehrle Jr., P. M., and Bergman, H. L., *Biomarkers Biochemical, and Histological Markers of Anthropogenic Stress*, Lewis Publishers, Boca Raton, FL, 1992.
3. Peakall, D. B., *Animal Biomarkers as Pollution Indicators*, Ecotoxicological series 1, Chapman & Hall, London, 1992.
4. Peakall, D. B. and Shugart, L. R., *Biomarker-Research and Application in the Assessment of Environmental Health*, NATO ASI Series, Ser. H, Cell Biology, Vol. 68, Springer-Verlag, Berlin, Heidelberg, 1993.
5. Walker, C. H., Wild birds as indicators of pesticide use and agricultural practice, in *Biological Indicators of Pollution*, Richardson, D. S., Ed., 1989.
6. Thompson, H. M., Walker, C. H., and Hardy, A.R., Esterases as indicators of avian exposure to insecticides, in: *Field Methods for the Study of Environmental Effects of Pesticides*, BCPC, Mono No. 40 Croydon, 39, 1988.
7. Fairbrother, A., Bennet, R. S., and Bennet, J. K., Sequential sampling of plasma cholinesterase in mallards *(Anas platyrhyncos)* as indicator of exposure to cholinesterase inhibitors, *Environ. Toxicol. Chem.*, 8, 117, 1989.
8. Lambertsen, R. H., A biopsy system for large whales and its use for cytogenetics, *J. Mamm.*, 68, 443, 1987.
9. Fossi, M. C., Marsili, L., Leonzio, C., Notarbartolo di Sciara, G., Zanardelli, M., and Focardi, S., The use of non-destructive biomarker in Mediterranean cetaceans: preliminary data on MFO of activity in skin biopsy, *Mar. Pollut. Bull.*, 24, 9, 459, 1992.

Acknowledgments

Financial support for the Workshop was provided by the U.S. Department of Energy (USDOE). Additional funding was obtained from Consiglio Nazionale delle Ricerche and Monte dei Paschi di Siena (MPS). Sincere thanks are due to Dr. Lee Shugart, Dr. John McCarthy, Dr. David Peakall, and Dr. Colin Walker for their precious scientific support, comments, and suggestions for the organization of the Workshop and the preparation of this volume. We also thank Prof. Aristeo Renzoni, Prof. Silvano Focardi, and Prof. Mauro Cresti, from the Dipartimento di Biologia Ambientale, Siena University, for their organizational assistance in the preparation of the Workshop. Particular thanks are due to the authors for their time and effort in the production of this book.

About the Editors

Maria Cristina Fossi is a Staff Member of the Department of Environmental Biology of Siena University. She graduated with a Ph.D. in Environmental Science from Genoa University and has carried on postdoctoral research in the Department of Environmental Biology of Siena University on the fate, bioaccumulation, and biological effects of xenobiotic compounds in the Mediterranean. Since 1981, she has worked on biochemical adaptation to polluted environments by pollution-tolerant species, toxicological effects of xenobiotics on sea-bird populations, the use of biochemical and metabolic biomarkers (MFO, esterases, porphyrins) for the evaluation of environmental pollution and MFO detoxication activity in Antarctic organisms. Her recent research has been concerned with the development of a methodological approach for the assessment of toxicological risk by means of nondestructive biomarkers in endangered vertebrate species. Dr. Fossi is author of more than 90 articles in the scientific literature.

She is on the editorial board of the international journal of *Ecotoxicology* and a member of the European Science Foundation Group for Biological Impact Assessment. She is a member of the American Chemical Society (ACS), the Society of Environmental Toxicology and Chemistry (SETAC), the European Society for Comparative Physiology and Biochemistry (ESCPB), the Italian Society of Ecology (SITE), and the Italian Society of Marine Biology (SIBM).

Claudio Leonzio is a Senior Research Assistant in the Department of Environmental Biology of Siena University. Since 1976, he has worked on the distribution of heavy metals and organochlorine compounds in the Mediterranean marine environment and on the dynamic accumulation of mercury in the trophic chain in relation to local geochemical anomalies, in particular. These studies were extended to birds in long-term monitoring studies of populations of marine species, based on residue analysis and biomarkers, especially monooxygenases. He has also been concerned with the relationship between mercury and selenium in processes of antagonism and detoxication. These studies were performed on birds, marine mammals, and experimental animals. He is author of more than 120 articles in the scientific literature.

Dr. Leonzio is a member of the Society of Environmental Toxicology and Chemistry (SETAC), the European Cetacean Society (ECS), the Italian Society of Ecology (SITE), and the Italian Society of Marine Ecology (SIBM).

Participants

NONDESTRUCTIVE BIOMARKERS IN VERTEBRATES

International Workshop: May 25–27, 1992, University of Siena, Certosa di Pontignano, Siena, Italy

Alex Aguilar, Department of Animal Biology, Faculty of Biology, University of Barcelona, Avenida Diagonal, 645, 08028 Barcelona, Spain

Asunción Borrell, Department of Animal Biology, Faculty of Biology, University of Barcelona, Avenida Diagonal, 645, 08028 Barcelona, Spain

John W. Bickham, Department of Wildlife and Fisheries Sciences,Texas A&M University, College Station, Texas 77843, U.S.A.

Michael H. Depledge, Institute of Biology, Odense University, Campusvej 55-DK5230, Odense M, Denmark

Francesco De Matteis, MRC Toxicology Unit, University of Leicester, Lancaster Road, Leicester LE1 9HN, United Kingdom

Anne Fairbrother, U.S. Environmental Protection Agency, Environmental Research Laboratory, Corvallis, Oregon 97333, U.S.A.

Silvano Focardi, Dipartimento Biologia Ambientale, Siena University, Via delle Cerchia 3, 53100 Siena, Italy

M. Cristina Fossi, Dipartimento Biologia Ambientale, Siena University, Via delle Cerchia 3, 53100 Siena, Italy

R. H. Lambertsen, Ecosystem Technology Transfer, Inc., University City Science Center, 3624 Market Street, Philadelphia, Pennsylvania 19104–6068, U.S.A.

Claudio Leonzio, Dipartimento Biologia Ambientale, Siena University, Via delle Cerchia 3, 53100 Siena, Italy

Chang K. Lim, MRC Toxicology Unit, University of Leicester, Lancaster Road, Leicester LE1 9HN, United Kingdom

John F. McCarthy, Environmental Sciences Division, Oak Ridge National Laboratory, P.O. Box 2008, Oak Ridge, Tennessee 37831–6036, U.S.A.

Michael Moore, Plymouth Marine Laboratory (NERC), Citadel Hill, Plymouth, PL1 2PB, United Kingdom

David Peakall, Monitoring and Assessment Research Centre, The Old Coach House, Campden Hill Road, London W8 7AD, United Kingdom

Aristeo Renzoni, Dipartimento Biologia Ambientale, Siena University, Via delle Cerchia 3, 53100 Siena, Italy

Lee R. Shugart, Environmental Sciences Division, Oak Ridge National Laboratory, P.O. Box 2008, Oak Ridge, Tennessee 37831–6036, U.S.A.

Helen Thompson, Central Science Laboratory, MAFF, London Road, Slough, Berkshire, SL3 7HJ, United Kingdom

Colin Walker, School of Animal and Microbial Sciences, Department of Biochemistry and Physiology, University of Reading, Whiteknights, P.O. Box 228, Reading RG6 2AJ, United Kingdom

Contributors

Alex Aguilar
Department of Animal Biology
Faculty of Biology
University of Barcelona
Avenida Diagonal, 645
08028 Barcelona, Spain

Asunción Borrell
Department of Animal Biology
Faculty of Biology
University of Barcelona
Avenida Diagonal, 645
08028 Barcelona, Spain

John W. Bickham
Department of Wildlife and Fisheries
   Sciences
Texas A&M University
College Station, Texas 77843, U.S.A.

C. Scott Baker
Pacific Biomedical Research
   Laboratory
University of Hawaii
41 Ahui Street
Honolulu, Hawaii 96813, U.S.A.

Michael H. Depledge
Institute of Biology
Odense University
Campusvej 55-DK5230
Odense M, Denmark

Francesco De Matteis
MRC Toxicology Unit
University of Leicester
Lancaster Road
Leicester, LE1 9HN,
United Kingdom

Anne Fairbrother
U.S. Environmental Protection
   Agency
Environmental Research Laboratory
Corvallis, Oregon 97333, U.S.A.

M. Cristina Fossi
Dipartimento Biologia Ambientale
Siena University
Via delle Cerchia 3
53100 Siena, Italy

Angela Köhler
Biologische Anstalt Helgoland/
   Zentrale
Notkestrasse 31
2000 Hamburg 52, FRG

R. H. Lambertsen
Ecosystem Technology Transfer, Inc.
University City Science Center
3624 Market Street
Philadelphia, Pennsylvania
19104–6068, U.S.A.

Claudio Leonzio
Dipartimento Biologia Ambientale
Siena University
Via delle Cerchia 3
53100 Siena, Italy

Chang K. Lim
MRC Toxicology Unit
University of Leicester
Lancaster Road
Leicester LE1 9HN, United Kingdom

David M. Lowe
Plymouth Marine Laboratory (NERC)
Citadel Hill, Plymouth, PL1 2PB
United Kingdom

John F. McCarthy
Environmental Sciences Division
Oak Ridge National Laboratory
P.O. Box 2008
Oak Ridge, Tennessee 37831–6036,
U.S.A.

William S. Modi
Biological Carcinogenesis and
  Development Program
Program Resources Inc./Dyncorp.
National Cancer Institute-FCRDC
Frederick, Maryland 21702, U.S.A.

Michael Moore
Plymouth Marine Laboratory (NERC)
Citadel Hill, Plymouth, PL1 2PB
United Kingdom

David Peakall
Monitoring and Assessment Research
  Centre
The Old Coach House
Campden Hill Road
London W8 7AD
United Kingdom

Lee R. Shugart
Environmental Sciences Division
Oak Ridge National Laboratory
P.O. Box 2008
Oak Ridge, Tennessee 37831–6036
U.S.A.

Michael G. Simpson
ZENECA, Central Toxicology
  Laboratory
Alderley Edge, Macclesfield,
Cheshire, SK 10 4TJ,
United Kingdom

Helen Thompson
Central Science Laboratory
MAFF
London Road
Slough, Berkshire, SL3 7HJ
United Kingdom

Colin Walker
School of Animal and Microbial
  Sciences
Department of Biochemistry and
  Physiology
University of Reading
Whiteknights, P.O. Box 228
Reading RG6 2AJ
United Kingdom

Mason Weinrich
Cetacean Research Unit
P.O. Box 159, 33 Bass Avenue
Gloucester, Massachusetts 01930,
U.S.A.

A portion of the royalties arising from this book will be donated to the American Foundation for AIDS Research.

Contents

SECTION ONE

Overview

---

1. The Use of Nondestructive Biomarkers in the
Hazard Assessment of Vertebrate Populations

CHAPTER 1

The Use of Nondestructive Biomarkers in the Hazard Assessments of Vertebrate Populations

M. Cristina Fossi, Claudio Leonzio, and David B. Peakall

TABLE OF CONTENTS

0-87371-648-5/94/$0.00+$.50
© 1994 by Lewis Publishers

I. GENERAL INTRODUCTION ON THE BIOMARKER CONCEPT

In the assessment of the impact of chemical pollution on biota, the main goal of the ecotoxicologist is to define the effect of contaminants on natural communities. The great diversity of physiological responses among living organisms and the enormous number of parameters that influence the uptake and pharmacological action of a chemical, or mixture of chemicals, make it very difficult to extrapolate biological effects from laboratory tests or theoretical models.

An important new component in biological monitoring programs is the use of biomarkers, generally defined by the National Academy of Science[1] as "a xenobiotically-induced variation in cellular or biochemical components or processes, structures, or functions that are measurable in a biological system or samples." Such "variations" can indicate the magnitude of the organism's response to contaminants as well as provide the causal link between the presence of a chemical and an ecological effect. The use of biomarkers to evaluate pollution hazards has noticeably increased in the past few years and has attracted considerable attention from regulatory agencies at an international level as a new and potentially powerful, informative tool for detecting and documenting exposure to, and effects of, environmental contamination.

The utilization of biochemical and physiological responses in the biological assessment of the environmental impact of pollution has a relatively brief history. The first researchers to use this approach in ecotoxicological investigations were Bayne and collaborators,[2] Payne,[3] Bend and collaborators,[4] James and collaborators,[5] Stegeman,[6] Moore,[7] Depledge,[8] and Livingstone and collaborators[9] for the marine environment; and Peakall,[10] Peakall and collaborators,[11] and Walker[12,13]

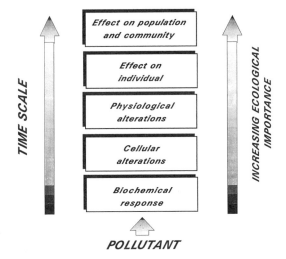

FIGURE 1. Relationship between responses to pollutants at different levels of biological organization including the continuum of these modifications along gradients of response time, toxicological and ecological relevance.

for terrestrial organisms. In 1984, Bayne defined the biochemical, cellular, and physiological responses of an organism to chemical insult as an "index of stress," defining stress as "the environmental stimulus (rather than the physiological state that constitutes the response) which when it exceeds a threshold value disturbs normal animal function." The term index of stress was widely used in the '80s[15-17] but was gradually replaced by a more concise and specific term: "biological marker" or "biomarker."[18-20] The advantages of biomarker measurements in biological monitoring were recently described in *Biomarkers of Environmental Contamination* by McCarthy and Shugart[19] and in *Animal Biomarkers as Pollution Indicators* by Peakall.[20] The strategy of using biomarkers in the assessment of environmental health has been examined at a recent NATO workshop.[21]

The relationship between responses at different levels of biological organization and their toxicological and ecological relevance taken from the book by Peakall[20] and the book by McCarthy and Shugart[19] is given in a modified form in Figure 1. It is based on the earlier work of Adams et al.[22] and had previously been foreshadowed by Stebbing.[23] All the effects of toxicants have their origins in chemical processes at a molecular level. Furthermore, if the impact of the toxicant at any level (biochemical, cytological, physiological, and autecological) is great enough to exceed the "compensatory" responses at that level, then its effect passes to successively higher levels of organization.[23] The temporal sequence is important in that indices at lower levels of organization make it possible to anticipate and predict effects that may occur later at higher levels of organization. In fact, biochemical responses measured at the lower levels of biological hierarchy, such as induction of mixed function oxidase (MFO) activity or formation of DNA adducts, often provide an early warning signal of the presence of a particular toxicant. These biomarkers can be a "measure of exposure," and may indicate the type of contaminant to which the organism is exposed. However, their biological significance at population and community levels is uncertain. On the other hand, responses at higher levels of biological organization such as changes in population structure or species diversity are "ecological effects," but by themselves cannot distinguish variations due to pollutants from effects due to natural ecological factors. An integrated monitoring program needs to consider responses at several levels of organization, ranging from biomarkers detected at the organismic level as well as population and community level indicators.[22,24]

From a practical point of view, the biomarker approach offers information that cannot be obtained from the measurements of chemical residues in environmental and biological media.[19] For example:

- Biomarkers make it possible to integrate the pharmacokinetic and toxicological interaction resulting from exposure to a complex mixture of chemicals in an exposed organism, providing the cumulative effect of toxicant interactions in molecular or cellular targets.
- Biomarkers can be used to integrate different episodes of exposure in time and space.
- Biomarkers represent rapid responses to toxicant exposure, providing an early warning signal of long-term effects.

On the other hand, the significance of chemical residues in tissues can only be assessed if detailed toxicological studies have been conducted on the target organism relating residue levels to effects. Even if this has been done for individual pollutants, this information is certainly not available for the complex mixtures of pollutants that occur in the real world. However, most biomarkers do not identify the casual agent; thus chemical and biomarker monitoring are complementary approaches.

II. NONDESTRUCTIVE BIOMARKERS

Most of the biomarkers commonly used in biomonitoring programs require the analysis of tissues and organs — such as the liver, kidney, or brain — involving the destruction of living organisms. Apart from ethical considerations, destructive testing may be undesirable in many situations; for example, the number of animals available at a site may be limited, it may be necessary to study an endangered species, or sequential samples from the same individual may be required for time course studies. Destructive approaches can not be used in hazard assessment of protected or threatened species.

The use of nondestructive techniques is not new. Certain traditional investigations (blood esterases and other enzymes, hormones and vitamin A in plasma, DNA alteration in blood cells, etc.) have been performed using noninvasive techniques for many years. However, the term "nondestructive biomarker" is relatively new in ecotoxicological research. Computerized bibliographic research in several data banks (Biological Abstracts, Science Citation Index, Chemical

Abstracts, Toxline, Zoological Records) covering the period from 1970 to 1991 revealed the term "biomarker" — or synonyms — in a total of 184,808 papers, but the term "nondestructive" or "noninvasive" combined with biomarker appeared in only 33 papers. Only a limited number of these concern ecotoxicological applications.[20,25-27] The remainder essentially belong to experimental toxicology, pharmacology, and veterinary science.

The concept of nondestructive testing has great merit and potential but progress in its development and application to environmental health assessment is hampered because no one has brought all the different methodologies and concepts of nondestructive testing together in an organized way. This book attempts a comprehensive review of the state of the art, establishing a consensus on the most useful and sensitive nondestructive biomarkers, and proposing research priorities for the development and validation of this promising methodology.

The limited amount of information available on this methodological approach makes it important to clarify certain important aspects. These include, for example:

1. the advantages of the use of nondestructive strategies in biomonitoring programs and the research fields in which nondestructive biomarkers can or should be applied
2. the biological materials suitable for nondestructive biomarkers and residue analysis
3. the knowledge of which biomarkers lend themselves to noninvasive techniques
4. the relationship between residue analysis on samples obtained by noninvasive methods and biomarkers
5. the validation and implementation strategy of the nondestructive biomarker approach

III. THE USE OF NONDESTRUCTIVE BIOMARKERS IN BIOMONITORING PROGRAMS: WHY AND WHEN?

First, the advantages of nondestructive method over conventional invasive or destructive techniques are to be clarified. In this section, we will discuss the specific cases of biomonitoring in which nondestructive biomarkers can or must replace destructive techniques.

A. Environmental Hazard Assessment

Nondestructive biomarkers can replace destructive techniques in environmental monitoring (Figure 2). With the aim of evaluating the "health" of a given environment (terrestrial, marine, or freshwater), a series of sentinel species can be tested with a suite of nondestructive biomarkers and the values obtained compared with results from the same species in a control area. This procedure has hitherto been used with destructive sampling in aquatic[3,6,16,22,24] and terrestrial[28-35] environments, based on the analysis of target tissues such as the liver, kidney, and brain. The use of noninvasive techniques in environmental monitoring would, however, have many important advantages:

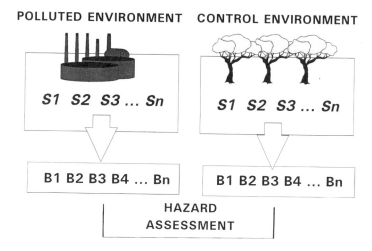

FIGURE 2. Application of nondestructive biomarkers in the environmental hazard assessment (see descriptions in the text). S = "Sentinel" Species; B = Nondestructive Biomarkers.

- Population decrements are avoided and legislative restrictions on the sacrifice of higher vertebrates (reptiles, birds, and mammals) can be overcome.
- Ecologically important species having reduced numbers can be analyzed. Such species could not be tested using invasive methods without further endangering the population.
- Analysis of blood samples can be more rapid than that of tissue samples since preparative time is reduced. In these cases, a larger number of individuals can be sampled per station which gives the data greater statistical weight.
- If animals can be recaptured, time series of measurements of the same biomarkers can be obtained from a given individual subjected to constant or variable chemical insult. The toxicological data thus acquired could otherwise only be found in laboratory experiments.
- In laboratory studies, the role of endogenous (sexual cycle, age, nutritional status, etc.) and exogenous (temperature, daylight, etc.) factors in variations in certain biochemical or physiological (biomarker) responses can be studied in the same individual (thus excluding intraspecific variation).

The application of the nondestructive biomarker strategy to environmental hazard assessment will be extensively discussed in the last chapters.

B. Hazard Assessment in Populations of Endangered Species

The principal ecotoxicological application of nondestructive biomarkers is in the hazard assessment of endangered species of vertebrates. Increasingly frequent incidents of drastic reductions in populations linked to the presence of contaminants of anthropic origin in populations of higher vertebrates (e.g., marine mammals, endangered bird species, etc.) make the development of techniques of biomonitoring and hazard assessment based on nondestructive methods indispensable. For

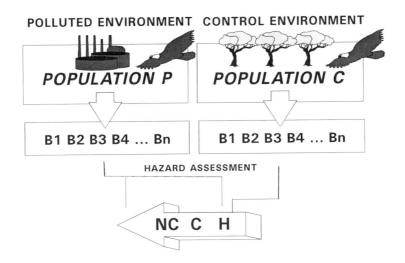

FIGURE 3. Application of nondestructive biomarkers in the hazard assessment of population of endangered species (see descriptions in the text). H = Homeostasis; C = Compensation; NC = Noncompensation.

example, noninvasive techniques have recently been successfully applied to the biomonitoring of populations of Mediterranean marine mammals, striped dolphins *(Stenella coeruleoalba)* and common whale *(Balenoptera physalus),* via skin biopsy using chemical analyses[36-37] and biochemical and cytochemical biomarkers.[38-39] Another good example of the application of the nondestructive strategy is the use of blood esterase assays in populations of wild birds accidentally exposed to organophosphorus insecticides. This approach has allowed the nondestructive assessment of toxicological risks.[26,27,40]

In the risk assessment of a population of endangered species suspected to be exposed to toxic substances of anthropic origin, nondestructive biomarkers may be applied in the following way. A series of nondestructive biomarkers may be tested in the population in question and compared with data from a control population (Figure 3). An estimate of how much the biomarker values of the endangered population differ from control population values (in the passage between homeostasis, compensation, and noncompensation) gives a measure of the risk of the population studied. In this case, the goal of biomarker research is to identify how biomarker responses correspond to different levels of departure from normal homeostasis, as extensively discussed by Depledge in Chapter 12.

C. Identification of "Species at Risk"

Nondestructive biomarkers can be used in the evaluation of the "species at risk" in a polluted environment. This research is based on the assumption that interspecific differences exist within one class of vertebrates in the susceptibility to contaminants. For example, it is well known that different species of wild birds have different levels of tolerance to liposoluble xenobiotics.[12,29,35] Such interspe-

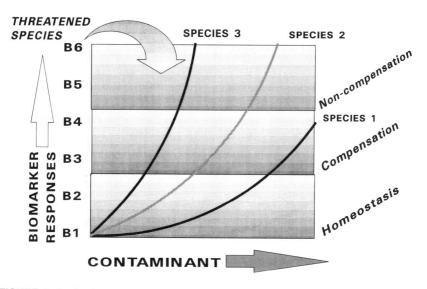

FIGURE 4. Application of nondestructive biomarkers in the identification of species at risk in a polluted environment (see descriptions in the text).

cific variation, which is mainly due to the differing ability of the mixed function oxidase detoxifying system, is expressed as a differing adaptability or nonadaptability to survival in a polluted environment.

If our goal is to identify the species at risk in a particular polluted environment, the use of noninvasive techniques (mainly biomarkers of the effects) should be directed toward the evaluation of interspecific differences in response to the sum, known or unknown, of the polluting agents. With this aim, a suite of nondestructive biomarkers may be tested in several species of the same class suspected to be threatened (for example, fish-eating birds, exposed to a biomagnification process in the marine food chain). An estimate of how much the biomarker values of the different species differ from each other in relation to the same level of environmental contamination (in the passage between homeostasis, compensation and non-compensation) permits the identification of the species at risk or "threatened species" in that particular environment (Figure 4).

IV. WHICH BIOLOGICAL MATERIALS ARE SUITABLE FOR NONDESTRUCTIVE BIOMARKERS AND RESIDUE ANALYSIS?

Once a program is decided upon, the most important consideration in hazard assessment is that techniques be valid. There is no merit in moving to nondestructive techniques unless they are clearly and definitively shown to reflect different degrees of pollutant stress. The most wasteful studies of all are those that do not produce meaningful results. Studies that can be considered under the heading nondestructive can be divided into four categories:

Table 1. The Application of Nondestructive Techniques to the Determination of Residue Levels of the Major Classes of Pollutants

Test system	Heavy metals	Organo-chlorines	Polyaromatic hydrocarbons	Museum specimens	Vertebrate groups
Blood	Yes	Yes	As adducts	No	All
Milk	Yes	Yes	No	No	Mammals
Skin biopsy	Yes	Yes	No	No	All
Liver biopsy	Yes	Yes	Yes	No	All
Hair	Yes	No	No	Yes	Most mammals
Antlers	Yes	No	No	Yes	Some mammals
Feathers	Yes	No	No	Yes	Birds
Wings	Yes	Yes	No	Yes	Birds
Feces	Yes	Yes	Yes	No	Colonial
Eggshell	Yes	Membrane	No	Yes	Birds
Egg contents	Some	Yes	Yes	No	No mammals

1. purely nondestructive methods, such as taking blood samples, after which the animal is released unharmed
2. invasive, but nonlethal techniques, such as liver and muscle biopsies
3. techniques that can be performed without harm to the animal, such as hair and feather samples, but which are generally collected after the animal has been killed for some other purpose, e.g., wings of ducks collected from hunters (the ability to use museum specimens which can greatly extend the temporal aspects of the study an important aspect of this approach)
4. studies on eggs which, while destructive to the egg, involve minimal harm to the species (these studies discussed by Peakall later in Chapter 9)

The techniques that have been used in the nondestructive measurement of residue levels are listed in Table 1. It is clearly important to be able to relate the degree of the change in a biomarker to a specific concentration of an environmental contaminant. The relationship between biomarkers and residue analysis can be considered in two broad categories: those in which the biomarker response and residue levels are measured in the same tissue and those in which the measurements are made in different tissues. The limitations of using the same tissue are, largely, those imposed by the type of tissue that is used for the biomarker measurement. Of the tissues considered here, blood, liver, and eggs have been used widely — and skin to a lesser extent — for biomarker studies. The other means of obtaining residue levels nondestructively — milk, hair, antlers, feathers, and excreta — do not lend themselves to biomarker measurements, except in some rare cases.

If residue levels are to be measured in a tissue other than that in which the biomarker is measured, a detailed knowledge of the pharmacodynamics of the contaminant is necessary. The critical word in the previous sentence is "if." For biochemical markers, which are the focus of this volume, there does not seem to be a justification for this approach. When, for example, liver samples are obtained

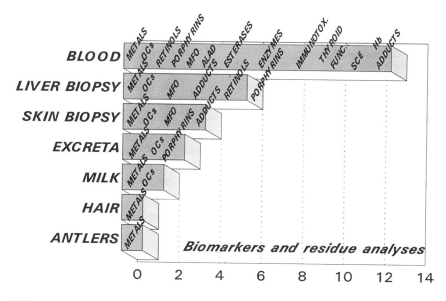

FIGURE 5. Biological material potentially obtainable by noninvasive techniques in mammals. The respective biomarkers and residue analyses for each material are reported in the figure.

for biomarker studies, the residue studies should be made on the same tissue. For biomarkers which are measured on the intact organism, such as reproductive capacity and behavioral changes, a wide range of possible tissues for nondestructive measurement of residue levels can be considered. Pharmacodynamics need to be considered in the context of both the relationship of the contaminant levels in the various tissues and the changes in these levels with time. With hair, antlers, and feathers the time aspect is particularly important. The contaminants are transferred from the body to these structures at the time they are laid down, and subsequently the residue levels do not alter. Thus, if one is studying reproductive parameters, one would want to look at feathers laid down during that period, instead of those laid down the previous year. With milk and feces, these time considerations do not apply.

In the following paragraphs, the principal biological materials available for biomarker studies and residue analyses are explored. The main research findings on the subject are reviewed, with special attention to proven techniques and theoretically more promising methods that require further validation. At the end of this major section, the biological materials potentially obtainable in a nondestructive way from the main classes of vertebrates are summarized graphically. The respective biomarkers and residue analyses for each material are also shown (Figures 5, 6, and 7). The tables will be useful in biomonitoring free-ranging vertebrates and in the hazard assessment of endangered species.

A. Blood

This is, in many ways, the tissue of choice for nondestructive biomarker/ residue work. It is readily obtained with minimal risk, and a wide range of both

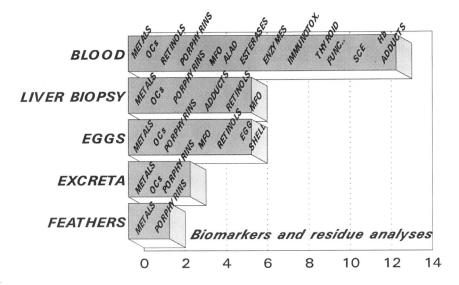

FIGURE 6. Biological material potentially obtainable by noninvasive techniques in birds. The respective biomarkers and residue analyses for each material are reported in the figure.

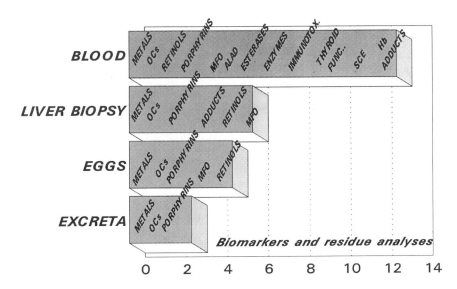

FIGURE 7. Biological material potentially obtainable by noninvasive techniques in reptiles, amphibians, and fish. The respective biomarkers and residue analyses for each material are reported in the figure.

biomarkers and residue levels can be measured in it. Its major limitations are the fact that residue levels may be temporarily high following the ingestion of a meal containing especially high levels of contaminants and that some biomarkers fluctuate more rapidly than in other organs. Trapping animals and taking blood samples is not risk free, although under good conditions the risk is low. Minimal effects on survival have been shown not only in captivity, but also in the wild.[41-42] Minimal effects have also been demonstrated on return rates and clutch desertion in shorebirds.[43] Nevertheless, the disturbance factor, especially during the breeding season, can be serious. This factor varies a great deal with the ecology of the species. For example, blood samples obtained from the burrow-nesting Puffin, *Fratercula arctica,* can be taken with minimum disturbance; scaling cliffs to obtain samples from other members of Alcidae, such as the Guillemot *(Uria aalge),* can cause wholesale loss of eggs because the birds leave the cliffs in fright.

Blood has been used to monitor the levels of organochlorines (OCs) in humans since the introduction of these compounds. In 1951, Laug et al.[44] demonstrated by analyzing human blood that widespread, nonoccupational exposure to dichlorodiphenyltrichloroethane (DDT) was occurring. Radomski et al.[45] found that the levels of several OCs in human blood were remarkably constant throughout both the day and the working week. In autopsy specimens, a good correlation was found between blood and fat levels. The authors conclude that the OCs in the blood are in equilibrium with other tissues and can be used as a reliable means of estimating the body burden. The work to diagnose death from organochlorines in both mammals and birds has focused on the levels in the brain and, to a lesser extent, in the liver. Nevertheless, some substantial work on the pharmacology dynamics of OCs in birds covering a wide range of tissues including blood has been conducted. Ecobichon and Saschenbrecker[46] found that in healthy chickens the levels of DDT and its metabolites were remarkably low in circulating plasma regardless of the levels in other tissues. When toxic signs were observed, the circulating levels were much higher in plasma as well as in other tissues. These workers found that the mean values (from the various dietary conditions) in the plasma showed a good correlation to the mean brain levels, but that this correlation disappeared when individual values were used. They conclude that it is questionable whether a positive diagnosis of poisoning can be made on the basis of residue levels in tissues other than the brain.

Capen and Leiker[47] conducted studies on white-faced ibis, *Plegadis chihi,* collected at three times during the nesting season to study the pharmacodynamics of dichlorodiphenyldichloroethylene (DDE). They found that DDE in blood serum was strongly correlated with levels in fat and breast muscle. Residues of DDE in the serum varied significantly throughout the breeding season, and this was correlated with changes in the mobilization of fat. Henny and Meeker,[48] in an experimental study feeding DDE at a wide range of dosages to American kestrels *(Falco sparverius)* found good correlations between DDE in the diet and in plasma, between DDE in plasma and in the brain, and finally a log-log relationship between tDDT in plasma and in eggs. Field studies on three other birds of prey confirmed the last correlation. However, these workers conclude that the ability to predict accurately the tDDT egg residues for the individual bird from the blood levels is tenuous, and consider that the blood plasma approach should be viewed as a supplement to existing procedures and not as an end in itself. A detailed series of equations for using blood levels to infer carcass levels are given by Hensler and

Stout.[49] Studies have been conducted on migrating peregrines trapped while migrating to and from Latin America[50] and in wintering bald eagles, *Haliaeetus leucocephalus,* in the central U.S.[51]

Blood has also been widely used as a means of measuring exposure to metals. Levels of heavy metals in blood, especially those of lead and mercury, have been widely used as indicators of human health problems. Standards for some metals have been set by the World Health Organization and by various national bodies. The pharmacodynamics of heavy metals have been well studied in wildlife. Serious problems are caused by the ingestion of lead shot pellets by wildfowl, which can also cause secondary poisoning of eagles. Dieter and Finley[52] found an excellent correlation between blood and brain levels of lead in mallard, *Anas platyrhychos,* dosed with a single lead shot pellet. Furthermore, these workers found good correlations between the inhibition of the enzyme aminolevulinic acid dehydratase (ALAD) and levels of lead in both blood and the brain. ALAD is an essential enzyme in the synthesis of hemoglobin, and measurements of its activity in blood have become the standard means of diagnosis of lead poisoning in humans[53] and waterfowl.[54] The current status of methods available has been reviewed by Scheuhammer.[55] In this case, as with the inhibition of cholinesterase, the measurement of a biomarker has been preferred over the measurement of the chemical itself. The correlations of mercury levels among the various tissues of mallard has been studied by Heinz.[56] Reasonable (r values of 0.5–0.7) correlations were found between blood and brain and between muscle and egg. Good correlations between blood and tissues have been found for other metals, namely, cadmium,[57] vanadium,[58] and nickel.[59] Of these metals, only mercury is transported into eggs in appreciable concentrations.

The available biomarkers in blood are summarized in Table 2. Blood is the best biological material for nondestructive biomarker analysis. The use and environmental applications of the most proven biomarkers (e.g. esterases, porphyrins, retinols, ALAD, Hb adducts, etc.) are explored in the following chapters of this book.

The use of adduct formation between pollutants and hemoglobin is a powerful tool that has, as yet, been used in only a few field situations. The covalent binding of pollutants to hemoglobin or DNA is a clear demonstration of exposure to these agents and an indication of possible adverse effects. The studies have largely been concerned with polynuclear aromatic hydrocarbons (PAHs). Despite the fact that petroleum products are — on a weight basis — the most common toxic substances released in to the oceans of the world, there are few good methods for the determination of exposure to these compounds. In the case of birds, it should be possible to study adduct formation with the DNA of the nucleated red blood cells. The use of adduct formation to hemoglobin to assess exposure has been reviewed by Neumann.[60] He found that adduct formation is proportional to the dose over a wide concentration range, accumulates on repeated exposure, and is stable throughout the life of the erythrocytes. The relationship between binding to hemoglobin and to DNA in other organs is constant in rats exposed to repeated doses of 4-acetylaminostilbenzene for 4 weeks, but after that point in time the

Table 2. Available Biomarkers in Blood

Biomarker	Blood	Tissue of Choice	Comment
AChE inhibition	+?	Brain	Effects in blood more transient
Neurotoxic esterases	–	Brain	Enzyme is limited to brain
Biogenic amines	–	Brain	Changes in blood too transient
DNA strand breakage	?	Wide range	Nucleated avian red blood cells
Adduct formation	+	Wide range	Hemoglobin is good substitute for DNA
SCE	+	Wide range	Blood lymphocytes can be used
Degree of methylation	?	Wide range	Nucleated avian red blood cells
MFO	–	Liver	Western blotting technique on leukocytes is possible
Thyroid	+	Thyroid	Circulating levels of T3 and T4 are sensitive
Retinols	+	Liver	Advances to use plasma are being made
Porphyrins	+	Liver	Advances to use plasma are likely
ALAD	+	Blood	Tissue of choice
Enzymes	+	Blood	Tissue of choice
Immunotoxicology	–	Lymphtic cells Bone marrow	Limited number of tests available for blood

turnover of red blood cells causes a plateau to be reached whereas levels in other tissues continue to rise. Detailed studies on the extent of binding of PAHs in fish and their relationship to adverse changes have been conducted in Puget Sound, in the state of Washington.[61,62] Monitoring of marine life for the determination of the extent of the formation of PAH adducts would give a clearer picture of the extent of oil contamination than is currently available.

B. Excreta

Levels in excreta, or guano, have been used to examine the overall pollution of colonial nesting animals. Petit and Altenbach[63] have analyzed the levels of mercury in stratified bat guano and related it to industrial activity. However, these workers are concerned with using the technique to establish a chronological record of environmental pollution, instead of considering the effects of these pollutants on organisms. The levels of three heavy metals in excrements of the great blue heron, *Ardea herodias,* have been examined by Fitzner et al.[64] They found that lead was the most abundant metal and that the amounts of cadmium and mercury were much lower. The levels of lead could be correlated with the overall contamination of the site, but no attempt was made to correlate with tissue levels. Clark and Prouty[65] determined the levels of OCs in guano from the little brown bat, *Myotis lucifugus*. Samples of 16–20 g were used and considered to have originated from 50 to 500 different bats feeding at different times of the year. When the ratio of the principal OCs in guano and carcasses is examined, the correspondence is not close. The ratios for DDE:dieldrin:PCB are 1:0.56:1.77 for the carcasses and 1:2.67:1.82 for the guano.

For colonial species, levels in guano can be used for intersite and temporal (assuming depth can be related to time) comparisons. It seems unlikely that a good correlation can be obtained with tissue levels and thus with biological effects.

Porphyrin profiles can be analyzed in excreta (see Chapter 4). This material is not currently being used in biomarker field studies. Due to its stability and ease of analysis, it could be an excellent tool for comparing populations of colony-forming species.

C. Liver Biopsy

Liver has been widely used for both residue levels and biomarkers. However, virtually all of this work has been done destructively. Although liver biopsies can be conducted without killing the animals, comparatively few studies have been made. While the technique has been used under laboratory conditions, it does not seem to have been used for monitoring purposes. Although the technique is more difficult than the others considered in this section, it should be possible to apply it in the field (John Cooper, private communication).

In view of the central role of the liver in biomarker studies, the possibilities of extending this technique to wildlife monitoring programs should be investigated.

D. Skin Biopsy

The use of skin biopsies seems to be limited to marine mammals. The reasons for this being the thick, well vascularized skin and the long-standing concern for the conservation of this group of animals. Techniques are available for collection of skin samples that do not require capture of the animal. Lambertsen and Aguilar will discuss these procedures in Chapters 10 and 11. Studies that have been made include those on striped dolphins and the common whale in the Mediterranean; chemical assay was performed on these via skin biopsy by Borrell and Aguilar[36] and Focardi et al.,[37] and on the beluga whale *(Delphinapterus leucas)* in Canada (Beland, personal communication).

Skin has rarely been used for biomarker studies, although some preliminary studies on adduct formation have been made on skin samples of the beluga whale (Shugart, unpublished). Preliminary research on mixed function oxidase activity in skin biopsy specimens of Mediterranean cetaceans has been performed by Fossi et al.[39] The sampling of marine mammals by means of a harpoon dart would extend our studies on these creatures which, until now, have been largely conducted on beached specimens. The possibility of extending these studies to other suitable groups of animals — amphibians and reptiles — should be examined.

E. Eggs

The justification for considering eggs in a book on nondestructive techniques is that the collection of eggs is a good deal less destructive than the collection of adults. In the case of birds, individuals will often re-lay if the eggs are collected early in the breeding cycle; and in the case of reptiles, amphibians, and fish the number of eggs laid per female is often very large.

In most studies of biomarkers in eggs, the relationship established is the straightforward one of biomarker response against the level in the egg. In some cases, calculations have been made to relate the levels in the egg to those in the female. The detailed pharmacodynamics of organochlorines in herring gulls *(Larus argentatus)* by Norstrom and co-workers[66] have included data relating egg contents to both total body burden and that of individual organs. The greatest difficulty with biomarker studies in embryos is the rapid change in the basal rate that frequently occurs during development. For example, the activity of cholinesterase increases fivefold from day 11 to day 26 in the mallard *(Anas platyrhynchos)*.[67] Thus any effect of pollutants on this activity has to be measured against a rapidly changing background. These problems are considered in more detail in Chapter 9.

1. Eggshells

Heavy metals can be laid down in the matrix of the eggshell. In mallard exposed to a dietary level of 100-ppm lead, Haegle et al.[68] determined the ratio of levels of lead in eggshells:whole body:bone to be 1:1.8:3.8. Despite the fact that lead is not as strongly concentrated in eggshells as in other tissues, the absolute level (2.5 ppm) is within a range that can be readily determined. Grandjean[69] has examined the lead levels in the eggshells of the European kestrel, *Falco tinnunculus,* and found an increase in recent shells compared to those collected during the period 1874–1953 and a higher level in urban than in rural sites. The levels of strontium 90 in the eggshells of Canada geese, *Branta canadensis,* have been related to the distance from a nuclear plant on the Columbia River.[70] A technique to extract OCs from membranes of eggshells was devised to demonstrate the presence of DDE in peregrine, *Falco peregrinus,* eggs collected soon after the introduction of DDT as a pesticide in the mid-1940s.[71] Subsequently, the technique was validated against egg contents[72] for a series of peregrine eggs. While the correlation was good, the limitation of the technique is the accuracy required in determining the small amount of lipid that can be extracted from the membranes.

2. Egg Contents

A great many studies have been made on the OC content of bird eggs. One of the major monitoring programs based on eggs has been that of the herring gull on the Great Lakes of North America. This study has been conducted on colonies throughout the Great Lakes from 1974 to the present time.[73] Another important study concerned the determination of levels of DDE in peregrine eggs and their relationship to eggshell thinning.[74] Mercury, because of its occurrence in an organic form, is readily transferred to eggs; other heavy metals occur only in small quantities. The nondestructive determination of cadmium is a particular problem because this metal is not transferred to any appreciable extent to either eggs or feathers.

F. Feathers/Wings

Feathers have been used to measure levels of metals in ways similar to those used on hair. Detailed pharmacodynamic studies have been conducted. Furness et al.[75] suggest that the amount of mercury stored in feathers reflects the overall body burden rather than the dietary levels at the time the feathers were laid down. The use of museum specimens is well illustrated in a study by Appelquist et al.;[76] these workers examined the mercury content of guillemot feathers over the period from 1835 to 1980, and substantial increases were found for some areas of the Baltic. Goede and de Bruin[77] examined the use of various parts of feathers for monitoring several metals. They concluded that mercury could be assayed using either the whole feather or parts, and that arsenic, lead, selenium, and zinc were best measured in the vane. The levels of cadmium found were low, often below detection limits. Mercury has been the metal most widely studied by this approach. Nationwide surveys based on duck wings (Heath,[78] and subsequent surveys) have been conducted in North America. The use of wings can be described as "no more destruction" rather than nondestructive. These wings are mailed in by hunters and are used to identify species and to determine age, sex, and distribution of the kill. Since these wings contain muscle, they can be used for the determination of OCs as well as heavy metals. In fact, the national U.S. surveys have determined only mercury in addition to OCs, although a recent survey has been conducted in Canada to determine the extent of lead contamination (Scheuhammer, unpublished data). The relationship of wing and body tissue levels of tDDT was examined by Dindal and Peterle.[79] The authors claim that the correlations were significant at the 0.01 level for 10 of the 11 tissues examined (the exception was breast feathers). Nevertheless, some of the correlation coefficients, especially those of the most widely used tissues, are rather low (0.55 for brain, 0.47 for liver). Furthermore, the absolute levels of tDDT in breast feathers are remarkably high (1.55 ppm compared to 1.35 ppm in the liver); this finding is unexpected. The analysis was based on the distribution of radioactivity following exposure to radiolabeled DDT. If these results are going to be used to calculate tissue levels and assess possible effects of these levels, then it seems necessary to reexamine the relationships using more precise analytical methods. The major use of the surveys — to examine geographic and temporal trends — is not affected by these criticisms.

In Chapter 4 of the book, De Matteis and Lim propose a validation procedure based on porphyrin studies in feathers.

G. Hair

The use of human hair to demonstrate exposure to heavy metals and to diagnose poisoning has been known for a long time. The most celebrated case has been the attempt to prove that Napoleon was poisoned by arsenic by analyzing his hair. An analysis of lead in human hair from 1871 to 1923 in comparison to hair from 1971 has shown an order of magnitude decrease in the U.S.[80]

Hair can be readily obtained for many mammalian species from live trapping, from hunters, or from museum specimens. The levels found reflect those at the time of formation of the hair rather than at the time of collection. Renzoni and Norstrom[81] found no correlation between the mercury levels in the hair and livers of polar bears. They ascribe this to the fact that the levels in liver represent long-term storage, whereas hair represents circulating levels over the period of its growth.

H. Antlers

While antlers are shed, most of the studies on antlers have been conducted on material from hunters or in collections. Antlers of roe deer, *Capreolus capreolus,* have been used to monitor heavy metal pollution.[82] During the 130 days of their growth, these appendices accumulate pollutants. Detailed pharmacodynamic studies relating levels in antlers to those in other body tissues do not appear to be available. Although measurements in antlers could be related to such parameters as breeding success, they do not seem to be of value in relationship to physiological/biochemical biomarkers. A technique for sampling without appreciable damage to specimens held in museums or as trophies has been devised by Jones and Samiullah.[83]

I. Milk

Residue levels in milk, particularly human milk, have been used to assess the importance of the transfer of pollutants from mother to offspring. As early as 1945, Woodard et al.[84] demonstrated the maternal transfer of DDT in dogs. Laug et al.[44] showed that DDT was widespread in human milk even when there was no occupational exposure. Unless biomarkers are developed in milk, there seems to be little point in determining residue levels in milk to relate them to other biomarker studies. The collection of milk can only be conducted on the female during a restricted time of the year and requires capture and a considerable amount of handling of the individual.

V. WHICH BIOMARKERS?

The choice of biomarker or series of biomarkers to use in a biomonitoring program should be guided by the information to be obtained; in other words, it has to be considered whether the study is aimed at evaluating the overall environmental contamination or at a more specific evaluation of risk to the population of a specific species.

Having decided which type of information is to be obtained, one should review the available methods which can be applied to the species in question (Figures 5, 6, and 7 summarize the available biological material for each vertebrate group) using biological material previously discussed. In this introductory part, the noninvasive techniques suitable for biomonitoring will be summarized. Suitable nondestructive biomarkers include old and new generation biomarkers. The former (MFO activity, porphyrins, DNA adducts, esterases, etc.) were traditionally used with invasive methods, but when modified can also be applied to material obtainable in a noninvasive way. The latter include methods conceived and standardized on noninvasive material (blood chemistry, vitamin A, micronuclei, etc.) normally used in clinical practice; in order to be applied in the field of ecotoxicology, they require validation and implementation.

In this section, some of the biomarkers applicable with invasive and noninvasive methods, more commonly used in biomonitoring, will be briefly examined. For each biomarker, we shall give a succinct ID consisting of (Table 3):

- biological response and relationship to environmental pollution
- use of invasive techniques and available nondestructive techniques
- temporal occurrence
- reliability index

In the following chapters of this book, individual authors will deal more systematically with the different biomarkers, giving details of current knowledge, methodology, field application, development, and validation.

A. Esterases

The esterases are members of a large and varied group of enzymes classified by Aldridge[85] as A esterases which hydrolyze organophosphates and B esterases that are inhibited by them. The biomarker most frequently used by toxicologists for diagnosing exposure of wildlife to organophosphate (OP) or carbamate (CB) compounds is the measurement of the inhibition of the B-type esterase, cholinesterase (ChE).[28,86-87] Most studies have involved the measurement of esterases in the brain. However, nondestructive sampling methods based on the detection of serum esterases can be conducted with high sensitivity on a small sample of serum. This gives the possibility of evaluating time changes, for example, in individual birds and can give an early warning signal of the presence and toxic effect of this class of insecticide. However, the application of the method has so far been limited by diurnal fluctuations of blood enzyme activities,[88] the rapid recovery of esterase levels in the serum as compared to the brain,[89] and the partial reactivation capacity of the activity during storage. Esterases are considered in more detail in Chapter 2.

B. Porphyrins

It is well known that some chemicals, such as chlorinated aromatics (PCBs, HCB, etc.) and heavy metals (Pb), may disturb porphyrin metabolism in mammals and birds. In chemically induced porphyria, these chemicals or their metabolites modify the activity of one or more of the enzymes involved in heme biosynthesis resulting in an alteration in the amount and/or composition of the porphyrin pool.[90] Patterns of porphyrin accumulation in tissues, blood, and excreta may be used to

Table 3. Biomarkers for Environmental Monitoring: Destructive and Nondestructive Use

Biomarker	Biological response	Pollutant	Invasive techniques	Nondestructive techniques	Temp.	Rel. index
Esterases	Enzyme inhibition	OPs and CBs	Brain	Blood	Early	S, D, P
Porphyrins	Metabolic disorder	Toxic metals, PHAHs	Liver	Blood, excreta, feathers?	Middle	S, D, P
MFO	Enzyme induction	PAHs, PHAHs	Liver	Skin, mucosa, blood?	Early	S, D
Blood chemisry	Various enzymes	Toxic metals, PHAHs, OPs	—	Blood	Middle	S, D
Retinols	Retinol changes	PHAHs	Liver	Blood	Early	S
Thyroid function	Thyroid function alteration	PHAHs	Thyroid	Blood	Middle	S
ALAD	Inhibition	Toxic metals	—	Blood	Early	S, D, P
Immunotoxicology	Various	Toxic metals, PAHs, PHAHs, OPs	Lymphatic cells	Blood	Middle late	S/
Hb adducts	Adducts	PAHs, PHAHs	—	Blood	Early	S, D
Stress proteins	Protein induction	Toxic metals, PHAHs	—	Blood	Early	S
DNA						
Strand breakage	Strand breaks	PAHs, PHAHs	Several tissues	Blood, skin	Early	S
Adducts	Adducts	PAHs, PHAHs	Several tissues	Blood, skin	Early	S, D, P
SCE	Chromosome	PAHs, PHAHs	Several tissues	Blood	Middle late	S, D, P

Note: CBs = Carbamates; OPs = organophosphates; PAHs = polynuclear aromatic hydrocarbons; PHAHs = polyhalogenated aromatic hydrocarbons; Temp. = temporal occurrence; (early — hours to days, middle — days to weeks/months, late — weeks/ months to years); Rel. index = reliability index (expanded from Shugart et al.[107]); S = signal of potential problem; D = definitive indicator of type or class of pollutant; P = predictive indicator of a long-term adverse effect.

predict the types of chemical insults and the sites of action of a pollutant within the pathway of heme biosynthesis.[91] In field studies, the most widely used tissue for porphyrin determinations is the liver. Fox et al.[32] showed that gulls from contaminated areas have considerably higher concentrations of highly carboxylated porphyrins in their livers than gulls from "clean" areas. Very few ecotoxicological studies have been based on the use of noninvasive techniques. Roscoe et al.[92] reported increased levels of protoporphyrin IX in a single drop of untreated blood following administration of lead shot pellets to mallard ducks. In Chapter 4, De Matteis and Lim propose the use of blood, excreta, and feathers as alternative materials to assess porphyrin disorders in wild species of higher vertebrates.

C. Mixed Function Oxidases

The cytochrome P-450 monooxygenase systems, mixed function oxidases (MFO), play a central role in the metabolism of a number of lipophilic organic contaminants. MFO-mediated reactions are considered important detoxification pathways. However, in a number of instances, this pathway serves to convert compounds to more reactive, toxic products. Polycyclic aromatic hydrocarbons (PAHs), for example, are metabolized to DNA-reactive carcinogens via the MFO activities. The MFO system has several important characteristics; for example, it provides indices of both exposure and effects, the biochemical responses of the different isoforms of cytochrome P-450 are relatively specific to the compounds or class of compounds evoking the response, and MFO is simple to detect and highly sensitive. For this reason induction of the MFO system is one of the most widely used biomarkers in field studies as an indicator of exposure to a variety of petroleum hydrocarbons (particularly PAHs) and halogenated hydrocarbons (dioxins, polychlorobiphenyls [PCBs], polybrominated biphenyls [PBBs], etc.).

Conventionally, the mixed function oxidase activity is detected in a destructive way, using mainly liver samples. The possibility of measuring MFO activities in the lymphocytes of mammals is well documented in the papers of Busbee et al.[93] and Bast et al.[94] However, the large amount of sample suggested in those papers makes it impossible to apply this analytical test to free-ranging animals. The possibility of using a more sensitive detection system, such as the Western blotting technique on leukocytes, could make the test applicable in environmental studies. The biopsy system, particularly on liver and skin, is rarely used for the detection of mixed function oxidase activity. Recently, Fossi et al.[39] presented preliminary results on mixed function oxidase activity in skin biopsy specimens of Mediterranean cetaceans. The theoretical background for this research was knowledge of the presence of MFO activity in mammalian skin.[95-96] Experimental confirmation that MFO activities (aryl hydrocarbon hydroxylase [AHH] and 7-Ethoxycoumarin deethylase [7-EC]) are induced in the whole skin of neonatal rats by topical application of Arochlor 1254,[95] suggests that this method might be used for the nondestructive testing of marine mammals.

D. Clinical Biochemistry

The evaluation of the health status of animals through the measurement of cellular, biochemical, and macromolecular constituents in blood is commonly referred to as clinical biochemistry and is a standard part of diagnostic protocols for investigating health problems in humans and domestic animals. Although clinical biochemistry has only rarely been applied to free-ranging vertebrates, research on captive wildlife indicates that such methods are now available.

The application in environmental studies of the several subdisciplines of clinical biochemistry — such as clinical enzymology, products of metabolism, hematology, clinical endocrinology, and diagnostic immunology — will be discussed extensively by Fairbrother in Chapter 3.

E. Thyroid Function and Retinols

The thyroid gland plays an important role in metabolic processes, particularly in those related to metabolism, development, and growth. The specific function of this gland is to produce the thyroid hormones thyroxine (T4) and triiodothyronine (T3), which convert plasma iodide to iodine. Investigations of the mechanism of PCB-induced toxicity in experimental animals have indicated an important role of vitamin A and thyroid hormones. The concomitant reductions in plasma retinol and thyroxine levels seem to be a consequence of interference of a PCB-metabolite with the plasma transport protein complex for both ligands.[97] A reduced plasma retinol concentration is an expression of vitamin A deficiency, which may result in increased susceptibility to viral infections.

Significant environmental studies on the thyroid include the investigations of Jefferies[98] on birds, Leatherland and Sonstegard[99] on fish, and Brouwer et al.[100] on seals. Jefferies' experimental work established that DDE, DDT, dieldrin, and PCBs cause an alteration to the thyroid gland for several avian species. Leatherland and Sonstegard[99] established that mirex and PCBs caused effects on thyroid activity in rainbow trout *(Salmo gairdneri)*, and Leatherland et al.[101] found alterations in wild populations of coho salmon *(Oncorhynchus kisutch)* in the North American Great Lakes that paralleled the degree of organochlorine pollution. In the case of the common seal *(Phoca vitulina)*, Brouwer and co-workers[100] used a nondestructive test to assess the reduction of plasma retinol in relation to contamination by PCBs and DDE.

F. ALAD

Aminolevulonic acid dehydratase (ALAD) inhibition represents a sensitive, dose-dependent measurement that is specific for a single environmental pollutant, lead.[102] ALAD inhibition has been widely used in detecting environmental lead exposure in humans. This specific biomarker is purely nondestructive, and in fact the blood is the tissue of choice.

G. Immunotoxicology

The immune system — in its capacity to destroy foreign material and protect an organism against diseases — can serve as a useful sentinel of the health status of environmentally stressed organisms, as well as provide regulatory agencies with additional means of assessing the extent of pollution. It has been shown only recently that various chemicals can actually affect specific components and functional activities of the immune system as has been reviewed by Wong and co-workers.[103] Biomarkers have been developed to evaluate many aspects of immune function and status. Among them are responses such as lymphocyte mitogenesis, antibody-producing cell formation, antibody production, and nonspecific macrophage activity. The immunologic biomarkers are usually performed in a destructive way in lymphatic cells and in a noninvasive way in blood samples.

Immunotoxicology is considered as a nondestructive biomarker in more detail in Chapter 3.

H. Hemoglobin Adducts

Shortly after an organism has been exposed to contaminants, the presence of an exogenous substance or its interactive product may be detected in the form of an adduct. Since evidence of DNA alteration (such as DNA adducts, for example) is sometimes difficult to obtain in a noninvasive way, damage to ancillary molecules (such as hemoglobin) may serve as a surrogate. Because it fulfills a number of essential requirements, hemoglobin has been proposed as a surrogate for DNA for the assessment of the *in vivo* dose of chemicals subsequent to exposure. Over 60 compounds have been shown to yield covalent reaction products with hemoglobin. Alteration of hemoglobin has been shown to give an indirect measure of the dose that the DNA was exposed to in cells which are potential targets for genotoxic agents.

The detection of adducts to hemoglobin has been largely investigated in humans who are occupationally exposed to hazardous chemicals. Only recently have hemoglobin adducts received attention as nondestructive biomarkers to assess environmental contamination (Shugart, Chapter 7).

I. DNA Alterations

Damage to DNA has been proposed as a useful parameter for evaluating the genotoxic properties of environmental pollutants.[104] The exposure of an organism to toxic pollutants may result in the initiation of a cascade of events from the generation of an initial structural alteration in DNA, to the processing of the damage and the successive expression of mutant gene products and finally to diseases resulting from genetic damage. The detection and quantitative evaluation of various events in this sequence may be employed as biomarkers. The environmental genotoxicology in fact is an approach in which the principles and techniques of genetic toxicology are applied in the assessment of the potential effect of pollution, in the form of genotoxic agents, on the health of living organisms. Most of the common types of structural changes that genotoxic agents can cause to DNA (i.e., adducts, strand breakage, modification to bases, sister chromatid exchange) will be discussed extensively in Chapters 6 and 7. The majority of these genotoxic events can be detected in a nondestructive way using blood or skin biopsy samples (see Table 3).

J. Stress Proteins

Stress proteins are a suite of proteins that are induced by a variety of compounds as well as by certain physical conditions. Some of proteins are believed to play a role in protection from various environmental perturbations, including toxic chemicals. While stress protein responses are generally considered to be nonspecific, basic research conducted on this set of particular proteins may reveal some specificities. Several techniques are employed for stress protein analyses.[105] These include metabolic labeling with radiolabeled amino acids followed by electrophoresis, cDNA probes for specific stress protein mRNAs, and immunochemical techniques for direct measurements of stress proteins (enzyme-linked immunosorbent assay [ELISA]). The usefulness of these proteins as biomarkers is largely unexplored.[106] The possibility of assessing stress proteins in blood samples appears to be a potential nondestructive tool for evaluating environmental stress.

VI. THE VALIDATION AND IMPLEMENTATION STRATEGY OF THE NONDESTRUCTIVE BIOMARKER APPROACH

In the world of biomarker research, significant efforts have recently been made to identify new potential biomarkers. The characteristics of the "ideal" biomarker are listed below:

- measurement in readily available tissues or biological products obtainable in a noninvasive way
- possibility of relating the measurement to exposure and/or degree of harm to the organism
- direct relation to the mechanism of action of the contaminants
- highly sensitive techniques that require a minimal amount of sample and are easy to perform and cost-effective
- suitability for different species

Many of these features may be found in the ordinary clinical tests used in medicine. Thoroughly tested in humans, these techniques promise to be a rich source of new methods for use in environmental studies. One of the obstacles to this approach, as underlined by Fairbrother in Chapter 3, is constituted by enormous interspecific differences even for the same biochemical process. Knowledge of species-specific basal levels of certain enzyme activities or metabolic processes is therefore essential if these techniques are to be extrapolated into the environmental field.

Before it can be used in the field, a new biomarker requires much basic research into dose-response relationships, and biological (age, sex, genetic stock, reproductive status, etc.) and environmental influences (temperature, salinity, light, etc.) on baseline values of measured responses. One of the most important aspects to investigate in the validation of new nondestructive biomarkers is the identification by laboratory experimentation of the relation between nondestructive and destructive biomarker responses in the tissue and target organ. Only after such a preliminary phase can a series of interpretative models permitting application in field studies be realized.

VII. CONCLUSIONS

The use of biomarkers in monitoring programs and the move to use nondestructive methods of measurements wherever possible are two concepts which have become opportune. In this book we review the means of wedding these two concepts.

Environmental monitoring usually requires the measurement of both chemical residues and biological effects. The use of biological effects, biomarkers, has been demonstrated to be able to stand alone when testing for the inhibition of esterases by organophosphorus compounds and the inhibition of ALAD by lead. In these two cases, the measurement of chemical residue is considered unnecessary. Normally, an effective monitoring program requires both biomarkers and chemical analyses. If the means of monitoring are to be nondestructive, then both sets of measurements must be accommodated.

So far the conservational concerns of programs to determine residue levels have concentrated on using "already destroyed" material instead of purely nondestructive techniques. Many major monitoring programs use material collected for other purposes: road kills, ducks killed by hunters, or fish caught for commercial purposes. The drawback is that such material is often not appropriate for the measurement of biomarkers. The two major difficulties are — as in the case of duck wings — that it is often not the right tissue or — as in the case of road kills — the tissue is not fresh enough for biomarker measurements.

There are two ways out of this dilemma. The first is to devise biomarkers that can be used on the material collected for residue analysis and the second is to conduct chemical analysis on nondestructively collected material used for biomarkers. These two approaches are, fortunately, not mutually exclusive.

As scientists interested in biomarkers, we are likely to favor the second approach. If one is designing a new monitoring program, this would seem to be the way to go. Nevertheless, there can be serious practical difficulties in catching the necessary specimens. For example, if one wanted to study the food chain that leads to high contaminant levels in the Inuit in northern Canada, it would be difficult to catch seals and polar bears and to take the necessary blood samples. It is much easier, and in this case more appropriate, to make measurements on these animals after they have been killed for food. In some cases, the damage caused by catching may outweigh the benefit. For example, it is likely that the prolonged disturbance of gull breeding colonies caused by trapping of adults to take blood samples causes greater losses than the collection of eggs. In contrast, the most effective way of monitoring the general level of contamination of hawks is to take blood samples when the hawks are caught for banding on their migration.

Adding a biomarker component to existing chemical monitoring program has considerable attractions. In some cases (e.g., when using commercially caught fish) it may be possible to obtain the correct tissue in the appropriate condition. In others (e.g., in the case of beached marine mammals) it may be necessary to develop more robust tests.

In environmental monitoring, especially in major programs, first it is important to set out the objectives and their justification clearly. This may sound obvious, but in our view a great deal of valuable resources have been wasted on analytical work which did not provide useful information. Second, the validity of the techniques used must be beyond reproach. Conservational and ethical considerations apply at both stages, but the most wasteful experiments of all are those which do not yield valid results. For biomarker studies, the nondestructive tissue of choice is blood; for developmental studies, the "semi-non-destructive" use of embryos is widely used. It is obviously important to relate residue levels to biological effects. The biggest problems here are that most experimental studies are presented in terms of dose and that residue data are given only rarely. This is a serious limitation because in field studies we cannot measure exposure in terms of intake, but only in terms of the final residues in the animals. Thus, in most cases, we cannot make direct comparisons between laboratory and field studies.

A great deal of work remains to be done before biomarkers, used nondestructively, become a regular part of environmental monitoring. We are going to need robust tests that are easy to perform and capable of straightforward interpretation. In this book we describe the current state of the art and, hopefully, point the way forward in this exciting field.

REFERENCES

1. NRC, Biologic markers in reproductive toxicology, National Academy Press, Washington, DC, 1989, pp. 395.
2. Bayne, B. L., Livingstone, D. R., Moore, M. N., and Widdows, J., A cytochemical and biochemical index of stress in *Mytilus edulis, Mar. Pollut. Bull.,* 7, 221, 1976.
3. Payne, J. F., Mixed function oxidases in marine organisms in relation to petroleum hydrocarbon metabolism and detection, *Mar. Pollut. Bull.,* 8, 112, 1977.
4. Bend, J. R., James, M. O., and Dansette, P. M., *In vitro* metabolism of xenobiotics in some marine animals, *Ann. N.Y. Acad. Sci.,* 298, 505, 1977.
5. James, M. O. and Bend, J. R., Polycyclic aromatic hydrocarbon induction of cytochrome P-450 dependent mixed-function oxidases in marine fish, *Toxicol. Appl. Pharmacol.,* 54, 117, 1980.
6. Stegeman, J. J., Mixed-function oxygenase studies in monitoring effects of organic pollution, *Rapp. P.V. Reun. Cons. Int. Explor. Mer,* 179, 33, 1980.
7. Moore, M. N., Cytochemical determination of cellular responses to environmental stressors in marine organisms, *Rapp. P.V. Reun. Cons. Int. Explor. Mer,* 179, 7, 1980.
8. Depledge, M. H., Disruption of endogenous rhythms in *Carcinus maenas* (L.) following exposure to mercury pollution, *Comp. Biochem. Physiol.,* 78A(2), 375, 1984.

9. Livingstone, D. R., Moore, M. N., Lowe, D. M., Nasci, C., and Farrar, S. V., Responses of cytochrome P-450 monooxygenase system to diesel oil in the common mussel, *Mytilus edulis,* and the periwinkler, *Littorina littorea, Aquat. Toxicol.,* 7, 79, 1985.

10. Peakall, D. B., Effects of toxaphene on hepatic enzyme induction and circulating steroid levels in the rat, *Environ. Health Perspect.,* 13, 117, 1976.

11. Peakall, D. B., Norstrom, R. J., Rahimtula, A. D., and Butler, R. D., Characterization of mixed-function oxidase systems of the nestling herring gull and its implications for bioeffects monitoring, *Environ. Toxicol. Chem.,* 5, 379, 1986.

12. Walker, C. H., Species differences in microsomal monooxygenase activity and their relationship to biological half-lives, *Drug Metabol. Rev.,* 7, 295, 1978.

13. Walker, C. H., Species variations in some hepatic microsomal enzymes that metabolize xenobiotics, *Prog. Drug Metabol.,* 5, 113, 1980.

14. Bayne, B. L., General introduction, in *The Effects of Stress and Pollution on Marine Animals,* Praeger Scientific, New York, 1985, xi.

15. Bayne, B. L., Brown, D. A., Burns, K., Dixon, D. R., Ivanovici, A., Livingstone, D. R., Lowe, D. M., Moore, M. N., Stebbing, A. R. D., and Widdows, J., *The Effects of Stress and Pollution on Marine Animals,* Praeger Scientific, New York, 1985.

16. Moore, M. N., Cellular responses to pollutants, *Mar. Pollut. Bull.,* 16(4), 134, 1985.

17. Viarengo, A., Pertica, M., Mancinelli, G., Palmero, S., Zanicchi, G., and Orunesu, M., Evaluation of general and specific stress indices in mussels collected from populations subjected to different levels of heavy metal pollution, *Mar. Environ. Res.,* 6, 235, 1982.

18. NRC (National Research Council), Committee on biological markers, *Environ. Health Persp.,* 74, 3, 1987.

19. McCarthy, J. F. and Shugart, L. R., *Biomarkers of Environmental Contamination,* Lewis Publishers, Boca Raton, FL, 1990.

20. Peakall, D. B., *Animal Biomarkers as Pollution Indicators,* Ecotoxicological Series 1, Chapman & Hall, London, 1992.

21. Peakall, D. B. and Shugart, L. R., *Biomarker-Research and Application in the Assessment of Environmental Health,* NATO ASI Series, Ser. H, *Cell Biology,* Vol. 68, Springer-Verlag, Berlin Heidelberg, 1993.

22. Adams, S. M., Shepard, K. L., Greeley, M. S., Jr., Jimenez, B. D., Ryon, M. G., Shugart, L. R., and McCarthy, J. F., The use of bioindicators for assessing the effects of pollutant stress on fish, *Mar. Environ. Res.,* 28, 459, 1989.

23. Stebbing, A. R. D., *The Effects of Stress and Pollution on Marine Animals,* Praeger Scientific, New York, 1985, chap. 12.

24. Adams, S. M., Shugart, L. R., and Southworth, G. R., Application of bioindicators in assessing the health of fish populations experiencing contaminant stress, in *Biomarkers of Environmental Contamination,* Lewis Publishers, Boca Raton, FL, 1990, chap. 19.

25. Walker, C. H., Wild birds as indicators of pesticides use and agricultural practice, in *Biological Indicators of Pollution,* Richardson, D. S., Ed., 1989.

26. Thompson, H. M., Walker, C. H., and Hardy, A. R., Esterases as indicators of avian exposure to insecticides, in *Field Methods for the Study of Environmental Effects of Pesticides,* BCPC, Mono No. 40, Croydon, 1988, p. 39.

27. Fairbrother, A., Bennet, R. S., and Bennet, J. K., Sequential sampling of plasma cholinesterase in mallards *(Anas platyrhyncos)* as indicator of exposure to cholinesterase inhibitors, *Environ. Toxicol. Chem.,* 8, 117, 1989.

28. Ludke J. L., Hill, E. F., and Dieter, M. P., Cholinesterase (ChE) response and related mortality among birds fed ChE inhibitors, *Arch. Environ. Contam. Toxicol.*, 1, 21, 1975.

29. Walker, C. H. and Knight, G. C., The hepatic microsomal enzymes of sea birds and their interation with liposoluble pollutants, *Aquat. Toxicol.*, 1, 343, 1981.

30. Hill, E. F. and Fleming, W. J., Anti-ChE poisoning of birds: field monitoring and diagnosis of acute poisoning, *Environ. Toxicol. Chem.*, 1, 27, 1982.

31. Peakall, D. B., Norstrom, R. J., Rahimtula, A. D., and Butler, R. D., Characterization of mixed-function oxidase systems of the nestling herring gull and its implications for bioeffects monitoring, *Environ. Toxicol. Chem.*, 5, 379, 1986.

32. Fox, G. A., Kennedy, S. W., Norstrom, R. J., and Wigfield, D. C., Porphyria in herring gulls: a biochemical response to chemical contamination of Great Lakes food chains, *Environ. Toxicol. Chem.*, 7, 831, 1988.

33. Fossi, M. C., Leonzio, C., and Focardi, S., Increase of organochlorine and MFO activity in water-birds wintering in an Italian lagoon. *Bull. Environm. Contam.*, 37, 538, 1986.

34. Fossi, M. C., Leonzio, C., and Focardi, S., Mixed function oxidase activity and cytochrome P-450 forms in the black headed gull feeding in different areas, *Mar. Pollut. Bull.*, 17 (12), 546, 1986.

35. Fossi, M. C., Leonzio, C., Focardi, S., Lari, L., and Renzoni, A., Modulation of MFO activity in Black-headed gulls living in anthropized environment: biochemical acclimatization or adaptation?, *Environ. Toxicol. Chem.*, 10, 1179, 1188.

36. Borrel, A. and Aguilar, A., Loss of organochlorine compounds in the tissues of a decomposing stranded dolphin, *Bull. Environ. Contam. Toxicol.*, 45, 46, 1990.

37. Focardi, S., Marsili, L., Leonzio, C., Zanardelli, M., and Notarbartolo di Sciara, G., Organoclorines and trace elements in subcutaneous blubber of *Balenoptera physalus* and *Stenella coeruleoalba,* Eur. Cetacean Soc., 6th Annu. Conf., San Remo, February 20–22, 230, 1992.

38. Lambertsen, R. H., A biopsy system for large whales and its use for cytogenetics, *J. Mammal.*, 68, 443, 1987.

39. Fossi, M. C., Marsili, L., Leonzio, C., Notarbartolo di Sciara, G., Zanardelli, M., and Focardi, S., The use of non-destructive biomarker in Mediterranean cetaceans: preliminary data on MFO activity in skin biopsy, *Mar. Pollut. Bull.*, 24(9), 459, 1992.

40. Walker, C. H. and Mackness, M. I., Esterases: problems of identification and classification. *Biochem. Pharmacol.*, 32, 3265, 1983.

41. Franks, E. C., Mortality of bled birds as indicated by recapture rate, *Bird Banding*, 38, 125, 1967.

42. Bigler W. J., Hoff, G. L., and Scribner, L. A., Survival of Mourning Doves unaffected by withdrawing blood samples, *Bird Banding*, 48, 168, 1977.

43. Colwell, M. A., Gratto, C. L., Oring, L. W., and Fivizzani, A. J., Effects of blood sampling on shorebirds: injuries, return rates and clutch desertions, *Condor*, 90, 942, 1988.

44. Laug, E. P., Kunze, F. M., and Prickett, C. S., Occurrence of DDT in human fat and milk, *AMA Ind. Hyg. Occup. Med.*, 3, 245, 1951.

45. Radomski, J. L., Deichmann, W. B., Rey, A. A., and Merkin, T., Human pesticide blood levels as a measure of body burden and pesticide exposure, *Toxicol. Appl. Pharmacol.*, 20, 175, 1971.

46. Ecobichon, D. J. and Saschenbrecker, P. W., Pharmacodynamic study of DDT in cockerels. *Can. J. Physiol. Pharmacol.*, 46, 785, 1968.

47. Capen, D. E. and Leiker, T. J., DDE residues in blood and other tissues of white-faced ibis, *Environ. Pollut.*, 19, 163, 1979.

48. Henny, C. J. and Meeker, D. L., An evaluation of blood plasma for monitoring DDE in birds of prey, *Environ. Pollut. Ser. A*, 25, 291, 1981.
49. Hensler, G. and Stout, W., Use of blood levels to infer carcass levels of contaminants, *Arch. Environ. Contam. Toxicol.*, 11, 235, 1982.
50. Henny, C. J., Ward, F. P., Riddle, K. E., and Prouty, R. M., Migratory Peregrine Falcons, *Falco peregrinus*, accumulate pesticides in Latin America during winter, *Can. Field Nat.*, 96, 333, 1982.
51. Henny, C. J., Griffin, C. R., Stahlecker, D. W., Harmata, A. R., and Cromartie, E., Low DDT residues in plasma of Bald Eagles *(Haliaeetus leucocephalus)* wintering in Colorado and Missouri, *Can. Field Nat.*, 95(3), 249, 1982.
52. Dieter, M. P. and Finley, M. T., Delta-aminolevulinic acid dehydratase enzyme activity in blood, brain, and liver of lead-dosed ducks, *Environ. Res.*, 9, 127, 1979.
53. Hernberg, S., Nikkanen, J., Mellin, G., and Lilius, H., δ-Aminolevulinic acid dehydrase as a measure of lead exposure, *Arch. Environ. Health*, 21, 140, 1970.
54. Dieter, M. P., Blood delta-aminolevulinic acid dehydratase (ALAD) to monitor lead contamination in Canvasback Ducks *(Aythya valisineria)*, in *Animals as Monitors of Environmental Pollutants*, Nielson, S. W., Migaki, G., Scarpelli, D. G., Eds., National Academy of Sciences, Washington, DC, 1979, 177.
55. Scheuhammer, A. M., Monitoring wild bird populations for lead exposure, *J. Wildl. Manage*, 53, 759, 1989.
56. Heinz, G. H., Comparison of game-farm and wild-strain mallard ducks in accumulation of methylmercury, *J. Environ. Pathol. Toxicol.*, 3, 379, 1980.
57. White, D. H. and Finley, M. T., Uptake and retention of dietary cadmium in mallard ducks, *Environ. Res.*, 17, 53, 1978.
58. White, D. H. and Dieter, M. P., Effects of dietary vanadium in mallard ducks, *J. Toxicol. Environ. Health*, 4, 43, 1978.
59. Eastin, W. C., Jr. and O'Shea, T. J., Effects of dietary nickel on mallards. *J. Toxicol. Environ.*, 7, 883, 1981.
60. Neumann, H. G., Analysis of hemoglobin as a dose monitor for alkylating and arylating agents, *Arch. Toxicol.*, 56, 1, 1984.
61. Myers, M. S., Landahl, J. T., Krahn, M. M., Johnson, L. L., and McCain, B. B., Overview of studies on liver carcinogenesis in English sole from Puget Sole; evidence for a xenobiotic chemical etiology. I. Pathology and epizootiology, *Sci. Total Environ.*, 94, 33, 1990.
62. Stein, J. E., Reichert, W. L., Nishimoto, M., and Varanasi, U., Overview of studies on liver carcinogenesis in English sole from Puget Sound; evidence for a xenobiotic chemical etiology. II. Biochemical studies, *Sci. Total Environ.*, 94, 51, 1990.
63. Petit, M. G. and Altenbach, J. S., A chronological record of environmental chemicals from analysis of stratified vertebrate excretion deposited in a sheltered environment, *Environ. Res.*, 6, 339, 1973.
64. Fitzner, R. E., Rickard, W. H., and Hinds, W. T., Excrement from heron colonies for environmental assessment of toxic elements. *Environ. Monit. Assess.*, 1, 383, 1982.
65. Clark, D. R., Jr. and Prouty, R. M., Organochlorine residues in three bat species from four localities in Maryland and West Virginia, 1973, *Pestic. Monit. J.*, 10(2), 44, 1976.
66. Norstrom, R. J., Clark, T. P., Jeffrey, D. A., Won, H. T., and Gilman, A. P., Dynamics of organochlorine compounds in herring gulls *(Larus argentatus)*. I. Distribution and clearance of [^{14}C] DDE in free-living herring gulls *(Larus argentatus)*, *Environ. Toxicol. Chem.*, 5, 41, 1986.

67. Hoffman, D. J. and Eastin, W. C., Jr., Effects of malathion, diazinon, and parathion on mallard embryo development and cholinesterase activity, *Environ. Res.*, 26, 472, 1981.

68. Haegle, M. A., Tugker, R. K., and Hudson, R. H., Effects of dietary mercury and lead on eggshell thickness in mallards, *Bull. Environ. Contam. Toxicol.*, 11(1), 5, 1974.

69. Grandjean, P., Possible effect of lead on egg-shell thickness in Kestrels 1874–1974, *Bull. Environ. Contam. Toxicol.*, 16(1), 101, 1976.

70. Rickard, W. H. and Price, K. R., Strontium-90 in Canada goose eggshells and reed canary grass from the Columbia River, Washington, *Environ. Monit. Assess.*, 14, 71, 1990.

71. Peakall, D. B., DDE: its presence in peregrine eggs in 1948, *Science*, 183, 673, 1974.

72. Peakall, D. B., Lew, T. S., Springer, A. M., Wayman, W., II, Risebrough, R. W., Monk, J. G., Jarman, W. M., Walton, B. J., Reynolds, L. M., Fyfe, R. W., and Kiff, L. F., Determination of the DDE and PCB contents of Peregrine Falcon eggs: a comparison of whole egg measurements and estimates derived from eggshell membranes, *Arch. Environ. Contam. Toxicol.*, 12, 523, 1983.

73. Environment Canada, The chemical pollution of the Great Lakes and associated effects. Synposis, *Environ. Can.*, 43, 1991.

74. Cade, T. J., Lincer, J. L., White, C. M., Roseneau, D. G., and Swartz, L. G., DDE residues and eggshell changes in Alaskan falcons and hawks, *Science*, 172, 955, 1971.

75. Furness, R. W., Muirhead, S. J., and Woodburn, M., Using bird feathers to measure mercury in the environment: relationships between mercury content and moult, *Mar. Pollut. Bull.*, 17(1), 27, 1986.

76. Appelquist, H., Drabaek, I., and Asbirk, S., Variation in mercury content of guillemot feathers over 150 years, *Mar. Pollut. Bull.*, 16, 244, 1985.

77. Goede, A. A. and de Bruin, M., The use of bird feather parts as a monitor for metal pollution, *Environ. Pollut. Ser. B.*, 8, 281, 1984.

78. Heath, R. G., Nationwide residues of organochlorine pesticides in wings of Mallard and Black Ducks, *Pest. Monit. J.*, 3(2), 115, 1969.

79. Dindal, D. L. and Peterle, T. J., Wing and tissue relationships of DDT and metabolite residues in mallard and lesser scaup ducks, *Bull Environ. Contam. Toxicol.*, 3, 37, 1968.

80. Weiss, D., Whitten, B., and Leddy, D., Lead content of human hair (1871–1971), *Science*, 178, 69, 1972.

81. Renzoni, A. and Norstrom, R. J., Mercury in the hairs of polar bears, *Ursus maritimus*, *Polar Rec.*, 26, 326, 1990.

82. Sawicka-Kapusta, K., Roe deer antlers as bioindicators of environmental pollution in southern Poland, *Environ. Pollut.*, 19, 283, 1979.

83. Jones, R. C. and Samiullah, Y., Deer antlers as pollution monitors, *Deer*, 6, 253, 1985.

84. Woodard, G., Ofner, R. R., and Montgomery, C. M., Accumulation of DDT in the body fat and its appearance in the milk of dogs, *Science*, 102, 177, 1945.

85. Aldridge, W. N., Serum esterases. I. Two types of esterase (A and B) hydrolysing p-nitrophenyl acetate, propionate and butyrate and method for their determination, *Biochem. J.*, 53, 110, 1953.

86. Bunyan, P. J. and Jennings, D. M., Organophosphorus poisoning; some properties of avian esterase, *J. Agric. Food Chem.*, 16, 326, 1968.

87. Venkateswara, G. P., Indira, K., and Rajendra, W., Inhibition of sheep brain acetyl-cholinesterase by hexachlorophene, *Bull. Environ. Contam. Toxicol.*, 38, 139, 1987.

88. Thompson, H. M., Walker, C. H., and Hardy, A. R., Avian esterases as indicators of exposure to insecticides: the factor of diurnal variation, *Bull. Environ. Contam. Toxicol.*, 41, 4, 1988.

89. Holmes, S. B. and Boag, P. T., Inhibition of brain and plasma cholinesterase activity in zebra fish orally dosed with fenitrothion, *Environ. Toxicol. Chem.*, 9, 323, 1990.

90. De Matteis, F., Drug-induced abnormalities of liver heme biosynthesis, in *Hepatotoxicology*, Meeks, G. G., Harrison, S. D., and Bull, R. J., Eds., CRC Press, Boca Raton, FL, 1991, p. 437.

91. Marks, G. S., Exposure to toxic agents: the heme biosynthetic pathway and hemoproteins as indicator, *CRC Crit. Rev. Toxicol.*, 15, 151, 1985.

92. Roscoe, D. E., Nielson, A. A., Lamola, A. A., and Zuckerman, D. A., Simple quantitative test for erythrocytic protophorphyrin in lead-poisoned ducks, *J. Wildl. Dis.*, 15, 127, 1979.

93. Busbee, D. L., Shaw, C. R., and Cantrell, E. T., Aryl hydrocarbon hydroxylase induction in human leukocytes, *Science*, 178, 315, 1972.

94. Bast, R. C., Jr., Okuda, T., Plotkin, E., Tarone, R., Rapp, H. J., and Gelboin, H. V., Development of an assay for aryl hydrocarbon (benzo(a)pyrene) hydroxylase in human peripheral blood monocytes, *Cancer Res.*, 36, 1967, 1976.

95. Bickers, D. R., Mukhtar, H., Dutta-Choudhury, T., Marcello, M. S. C., and Voorhees, M. D., Aryl hydrocarbon hydroxylase, epoxide hydrolase, and benzo(a)pyrene metabolism in human epidermis: comparative studies in normal subjects and patients with psoriasis, *J. Invest. Dermatol.*, 83, 51, 1984.

96. Mukhtar, H. and Khan, W. A., Cutaneous cytochrome P-450 *Drug Metab. Rev.*, 20 (2–4), 657, 1989.

97. Brouwer, A. and van den Berg, K. J., Binding of a metabolite of 3,4,3',4'-tetrachlorobiphenyl to transthyretin reduces serum vitamin A transport by inhibiting the formation of the protein complex carrying both retinol and thyroxin, *Toxicol. Appl. Pharmacol.*, 85, 301, 1986.

98. Jefferies, D. J., The role of the thyroid in the production of sublethal effects by organochlorine insecticides and polychlorinated biphenyls, in *Organochlorine Insecticides: Persistent Organic Pollutants*, Moriarty, F., Ed., Academic Press, London, 1975, p. 132.

99. Leatherland, J. F. and Sonstegard, R. A., Effects of dietary mirex and PCB (Aroclor 1254) on thyroid activity and lipid reserves in rainbow trout *Salmo gairdneri* Richardson, *J. Fish. Dis.*, 3, 115, 1979.

100. Brouwer, A., Reijnders, P. J. H., and Koeman, J. H., Polychlorinated biphenyl (PCB)-contaminated fish induces vitamin A and thyroid hormone deficiency in the common seal *(Phoca vitulina)*, *Aquat. Toxicol.*, 15, 99, 1989.

101. Leatherland, J. F., Sonstegard, R. A., and Moccia, R. D., Interlake differences in body and gonadal weights and serum constituents of Great Lake coho salmon *(Oncorhynchus kisutch)*, *Comp. Biochem. Physiol.*, 69A, 701, 1981.

102. Scheuhammer, A. M., Monitoring wild bird populations for lead exposure, *J. Wildl. Manage.*, 53, 759, 1989.

103. Wong, S., Fournier, M., Coderre, D., Banska, W., and Krzystyniak, K., Environmental immunotoxicology, in *Animal Biomarkers as Pollution Indicators*, Ecotoxicological Series 1, Peakall, D. B., Ed., Chapman & Hall, London, 1992, chap. 8.

104. Kohn, H. W., The significance of DNA-damaging assay in toxicity and carcinogeneicity assessment, *Ann. N.Y. Acad. Sci.,* 407, 106, 1983.
105. Lindquist, S., The heat shock response, *Annu. Rev. Biochem.,* 55, 1151, 1988.
106. Sanders, B. M., Stress protein: potential as multitiered biomarkers, in *Biomarkers of Environmental Contamination,* McCarthy, J. F. and Shugart, L. R., Eds., Lewis Publishers, Boca Raton, FL, 1990, chap. 9.
107. Shugart, L. R., Adams, S. M., Jiminez, B. D., Talmage, S. S., and McCarthy, J. F., Biological markers to study exposure in animals and bioavailability of environmental contaminations, in *Biological Monitoring for Pesticide Exposure: Measurement, Estimation and Risk Reduction,* ACS Symposium Series No. 382, 86, 1989.

SECTION TWO

Enzymatic Biomarkers

2. Blood Esterases as Indicators of Exposure to
Organophosphorous and Carbamate Insecticides

3. Clinical Biochemistry

CHAPTER 2

Blood Esterases as Indicators of Exposure to Organophosphorus and Carbamate Insecticides

Helen M. Thompson and Colin H. Walker

TABLE OF CONTENTS

0-87371-648-5/94/$0.00+$.50
© 1994 by Lewis Publishers

I. INTRODUCTION

Long before the term "biomarker" came into common usage, the inhibition of cholinesterases (ChE) was recognized as a characteristic indicator of poisoning by organophosphorus (OP) and carbamate (CB) compounds. The toxicity of certain nerve gases synthesized during the Second World War was based on their ability to phosphorylate and thus inhibit cholinesterases of the nervous system. Inhibition of acetylcholinesterase (AChE) in the central and peripheral nervous system can lead to a buildup of acetylcholine at the synapse resulting in overstimulation and depolarization of the postsynaptic membrane. Such overstimulation results in muscle tetani, and death is normally due to asphyxiation caused by tetanus of the diaphragm.

In the 1940s Schrader synthesized the first organophosphorus insecticide to have widespread use, parathion; and in the late 1950s the first commercial carbamate insecticide, carbaryl, was synthesized. The relatively low persistence of organophosphorus and carbamate pesticides, compared to organochlorine pesticides, together with their effectiveness has led to wide usage of these compounds. However, identifying nontarget exposure by residue analysis, the first step in assessing the effects of these compounds on wildlife, is difficult due to their high acute toxicity (less than 1 mg/kg for paraoxon) and rapid metabolism and elimination. This has led to increasing interest in the use of biological markers of poisoning by these compounds.

Serum cholinesterases are routinely used to monitor exposure of spray operators to organophosphorus and carbamate compounds,[1] and these methods have been further developed to study the effects of these compounds on wildlife. Significant (>50–80%) inhibition of brain acetylcholinesterase activity in the brain has been used routinely to provide evidence of the involvement of organophosphorus and carbamate pesticides in the poisoning of birds and mammals[2] and has been investigated for use in honeybees.[3] More recently interest has grown in

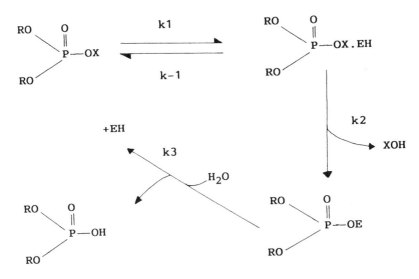

FIGURE 1. Interaction of B esterases with substrates and organophosphorus and carbamate compounds. R = alkyl group; E = enzyme.

the use of nondestructive blood sampling of nontarget vertebrates to monitor cholinesterase activity following agricultural use of organophosphorus or carbamate pesticides. Although interest has primarily centered on the use of serum cholinesterases, serum carboxylesterase has also been investigated as a monitor of exposure of birds to these compounds.

In the following account, the properties and classification of blood esterases will be discussed before considering the techniques involved in monitoring esterase activity and its use in the field.

II. THE BIOCHEMISTRY OF THE B ESTERASES

The term B esterase was introduced by Aldridge[4] to describe a large group of serine hydrolases which are inhibited by OPs. A esterases hydrolyze OPs and are virtually absent from the serum/plasma of avian species.[5] The way in which inhibition of B esterases occurs is shown in Figure 1. Organophosphates act as suicide substrate for this type of enzyme, i.e., hydrolysis of the substrate results in inhibition of the enzyme. Following formation of the enzyme-substrate complex, phosphorylation of the enzyme occurs — with the splitting of a phosphoroester bond — and the consequent loss of the leaving group X. In the case of cholinesterase, inhibition is due to phosphorylation (or with carbamates, carbamylation) of a serine residue at the active site; and the same is probably true of other B esterases. The phosphoryl moiety is very stable, with the consequence that the final stage of hydrolysis, the release of the phosphoryl moiety, and the reactivation of the enzyme proceed very slowly. Thus, although the organophosphate is hydrolyzed, turnover is very slow because the phosphorylated enzyme is stable.

This interaction occurs with organophosphates (also termed oxons) but not with organophosphorothionates (also called thions). This is important to recognize, because many OP insecticides are thions and need to be converted to their oxon forms by monooxygenases before they have toxic action. The inhibition of B esterases occurs readily at very low concentrations of organophosphates (I_{50}s are frequently in the range 10^{-8}–$10^{-10}M$) and should not be confused with the competitive inhibition that these compounds may show with other types of esterases at much higher concentrations.

After phosphorylation of the B esterase has occurred, one of the alkyl groups may be lost (Figure 1). This process is termed aging, and once it has occurred the enzyme cannot be reactivated, either spontaneously or by the action of reactivating agents such as oximes. Aging is particularly important in the case of neuropathy target esterase (NTE) where aging of the enzyme coincides with the appearance of the symptoms of delayed neuropathy.[6]

The rates of the different stages of the reaction between B esterases and organophosphates or carbamates, indicated by the rate constants shown in Figure 1, vary between the forms of the enzyme and the inhibitory compound. In rats, reactivation of diethyl phosphorylated esterases tends to be more rapid than the dimethyl phosphorylated form but this rate may also be species dependent.[7] Carbamylated cholinesterases spontaneously reactivate more readily than phosphorylated enzymes and inhibition may be reversed as a consequence of dilution or substrate competition, presenting problems in the measurement of inhibition. Phosphorylated cholinesterase can be reactivated by certain oximes, e.g., pyridine-2-aldoxime methiodide (2PAM); and these agents act as nucleophiles which interact with the electrophilic phosphorus atom of the phosphoryl moiety. This results in the phosphoryl moiety leaving the serine residue at the active site and consequent restoration of the activity of the enzyme. These oximes may also interact with, and thus deactivate, the unbound molecules of organophosphate.

A. Classification of B Esterases

As noted earlier, esterases which are inhibited by organophosphates are classified as B esterases. On the basis of studies with purified enzymes, two main classes of B esterases have been established (Figure 2 and Tables 1 and 2): cholinesterases (ChE) and carboxylesterases (CBE).

B. Properties of B Esterases

1. Cholinesterases (ChE)

This category of B esterase represents a subclass characterized by the presence of a binding site for the choline head as well as an esteratic site. ChEs are distinguished from other B esterases by their specific properties to hydrolyze choline esters in preference to other carboxylic esters and are inhibited by eserine salts at low concentrations ($10^{-6}M$).[8] Two types of ChE are recognized:

Acetylcholinesterase (AChE) (EC 3.1.1.7) — This esterase has high specificity for acetylcholine and is found primarily in nervous tissue, plasma of some mammalian, and other vertebrate erythrocytes. Within certain areas of nervous

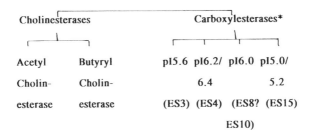

B Esterases (Inhibited by organophosphates)

Cholinesterases   Carboxylesterases*

Acetyl   Butyryl   pI5.6   pI6.2/   pI6.0   pI5.0/

Cholin-   Cholin-   6.4   5.2

esterase   esterase   (ES3)   (ES4)   (ES8?   (ES15)

                               ES10)

FIGURE 2. Classification of B esterases into cholinesterases and carboxylesterases. *pI values refer to classification of isoelectric focusing. (From Meintlein, R., Ronai, R., Kobbi, M., Heymann, E., and von Deimling, O., *Biochim. Biophys. Acta.* 913, 27, 1987. With permission.) ES values refer to genetic classification. (From Van Zutphen, L.F.M., *Transplant Proc.,* 15, 1687, 1983. With permission.)

TABLE 1. Types of Carboxylesterase from Rat Liver Microsomes

pI	Substrate	Comments
5.6	Simple aromatic esters, acetanilide, lysophospholipids, monoglycerides, long-chain acyl carnitines	Sometimes called lysophospholipase to distinguish it from other esterases
6.2/6.4	Aspirin, malathion, pyrethroids, palmitoyl CoA, monoacylglycerol cholesterol esters	May correspond to EC 3.1.2.2 and EC 3.1.1.23
6.0	Short-chain aliphatic esters, medium-chain length acylglycerols, clofibrate, procaine	ES8 may be a monomer, ES10 a dimer
5.0/5.2	Mono- and diacylglycerols, acetyl carnitine, phorbol diesters	Correspond to acetyl carnitine hydrolase EC 3.1.1.28

Source: Mentlein et al.[11]

tissue AChE is fundamental to the regulation of neurotransmission by the hydrolysis of acetylcholine. Its physiological function outside nervous tissue is not known. No ions are known to be essential for activity, although calcium and magnesium ions can enhance the activity. Inhibition of AChE occurs at high substrate concentrations, which is assumed to be due to the binding of the cationic head of one acetylcholine molecule to its binding site and the ester head of another molecule to the esteratic site. The structure of AChE basically consists of two main forms — globular and asymmetric — each of which has monomer, dimer, and tetramer forms.[9,10] The structure of the monomeric form has recently been further elucidated following X-ray analysis of the purified enzyme.[10]

Butyrylcholinesterase (BChE) (EC 3.1.1.8) — This is a relatively nonspecific esterase which readily hydrolyzes a number of cholinesters in addition to acetylcholine. It hydrolyzes butyrylcholine at approximately four times the rate of acetylcholine and is not inhibited at high substrate concentrations. BChE is found

Table 2. Substrate Specificities of BChE from Different Species

Source	PrCh	BCh	BzCh
Dog	160	270	60
Horse	170	250	30
Man	160	230	30
Pigeon	120	125	15
Rat	220	130	30
Chicken	150	90	10
Mouse	140	120	15
Pig	90	120	15

Source: Walker and Thompson.[43]
Note: Activities expressed relative to that of acetylcholine, which is ascribed a value of 100. PrCh = Propionylcholine, BCh = Butyrylcholine, BzCh = Benzoylcholine.

in a number of tissues including liver, muscle, and pancreas and is found at relatively high concentrations in the plasma of many vertebrate species although its physiological function is not known. In human plasma it is found as a monomer, dimer, and tetramer.[9] No ions are known to be required for activity.

2. "Unspecific" Carboxylesterase

A considerable number of enzymes which can hydrolyze a wide range of esters belong to this group. They differ from ChEs in that they do not appear to possess a binding site for the cationic head of cholinesters. Recent work on four of these esterases isolated from rat liver microsomes has rendered the term unspecific obsolete.[11] Each of the esterases has wide-ranging, yet characteristic substrate specificity (Table 1). All have molecular weights in the range 58,000–61,000 Da in the monomeric form. Each of these enzymes can hydrolyze both endogenous and xenobiotic esters.[11] In some cases the esterases clearly correspond to those previously given other designations in the International Union of Biochemistry (IUB) classification. The esterases are identified by both their isoelectric point (pI) on isoelectric focusing and an esterase (ES) number according to the genetic classification of van Zutphen.[12] The pI values are of limited usefulness in classification because they can vary between species and even between strains of the same species. Several aspects of the classification need further clarification in that monomers and dimers have been classified separately and several authors have attributed slightly different pI values to the same enzyme.[11]

Although these enzymes are wide ranging in their substrate specificity, it is still uncertain to what extent particular reactions are catalyzed in vivo; in vitro assays usually employ substrate concentrations far higher than would be found in vivo. Thus, the role of these enzymes in the detoxification of xenobiotics and in normal endogenous metabolism awaits clarification.

C. Variations in Forms and Activities of Serum B Esterases

In considering the use of B esterases as nondestructive biomarkers, account needs to be taken of the different forms and activity levels of the enzymes that occur. There are considerable differences in the levels of esterase activity in sera from different species, interindividual variations within a species, and temporal and developmental changes within individuals.

1. Acetylcholinesterase (AChE)

AChE is found in the plasma membrane of mammalian erythrocytes, but not to any significant extent in avian erythrocytes although it has been reported in the stroma of chicken erythrocytes.[13]

A number of studies have reported AChE activity in serum or plasma from different species toward acetylcholine or acetylthiocholine. Unfortunately, this assay does not distinguish between AChE and BChE, which will both hydrolyze these substrates. Thus many of these studies will overestimate the true activity of AChE. Studies which employ iso-OMPA to specifically inhibit BChE[4] give a more reliable measure of AChE activity of plasma and in birds have shown wide species differences in the levels of AChE present.

2. Butyrylcholinesterase (BChE)

BChEs occur throughout the animal kingdom and are found in a wide range of tissues. In mammalian blood high levels of activity occur in serum, particularly in the horse and dog, but there is little or no activity in erythrocytes. BChE activity is also found in avian and piscine sera and in insects.

Some data on the substrate specificities of BChEs from different species are given in Table 2. While there are no qualitative differences between enzymes from different species, there are contrasting substrate specificities for acetylcholine and the other three substrates.

BChEs are well represented in the serum of birds, where there are marked interspecies differences in activity.[14] No clear relationships were evident between BChE activity and body size, diet, or taxonomic group. However, there were differences between species in regard to the separation of BChEs by chromatofocusing (pI). The serum of the starling *(Sturnus vulgaris),* house sparrow *(Passer domesticus),* tree sparrow *(Passer montanus),* pigeon *(Columba livia),* and partridge *(Alectoris rufa)* were shown to contain several differing forms with BChE activity after chromatofocusing.[14]

3. Carboxylesterases (CBE)

CBE activity has been found in the serum or plasma of all vertebrate species investigated so far. Typically, the activity has been measured using naphthyl acetate or N-methyl indoxyl acetate although triacetin has also been used.

In mammals, several different serum CBEs have been identified. In mice, for example, four different serum CBEs have been identified and classified genetically.[15] This compares with 10 different esterases that have been recognized in livers from the same species.

A study of CBE activity in liver homogenates from 55 species of birds representing eight orders, all from North America,[16] employed triacetin and phenylacetate as substrates. Much of the activity toward triacetin appeared to be due to CBE activity since low levels of dichlorvos inhibited most of it. However, this esterase was more sensitive to eserine salts than mammalian CBE, raising doubts over its classification (ChEs are inhibited by low levels of eserine). There was no clear relationship between esterase activity and diet; although within the passerines, most warblers (which are insectivores) had lower esterase activity toward both substrates than did herbivores or omnivores of the same order. Larger birds tended to have higher levels of hepatic esterases than did smaller birds within the same order.

There is some evidence of differences in the quantity and nature of forms of serum CBE between different species of birds. In a study of 10 species the highest activities were found in 3 species of passerines, which were also the smallest in body size.[14]

The herbivorous and omnivorous species tended to show more complex patterns of CBE activity following chromatofocusing than did the four predatory species.Thus, the highest activity and/or greatest diversity of forms are associated with the smallest birds and the most diverse diets. These observations are consistent with the hypothesis that these esterases have a role in the detoxification of lipophilic esters in the diet. The smallest birds ingest food, and consequently xenobiotics in food, more quickly per unit body weight than larger birds; herbivores and omnivores are likely to have a greater quantity and variety of xenobiotics in their diet than are carnivores. It is interesting to note that the well-characterized CBEs of rat liver are able to hydrolyze a wide range of lipophilic xenobiotic esters, indicating the broad substrate specificity that would be required for an enzyme with such a dietary role. In a study on a wide range of avian and mammalian species, esterase activity toward nitrophenyl acetate was measured in samples of plasma, brain, and liver.[17] This is a general esterase substrate, but it would appear that much of the activity measured in birds — especially in plasma — was due to B esterases. Interestingly, the plasma activity toward this substrate was lower in avian and mammalian carnivores than it was in omnivores or herbivores. This lends further support to the suggestion that at least some of the enzymes of this type have a role in detoxifying xenobiotic esters in the food.

III. TECHNIQUES FOR MEASURING BLOOD ESTERASE INHIBITION

A. Sampling from Populations

The first decision to be made in the design of a sampling program is the species to be studied. This is dependent on the formulation of the pesticide and the reasons for the study, i.e., whether a particular rare species is thought to be at risk or an indicator species is required. The problems in the selection of a study species are thoroughly reviewed in Somerville and Walker.[18]

Sampling individuals from an exposed population following a pesticide application poses a major problem in its design. This is to ensure that sampled individuals are representative of the exposed population. Biases may occur if unexposed or less exposed individuals from outside the area are sampled, or if there are differences between exposed individuals in their susceptibility to capture. Birds can normally be readily captured by mist netting, and individuals may be marked (and possibly radiotracked) to enable time course changes to be established following exposure to pesticides. This approach also allows problems posed by interindividual variations to be overcome by using each individual preexposure as its own control in esterase analysis. In addition, concurrent controls from outside the study area should be sampled to reduce the possibility of ascribing changes due to temporal or environmental effects to the pesticide.

In some birds, nestlings can be used to reduce bias due to sampling problems. However, only readily accessible nests will be found, unless nest boxes are erected; and these may be more vulnerable to pesticide exposure. Other restrictions when using nestlings to assess exposure include the size of the individual, which may limit the number of times each can be sampled, and the time limit imposed by fledging time.

Small mammals, such as field mice *(Apodemus sylvaticus)* and rabbits can be livetrapped (e.g., mice in Longworth traps) marked by fur clipping or tagging, and released. As in the case of birds, such marking allows the time course of changes in esterase activity in individuals to be followed.

In all cases, as much information as practically possible should be collected about sampled individuals, e.g., age, sex, weight, condition, and time of day. This allows baseline population data on ChE variability to be collated as well as data on individuals. Such data are important to establish the reasons for inter- and intraindividual variations (see later).

B. Blood Sampling

The method for nondestructive blood sampling depends on the species under study. In birds, both adults and nestlings, blood samples can be readily obtained by using a fine gauge hypodermic needle to puncture the brachial vein and either a syringe or capillary tube to draw off blood samples for transfer to tubes. Samples should be kept at 4°C before centrifugation to prevent further inhibition or reactivation. No deleterious effects on sampled individuals have been observed in a number of species following this method of sampling.

In rabbits, blood samples can be taken directly from the ear vein into tubes for centrifuging, although care should be taken not to contaminate the sample with residues present on the fur of exposed animals. Ocular bleeding has been used to obtain samples from wild mice without apparent ill effect.[19] Whichever method is employed obviously the volume of blood sampled must be carefully monitored, particularly when multiple samples are taken, in order to ensure that any adverse effects are minimized. In addition, care should be taken not to allow samples to become contaminated by residues present on the surface of the animal although the effect of any compound used to clean the surface on esterase activity should also be taken into account.

C. Sample Preparation

Sample preparation for analysis is dependent on the source of the B esterase activity being monitored. In mammals this may be red blood cells or serum/plasma, whereas most birds have no erythrocyte AChE activity and therefore only serum/plasma is used. The decision to use serum or plasma depends primarily on the ease of preparation, since heparin does not appear to affect B esterase activity. Avian blood clots very readily, and therefore serum is usually more readily prepared.

Serum is isolated by allowing blood samples to clot, keeping at 4°C, and centrifuging at approximately 10,000 g for 5–10 min. Serum can then be readily aspirated from the surface and either analyzed immediately (short-term storage 4°C) or frozen at –20°C until analyzed (see Section III.E. for discussion of storage effects).

Erythrocytes can be prepared by centrifuging heparinised blood samples at 4000 g for 3 min. Blood cells should be washed with isotonic (0.9% w/v) saline, centrifuging at 4000 g for 3 min between washes. The cells can then be lysed by addition of 9 vol 10 mM Tris HCl (pH 7.35), and the membranes can be isolated by centrifuging at 30,000 g for 25 min.

D. Assays

Cholinesterase assays are normally based on the method of Ellman et al.[20] using acetylthiocholine and butyrylthiocholine as substrates for AChE and BChE assays, respectively.

Serum AChE and BChE activity is readily monitored by recording the change in absorbance at 415 nm during the hydrolysis of acetylthiocholine or butyrylthiocholine, respectively.[21] Serum (5 µL) is added to 0.1 mL 0.01 M 5,5 dithiobis 2-nitrobenzoic acid (DTNB) and 2.875 mL 0.1 M Tris HCl (pH 7.6) in a 3 mL cuvette, and the change in absorbance is monitored at 415 nm following addition of 20 µL 0.078 M substrate. The reaction should be incubated at the physiological temperature of the species source. However, since DTNB is relatively unstable above 37°C,[20] avian sera are incubated at this temperature. The hydrolysis rate is calculated using a millimolar extinction coefficient of 13.6 for the thiocholine-DTNB conjugate. The hydrolysis of cholinesterase is pH dependent.[22] These ChE assays may be scaled down to a final assay volume of 0.3 mL to allow use of microtiter plates and plate readers for analysis.[22]

A method has been established to quantify the contribution of serum BChE toward acetylthiocholine. This involves preincubation of serum samples with iso-OMPA which specifically inhibits BChE.[4] The concentration required to inhibit BChE is species dependent and should be established, using an inhibition curve, for the species under investigation. Serum AChE activity is assayed in the presence and absence of iso-OMPA. Serum AChE activity is calculated as the activity toward acetylthiocholine in the presence of iso-OMPA, and BChE activity is determined as the difference between activity in the presence and absence of iso-OMPA.[22]

Serum CBE activity is most readily assayed by monitoring hydrolysis of 1-naphthyl acetate.[23] Serum (20 µL) is added to 2.44 mL 0.1M Tris HCl (pH 7.6) and incubated for 10 min at 37°C, following addition of 40 µL 1-naphthyl acetate (3.46 mg/10 mL acetone). The reaction is stopped by addition of 0.5 mL sodium dodecyl sulfate (2.5% w/v in water) and 0.5 mL Fast Red ITR (0.1%w/v) in Triton X-100 (2.5% w/v in water). After mixing, the assay is allowed to stabilize at room temperature for 30 min and then read at 530 nm. The hydrolysis rate is calculated using a millimolar extinction coefficient of 33,225 for the naphthol-ITR complex. Again this assay may be scaled down for use in microtiter plates. Starling serum CBE activity is pH dependent.[24]

Assay of membrane-bound erythrocyte AChE requires the use of a titrometric method due to the presence of hemoglobin. The method is based on the release of acetic acid during hydrolysis of acetylcholine and is reviewed by Fairbrother et al.[22]

E. Effects of Storage

During laboratory studies it is usually relatively simple to assay samples immediately. However, in field studies it is not usually possible, and the effects of storage conditions must be taken into account when analyzing data. Several studies have shown spontaneous reactivation of ChE on storage following dosing with organophosphorus and carbamate compounds.[25,26] Hill[25] observed sex differences in the spontaneous reactivation on storage of quail plasma ChE inhibited by dicrotophos and carbofuran and also showed changes in the activity of uninhibited plasma ChE. Thompson et al.[26] showed that the effects of storage on inhibited serum BChE and CBE from starlings was dependent on the compound. In summary, the simplest approach is to analyze samples immediately after collection without freezing; however, if this is not possible, a laboratory study should be undertaken under the same storage conditions, pre-freezing and freezing times, as in the case of samples taken in the field.

F. Use of Reactivation Techniques

Martin et al.[27] reported that phosphorylated brain AChE is readily chemically reactivated by pyridine-2-aldoxime methiodide (2PAM) and carbamylated brain AChE is spontaneously reactivated following dilution. When applied to serum esterases, this approach may allow differences in activity to be attributed to inhibition by organophosphorus or carbamate pesticides, rather than interindividual or other sources of variability. However, there are problems with this approach in that reactivation depends on the esterase, the compound, and the time after exposure.[28] Figure 3 shows the dependence of in vitro reactivation of serum and brain esterases on the time since exposure. The time dependence of this process is due to progressive aging of the phosphorylated enzyme, which is obviously dependent on the compound. Even with these limitations, such an approach may

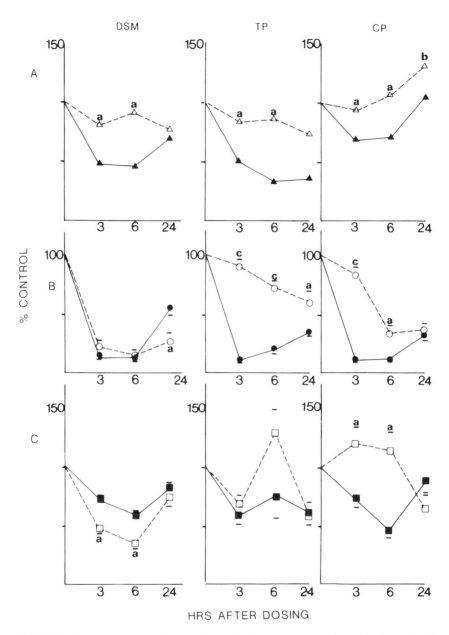

FIGURE 3. Time dependence of chemical reactivation (open symbols) of (A) brain AChE, (B) serum BChE, and (C) serum CBE activity following dosing (closed symbols) with DSM, demeton-S-methyl; TP, triazophos; and CP, chlorpyriphos. (Student's *t*-test, significant reactivation, a = $p < 0.1$, b = $p < 0.05$, c = $p < 0.001$.) (From Thompson, H. M., Walker, C. H., and Hardy, A. R., *Arch. Environ. Contam. Toxicol.*, 20, 509, 1991.)

help to confirm that inhibition is due to an organophosphorus or carbamate compound rather than heavy metals or a pathological condition.[29]

G. Interpretation of Activity Levels

The interpretation of serum esterase activity data is dependent on the collection and use of valid control data. Ideally two types of control should be used as already discussed: samples collected at the same time as samples from exposed individuals but from control areas and samples from the same individuals before and after pesticide application.

The interpretation of field data is greatly assisted if laboratory data is available on the effects of the compound under investigation on esterase activity, both inhibition and recovery. However, the route of exposure, dermal, ingestion of contaminated food, preening or grooming, etc. may affect the dose-response relationship; and therefore extrapolation of dose-response data from the laboratory to the field should not normally be attempted. In addition, inhibition of serum esterases to a particular class of compound should ideally be supported by further evidence, e.g., alkylphosphate residues in fecal samples,[30] or chemical reactivation.

There are several factors which need to be taken into consideration when comparing data from samples taken from exposed and unexposed individuals.

1. Species Differences

There are wide species differences in serum B esterase activity of both birds and mammals[17] (see Section II.C.3). This means that serum esterase activities from exposed individuals can only be compared with controls from the same species if large under- or overestimation of exposure is to be avoided.

2. Interindividual Differences

Wide interindividual differences in serum B esterase activity have been reported, including sex differences in some species.[29,31,32] This means that a relatively large number of control samples are required in order to establish the true level of inhibition in exposed individuals. An alternative approach is to sample the same individual sequentially before and after exposure in order to circumvent interindividual differences. This allows the time course of inhibition, i.e., exposure and recovery, to be established.

3. Diurnal Variations

Even when the same individual is sampled sequentially, diurnal changes need to be taken into account. Diurnal changes in serum ChE and CBE activity have been reported in buzzards *(Buteo buteo)* and starlings, respectively.[33,34] Increases of 150% were observed in starling serum CBE activity over a 12-hr period.[34] Obviously, such changes would severely affect the interpretation of data collected at different times of day. Therefore, if samples collected at varying times of day

SERUM BChE ACTIVITY

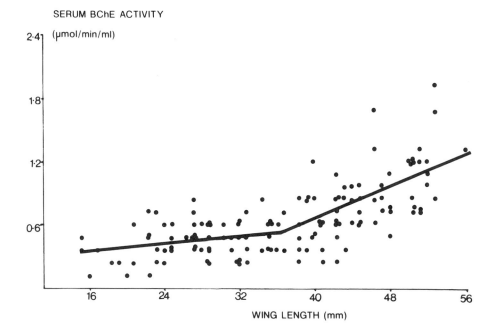

FIGURE 4. Developmental increase in tree sparrow nestling serum BChE activity. Each point represents a single measurement before exposure to demeton-S-methyl; several measurements are shown for each individual. (From Thompson, H. M., in *Cholinesterase Inhibiting Insecticides, Their Impact on Wildlife and the Environment,* Mineau, P., Ed., Elsevier, Amsterdam, 1991, p. 109.)

are to be compared, the diurnal variation in enzyme activity must be established in unexposed individuals. An alternative approach is to ensure that all samples are taken at around the same time of day; this is usually possible only in individuals known to be available for sampling at a specific time of day, e.g., nestling birds and field mice trapped overnight, and is impractical with free-flying birds.

4. Developmental Changes

If young animals are sampled, it is obviously important to establish whether there are any developmentally related changes in esterase activity. This is necessary for both sequentially and concurrently sampled birds as there may be substantial day-to-day changes in activity with growth. Such changes have been shown in tree sparrow nestlings with an approximately twofold increase in serum BChE activity during growth from 16 to 50-mm wing length (Figure 4). The effects of these changes on the interpretation of data have been reviewed.[35]

5. Seasonal Variations

Seasonal variations have also been reported in avian species.[32] However, these need not be taken into account if concurrent control or sequential sampling, over a short time period, is undertaken.

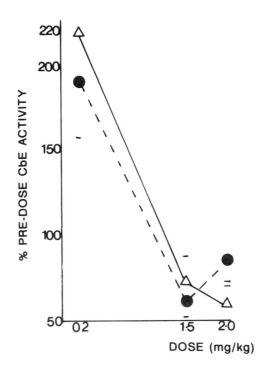

FIGURE 5. Effect on starling serum CBE activity of oral dosing with demeton-S-methyl (0.2 mg/kg = 1% oral LD_{50}) Δ 6 hr and • 24 hr after dosing. (From Thompson, H. M., Walker, C. H., and Hardy, A. R., *Arch. Environ. Contam. Toxicol.*, 20, 514, 1991.)

6. Increases in Esterase Activity

A complicating factor in the use of serum esterases in assessing the exposure of individuals to pesticides is an increase in activity which may occur following low-level exposure. An increase in serum CBE activity has been observed following dosing of starlings with 1% LD_{50} of demeton-S-methyl, whereas at higher doses inhibition was observed[28] (Figure 5). It is presumed that the increase in activity occurs due to release of enzyme from the liver. Such an increase in esterase activity directly counteracts any inhibition that may also be occurring, resulting in an underestimation of the actual level of inhibition. Thus if a 50% increase in the original level of esterase activity occurs at the same time as 100% decrease in the level of initial activity, only 50% inhibition would be observed.

IV. FIELD STUDIES

The usefulness of serum B esterase inhibition as a biomarker is dependent on the reason for which it is monitored. As blood B esterases have no known physiological function, the inhibition of these enzymes is purely a monitor of

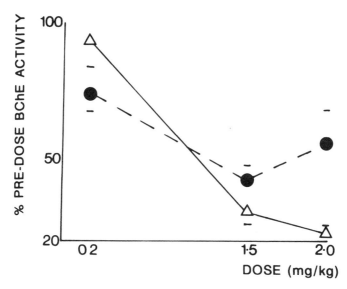

FIGURE 6. Inhibition of starling serum BChE activity following oral dosing with demeton-S-methyl △ 6 hr and • 24 hr after dosing. (From Thompson, H. M., Walker, C. H., and Hardy, A. R., *Arch. Environ. Contam. Toxicol.,* 20, 514, 1991.)

exposure, not of effect. Attempts to correlate serum ChE inhibition with that of brain AChE have so far been unsuccessful. Erythrocyte AChE inhibition has been reported to correlate more closely with that of brain; but this enzyme appears to be restricted to mammalian species and therefore is limited in its applicability.[36] However, the inhibition of serum B esterases is dose related[28] (Figures 5 and 6) and as such offers a sensitive method for monitoring levels of exposure to anticholinesterase chemicals and thus relating changes in other parameters, e.g., behavior to this exposure. The next section reviews examples of the use of serum B esterase inhibition in field studies in order to demonstrate their value and limitations in practical situations.

A. Exposure of Tree Sparrow Nestlings to Organophosphorus Pesticides

The exposure of tree sparrow nestlings was investigated as part of a large-scale project to investigate the economic and ecological effects of pesticides used on winter wheat.[37] The study on tree sparrows investigated whether nestlings were exposed to organophosphorus pesticides applied to an adjacent winter wheat crop in which the adult birds were feeding. The approach was to take blood samples and fecal samples for diet analysis from nestlings before and after application of demeton-S-methyl on the treated field. Additional samples were collected from nestlings around untreated fields.

Blood samples were taken by puncture of the brachial vein using a 25 G (0.5 × 16 mm) hypodermic needle and transferred to a microcentrifuge tube using a 50 μL capillary tube. The samples were centrifuged and assayed for serum BChE activity within 2 hr of collection.

Serum BChE activity of samples collected before pesticide application and in control areas was plotted against wing length of the nestlings, as a measure of growth stage. Serum BChE activity increased slowly and linearly between 16-mm and 36-mm wing length and then more rapidly between 36-mm and 54-mm wing length[35] (Figure 4). Sampling was stopped at around 50-mm wing length to prevent premature fledging caused by disturbance. Serum CBE activity also increased during the development of the nestlings, but far more scatter was observed than in the case of serum BChE.

Serum BChE activity of samples collected from nestlings after the demeton-S-methyl application was analyzed in relation to the activity of the same bird preexposure and in relation to the growth curve for serum BChE activity in control samples. Comparison of serum BChE activity in individual bird pre- and postexposure indicated that inhibition occurred for up to 1 day after demeton-S-methyl application (Figure 7). However, when the activities were compared to the growth curve, inhibition was found to continue throughout the 4-day sampling period postapplication[35] (Figure 8). This underlines the need for adequate data on underlying changes in serum B esterase activity in order to fully interpret data collected from exposed birds.

The ideal extension to this study would have been to collect blood samples from the parent as well as nestling birds, in order to assess and compare the level of parental and nestling exposure. However, the limiting factor in such a study is the effect of trapping the parent birds near the nest, which would be important to ensure that the adults could be identified with each nest. Adult tree sparrows appear to be particularly likely to abandon their nests when disturbed. In order to collect comparable data to that for the nestlings, parent birds would need to be sampled sequentially at the same time intervals as their brood.

Laboratory studies may be useful in aiding the interpretation of field data, i.e., the nature of the response to a chemical and its time dependency. Ideally studies should be undertaken on the same, or a closely related, species to allow comparison of the inhibition of serum esterases with that of the brain. These studies would allow at least some confounding factors to be investigated, such as the effects of reduced food availability on levels of B esterases. However, the utility of dose-response curves is limited since such relationships are dependent on the route of exposure, which may be multiple in individuals exposed in the field. Therefore, field data tend to be indicative of differences in exposure between individuals, rather than absolute in terms of dose.

B. Exposure of Mice to Organophosphorus Seed Treatments

Field mice are resident on cereal fields and are readily livetrapped. Thus, the field mouse is a useful indicator species for the exposure of wildlife to seed treatments. As a preliminary to field studies of OP seed treatments, laboratory studies undertaken with captive mice confirmed that plasma AChE was inhibited following exposure to chlorfenvinphos and carbophenothion.[21,38] In both studies plasma AChE inhibition was greater than that in the brain. Inhibition of plasma AChE reached a maximum level of inhibition 1 day after exposure to treated food and recovered within 7 days after removal of the treated diet.

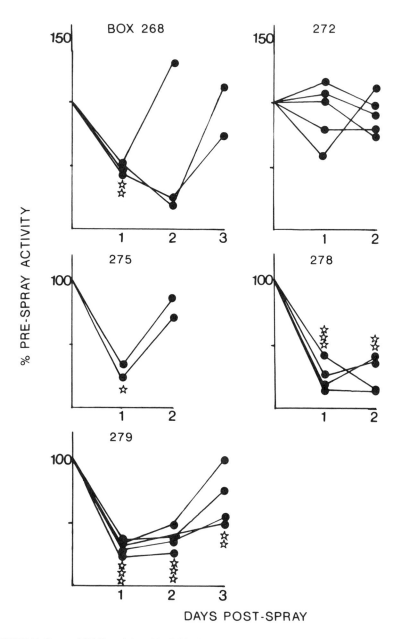

FIGURE 7. Serum BChE activity of individual tree sparrow nestlings in different nests 1–3 days after demeton-S-methyl application. Each point joined by a line represents data from the same individual (Student's t-test, significant inhibition, * $p < 0.02$, ** $p < 0.01$, *** $p < 0.001$).

In the field studies mice were livetrapped on fields treated with chlorfenvinphos or carbophenothion[21,38] and in untreated woodland. Brain and plasma AChE were monitored together with residue levels in the gut.

% SERUM BChE ACTIVITY

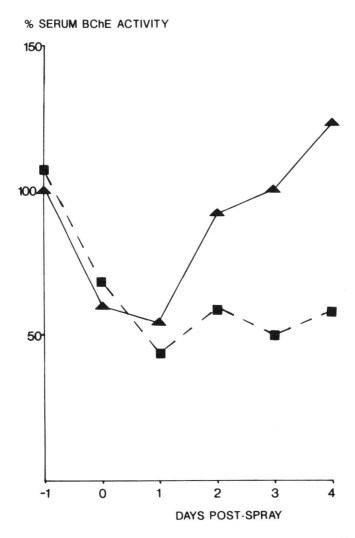

FIGURE 8. Comparison of nestling tree sparrow mean serum BChE activity inhibition in birds in one nest following demeton-S-methyl application. Activity is expressed as percentage of prespray values for the same birds (▲) and as a percentage of the expected value at that growth stage from the developmental control curve (■) (Figure 4). (From Thompson, H. M., in *Cholinesterase Inhibiting Insecticides, Their Impact on Wildlife and the Environment,* Mineau, P., Ed., Elsevier, Amsterdam, 1991, p. 109.)

In the field trial plasma AChE was significantly inhibited in trapped mice for 10 days after drilling with chlorfenvinphos treated grain, but for 6 months following drilling with the carbophenothion-treated seed. In both the trials pesticide residues in the gut were correlated with the levels of plasma AChE inhibition. Thus, plasma AChE inhibition was dose related and could be used to show significant differences in the persistence of the two organophosphorus pesticides. Although this study included the analysis of brain AChE and residues of organophosphorus compounds as well as analysis of plasma AChE in mice that were sacrificed, it provides a useful illustration of the potential utility of serum esterases if used nondestructively.

C. Exposure of Adult Birds to Organophosphorus Pesticides

In this study[39] serum BChE inhibition was used to indicate that histological damage in the liver was related to exposure to the organophosphorus pesticide demeton-S-methyl. Serum BChE inhibition was investigated since although the birds were sacrificed for histological studies, the liver is far more sensitive to inhibition by low levels of demeton-S-methyl than is the brain AChE.[28]

Birds in which serum BChE was inhibited, showing that exposure had occurred, also showed decreases in hepatocyte size and a significant increase in hepatocyte binucleation. This suggests that there was an induction of replication in order to increase liver cell numbers as a toxicological response to exposure to the pesticide. This study illustrates the usefulness of serum esterases in aiding the attribution of effects of pesticides to exposure.

D. Exposure of Red-Tailed Hawks to Organophosphorus Pesticides in Orchards

Hooper et al.[40] and Wilson et al.[41] have investigated the use of serum ChE in evaluating the cause of poisoning of red-tailed hawks in relation to the use of pesticides in almond orchards. Plasma samples were taken from hawks trapped in orchards and, in addition, from suspected cases of poisoning in birds submitted to wildlife rehabilitation centers. Plasma samples were assayed for serum ChE activity, and in inhibited samples the level of chemical reactivation which could be achieved was assessed. In addition to the ChE assays, alkylphosphate residues were monitored in foot and feather washes and in fecal samples from trapped birds.

This approach showed that red-tailed hawks were significantly exposed to organophosphorus pesticides used in almond orchards. Some of the applications of these pesticides are in the form of oil mists which are relatively stable and coat the almond trees in which the birds perch and roost. The class of chemical involved was established from the significant chemical reactivation of many ChE samples and the presence of alklylphosphate residues in foot and feather washes and in fecal samples.

V. DISCUSSION

The employment of blood esterases as biomarkers needs to be viewed in a wider context. As with other biomarkers, they have their strengths and limitations; and the measurement of the effects of chemicals on individuals and populations in the field is best accomplished by the use of a suite of biomarkers, each of which will give different, yet complementary information. Changes in blood esterases can give a measure of exposure to OP and carbamate pesticides in the field, following nondestructive sampling. The methods currently available do not give any reliable measure of the degree of inhibition of AChE of the central nervous system or of toxic manifestations, although some tentative correlations may be possible for particular compounds where there are good laboratory data (see, however, later discussion). The inhibition of blood esterases represents a nonspecific response to a wide range of OPs and carbamates; thus the identification of a particular chemical as a causal agent requires other evidence.

The selection of biomarkers and other techniques to be used alongside changes in blood esterases depends upon the purpose of the study. Critical questions are

1. Is it necessary to identify the particular chemical causing the response?
2. Is it necessary to measure the degree of inhibition of brain AChE?
3. Are there toxic interactions other than with brain AChE of interest? (e.g., interaction with neuropathy target esterase or inhibition of cytochrome P450)
4. Are there effects at other levels of the organism of interest (e.g., changes in physiological function or population parameter)?

The identification of a particular chemical will require chemical analysis. This may be the original chemical in the food, feathers, or fur; or in disgorged pellets in the case of owls and some other birds. Analysis of feces can lead to the identification of metabolites, although this does not necessarily enable identification of a specific chemical. Direct measurement of brain AChE requires a destructive assay and therefore falls outside the present discussion; thus the difficulty of predicting this from nondestructive assays remains. This will be discussed further in the next section. Other types of interaction may be of interest. Some birds, for example, are very susceptible to the action of certain OPs (e.g., mipafox, leptophos, methamidophos) which cause delayed neuropathy.[6] Unfortunately, a nondestructive assay is not available for this. Inhibition of cytochrome P-450 occurs when OP compounds which contain thion groups undergo oxidative desulfuration, and this may lead to toxic effects resulting from the reduced detoxication of other xenobiotics. Again, a nondestructive assay is not available to measure this. Biomarker responses at other levels are dealt with in other chapters of this book. Most of them do not relate specifically to compounds that inhibit esterases. The extent to which they may be of value will depend upon the objectives of the study, e.g., whether effects on behavior, reproduction, population numbers, and gene frequency are relevant.

The underpinning of field studies by laboratory studies and the maintenance of a close relationship between the two are very important with blood esterases, as with other systems used as biochemical biomarkers. In the first place, the development of methods, along with the refinement of techniques, needs to be conducted in the laboratory ahead of field studies. Once field studies are initiated it is important to keep laboratory studies going, and to develop and improve them in the light of field experience, including the extension of methods to species or groups not previously investigated. For example, laboratory studies have highlighted a basic problem with some blood esterases in that the levels can rise rapidly after exposure to low levels of OPs.[28] Such a tendency makes it almost impossible to measure low levels of exposure in the field with current methods.

A potential solution to this dilemma lies in the production of specific antibodies to blood esterases, which can be used to measure (1) levels of esterase protein (e.g., by enzyme-linked immunosorbent assay [ELISA]) and (2) specific activities following immunoprecipitation of the antigenic esterase. This approach is now being followed.[24] Such an approach can also resolve another fundamental problem with blood esterases — the use of reliable control values for activity. Because of interindividual and temporal variations, truly reliable control values are difficult to obtain using currently available techniques. By contrast, the specific activity of a particular esterase should remain constant, unless it is subject to activation or inhibition.

Laboratory studies can have a number of important functions in regard to supporting the interpretation of field studies which employ nondestructive biomarker systems such as blood esterases. In the first place, they can aid the identification of appropriate species to be used in field studies. This may be because of sensitivity to OPs or carbamates due to poor detoxication or sensitive brain AChE, or it may be because of only certain species possessing an esterase form for which an antibody is available when using an ELISA kit or similar test. Laboratory studies may also be of value in the identification of other environmental factors which may influence the interpretation of esterase changes. For instance, interactions with other compounds may lead to the enhanced toxicity of OPs or carbamates.[42] One compound may increase the toxicity of another, by inhibiting its detoxication or by increasing its activation; and such effects may be studied in the laboratory. In this situation, the presence of another pesticide or other environmental chemicals may lead to greater inhibition of a blood esterase by the application of an OP or carbamate in the field than would have been expected if the pesticide had been applied in the absence of this chemical. Again, laboratory studies may identify other factors (such as diet or ambient temperature) which influence the degree of blood esterase inhibition.

Finally, it is worth restating the point made earlier. There are many potential problems with regard to sampling and storage of samples, which — in some cases at least — differ between species. In particular, changes in activity during storage, due to reactivation of inhibited enzymes or deactivation of esterases, need to be closely observed. Supporting laboratory studies are essential to take account of these factors.

The selection of controls for field studies is vitally important because it is from these data that the levels of inhibition in exposed animals are assessed. Control data must be obtained from the same species because of wide interspecies variation. Use of controls from an untreated site allows the effects of environmental factors apart from the pesticide to be removed from the comparison (e.g., temperature, seasonal variations) and also allows diurnal and developmental changes to be investigated. The time course of exposure can be assessed only when laboratory data are available for the chemical because there are intercompound differences in the rates of inhibition and recovery.

The collection of exposed individuals that are representative of the exposed population is also vitally important, and the techniques available have been discussed above. Without the assurance that representative individuals have been sampled, extrapolation of the level of exposure and any observed effects from the individual to the population level cannot be made. The effects of pesticide exposure can substantially affect the chances of catching more exposed individuals and thus considerably bias collected data toward less affected individuals.[44] The level of exposure cannot be accurately determined from laboratory dose-response curves because the routes of exposure under laboratory and field conditions vary widely. Thus, the level of inhibition of serum esterases in field data cannot be used indirectly to predict either the dose received or the level of brain AChE inhibition from laboratory derived dose-response curves. Therefore, the utility of serum esterase inhibition is entirely dependent on the reason for their use. Serum B esterases are generally far more sensitive than brain AChE to inhibition by anticholinesterase pesticides and thus, given good control data, are ideal monitors of exposure to a known pesticide.

A more interesting use of blood esterase inhibition is in attributing the effects of pesticides to exposure (e.g., behavioral effects, reduced reproductive success, histological changes) when nondestructive methods are invaluable in monitoring individuals. For example, studies on the survival of fledged birds exposed as nestlings to pesticides are dependent on nondestructive methods to determine that exposure has occurred.

In summary, the use of blood esterases in monitoring exposure of animals to field applications of OP and carbamate pesticides can be invaluable, particularly in the interpretation of effects. However, caution must be taken in their use and interpretation since there are a large number of variables, apart from anticholinesterase pesticides, which affect their activity. In addition it must be noted that although laboratory studies are valuable in investigating the time course of inhibition and recovery following pesticide exposure, such data should be extrapolated to field data only with great care due to the widely differing routes and time course of exposure.

REFERENCES

1. Vandekar, M., Minimising occupational exposure to pesticides — ChE determination and OP poisoning, *Residue Rev.* 75, 67, 1980.
2. Greig-Smith, P. W., Use of cholinesterase measurements in surveillance of wildlife poisoning in farmland, in *Cholinesterase Inhibiting Insecticides, Their Impact on Wildlife and the Environment,* Mineau, P., Ed., Elsevier, Amsterdam, 1991, p. 127.
3. Westlake, G. E., Hardy, A. R., and Stevenson, J., Effects of storage and pesticide treatments on honey bee brain acetyl cholinesterase activities, *Bull. Environ. Contam. Toxicol.,* 34, 668, 1985.
4. Aldridge, W. N., The differentiation of true and pseudo cholinesterase by organophosphorus compounds, *Biochem. J.,* 53, 62, 1953.
5. Mackness, M. I., Thompson, H. M., Hardy, A. R., and Walker, C. H., Distinction between A esterases and arylesterases: implications for esterase classification, *Biochem. J.,* 245, 293, 1987.

6. Johnson, M. K., Neurotoxic esterases, *Crit. Rev. Toxicol.,* 3, 289, 1975.
7. Hollingworth, R. M., The dealkylation of organophosphorus triesters by liver enzymes, in *Biochemical Toxicology,* O'Brien, R. D. and Yamamoto, I., Eds., Academic Press, New York, 1970, p. 75.
8. Augustinsson, K. B., Multiple forms of esterases in vertebrate blood plasma, *Ann. N.Y. Acad. Sci.,* 94, 844, 1961.
9. Massoulie, J., The polymorphism of cholinesterase and its physiological significance, *TIBS,* 5, 160, 1980.
10. Sussman, J. L., Harel, M., and Silman, I., Three-dimensional structure of acetylcholinesterase and its complexes with anticholinesterase drugs, *Chem. Biol. Interactions,* in press.
11. Mentlein, R., Ronai, R., Kobbi, M., Heymann, E., and von Deimling, O., Genetic identification of rat liver carboxylesterases isolated in different laboratories, *Biochim. Biophys. Acta,* 913, 27, 1987.
12. van Zutphen, L. F. M., Review of the genetic nomenclature of esterase loci in the rat *(Rattus norvegicus), Transplant. Proc.,* 15, 1687, 1983.
13. Pickering, C. E. and Pickering, R. G., The interference by erythrocyte acetylthiocholinesterase in the estimation of blood cholinesterase activity of the chicken, *Toxicol. Appl. Pharmacol.,* 39, 229, 1977.
14. Thompson, H. M., Mackness, M. I., Walker, C. H., and Hardy, A. R., Species differences in avian serum B esterases revealed by chromatofocusing and possible relationships of esterase activity to pesticide toxicity, *Biochem. Pharmacol.,* 41, 1235, 1991.
15. Ronai, A., Wassmer, B., DeLoose, S., and von Deimling, O., Immunochemical interrelationships between carboxylesterase isozymes (EC 3.1.1.1) of the house mouse *(Mus musculus), Comp. Biochem. Physiol.,* 82B, 201, 1985.
16. Bush, F. M., Price, J. R., and Townsend, J. I., Avian hepatic esterases, pesticides and diet, *Comp. Biochem. Physiol.,* 44B, 1137, 1973.
17. Westlake, G. E., Martin, A. D., Stanley, P. I., and Walker, C. H., Control enzyme levels in the plasma, brain and liver from wild birds and mammals in Britain, *Comp. Biochem. Physiol.,* 76C, 15, 1983.
18. Somerville, L. and Walker, C. H., Eds., *Pesticide Effects on Terrestrial Wildlife,* Taylor & Francis, London, 1990.
19. Fairbrother, A., personal communication, 1989.
20. Ellman, G. L., Courtney, K. D., Andreas, V., Jr., and Featherstone, R. M., A new and rapid colorimetric determination of acetylcholinesterase activity, *Biochem. Pharmacol.,* 7, 88, 1961.
21. Westlake, G. E., Blunden, C. A., Brown, P. M., Bunyan, P. J., Martin, A. D., Sayers, P. E., Stanley, P. I., and Tarrant, K. A., Residues and effects in mice after drilling wheat treated wuth chlorfenvinphos and an organomercurial fungicide, *Ecotoxicol. Environ. Health,* 4, 1, 1980.
22. Fairbrother, A., Marden, B. T., Bennett, J. K., and Hooper, M. J., Methods used in the determination of cholinesterase activity, in *Cholinesterase Inhibiting Insecticides, Their Impact on Wildlife and the Environment,* Mineau, P., Ed., Elsevier, Amsterdam, 1991, p. 72.
23. Gomori, G., Human esterases, *J. Lab. Clin. Med.,* 42, 445, 1953.
24. Thompson, H. M., Esterases as Indicators of Exposure of Birds to Pesticides, Ph.D. thesis, University of Reading, U.K., 1988.

25. Hill, E. F., Divergent effects of post-mortem ambient temperature on organophosphorus and carbamate-inhibited brain cholinesterase activity in birds, *Pestic. Biochem. Physiol.,* 33, 264, 1989.

26. Thompson, H. M., Walker, C. H., and Hardy, A. R., Inhibition of avian esterases by organophosphorus insecticides: problems of reactivation and storage, *Arch. Environ. Contam. Toxicol.,* 20, 509, 1991.

27. Martin, A. D., Norman, G., Stanley, P. I., and Westlake, G. E., Use of reactivation techniques for the differential diagnosis of organophosphorus and carbamate pesticide poisoning in birds, *Bull. Environ. Contam. Toxicol.,* 26, 775, 1981.

28. Thompson, H. M., Walker, C. H., and Hardy, A. R., Changes in activity of avian serum esterases following exposure to organophosphorus insecticides, *Arch. Environ. Contam. Toxicol.,* 20, 514, 1991.

29. Rattner, B. A. and Fairbrother, A., Biological variability and the influence of stress on cholinesterase activity, in *Cholinesterase Inhibiting Insecticides, Their Impact on Wildlife and the Environment,* Mineau, P., Ed., Elsevier, Amsterdam, 1991, p. 89.

30. Hooper, M. J., Brewer, L. W., Cobb, G. P., and Kendall, R. J., An integrated laboratory and field approach for assessing hazards of pesticide exposure to wildlife, in *Pesticide Effects on Terrestrial Wildlife,* Somerville, L. and Walker, C. H., Eds., Taylor & Francis, London, 1990, p. 271.

31. Greig-Smith, P. W., Walker, C. H., and Thompson, H. M., Ecotoxicological consequences of the interactions between avian esterases and organophosphorus compounds, in *Clinical and Experimental Toxicology of Organophosphates and Carbamates,* Ballantyne, R. M. and Marrs, T., Eds., Butterworth-Heinemann, London, 1992, p. 295.

32. Hill, E. F., and Murray, H. C., Seasonal variation in diagnostic enzymes and biochemical constituents of captive northern bobwhites and passerines, *Comp. Biochem. Physiol.,* 87B, 933, 1987.

33. Garcia-Rodriguez, T., Ferrer, M., Recio, F., and Castroviejo, J., Circadian rhythms of determined blood chemistry values in buzzards and eagle owls, *Comp. Biochem. Physiol.,* 88A, 663, 1987.

34. Thompson, H. M., Walker, C. H., and Hardy, A. R., Avian esterases as indicators of exposure to pesticides — the factor of diurnal variation, *Bull. Environ. Contam. Toxicol.,* 41, 4, 1988.

35. Thompson, H. M., Serum "B" esterases as indicators of exposure to pesticides, in *Cholinesterase Inhibiting Insecticides, Their Impact on Wildlife and the Environment,* Mineau, P., Ed., Elsevier, Amsterdam, 1991, p. 109.

36. Lawson, A. A. and Barr, R. D., AChE in red blood cells, *Am. J. Hematol.,* 26, 101, 1987.

37. Greig-Smith, P. W., The Boxworth Project — environmental effects of cereal pesticides, *J. R. Agric. Soc. Engl.,* 150, 171, 1989.

38. Westlake, G. E., Bunyan, P. J., Johnson, J. A., Martin, A. D., and Stanley, P. I., Biochemical effects in mice following exposure to wheat treated with chlorfenvinphos and carbophenothion under laboratory and field conditions, *Pestic., Biochem. Physiol.,* 18, 49, 1982.

39. Tarrant, K. A., Thompson, H. M., and Hardy, A. R., Biochemical and histological effects of the aphicide demeton-S-methyl on house sparrows *(Passer domesticus)* under field conditions, *Bull. Environ. Contam. Toxicol.,* 48, 360, 1992.

40. Hooper, M. J., Detrich, F. J., Weisskopf, C. P., and Wilson, B. W., Organophosphate exposure in hawks inhabiting orchards during winter dormant-spraying, *Bull. Environ. Contam. Toxicol.*, 42, 651, 1989.
41. Wilson, B. W., Hooper, M. J., Littrell, E. E., Detrich, F. J., Hansen, M. E., Weisskopf, C. P., and Seiber, J. N., Orchard dormant sprays and exposure of red-tailed hawks to organophosphates, *Bull. Environ. Contam. Toxicol.*, 47, 717, 1991.
42. Johnson, G., Walker, C. H., Dawson, A., and Furnell, A., Interactive effects of pesticides in the red-legged partridge, *Functional Ecol.*, 4, 309, 1990.
43. Walker, C. H. and Thompson, H. M., Phylogenetic distribution of cholinesterases and related esterases, in *Cholinesterase Inhibiting Insecticides, Their Impact on Wildlife and the Environment*, Mineau, P., Ed., Elsevier, Amsterdam, 1991, p. 1.
44. Mineau, P. and Peakall, D. B., An evaluation of avian impact assessment techniques following broad-scale forest insecticide sprays, *Environ. Toxicol. Chem.*, 6, 781, 1987.

CHAPTER 3

Clinical Biochemistry

Anne Fairbrother

TABLE OF CONTENTS

0-87371-648-5/94/$0.00+$.50
© 1994 by Lewis Publishers

I. INTRODUCTION

Assessment of the health status of animals through measurement of cellular, biochemical, and macromolecular constituents in blood, secretions, and excretions has been variously referred to as clinical chemistry, clinical biochemistry, or clinical pathology. The genesis of this discipline occurred in the mid-1800s, although the applications to medical and veterinary practices did not blossom until after the Second World War when automated equipment required for processing large numbers of samples became available.[1] Clinical biochemistry (as it will be referred to here) has now become a standard part of all diagnostic protocols for investigating health problems in humans and domestic or captive animals. A suitable group of tests (commonly called a "panel") can be defined to evaluate the function of most major organs, the endocrine system, the immune system, and the nervous system. By using this approach, the affected organs and, potentially, the processes responsible for the observed disease syndrome can be identified and further diagnostic tests (e.g., radiography) can be called into play to make a precise diagnosis. While clinical biochemistry panels have been applied to free-ranging vertebrates only infrequently, this chapter will show that studies of captive animals indicate that such methods are now available. Application of clinical biochemistry procedures as nondestructive biomarkers for assessing the health status of free-living animals is limited only by our ability to validate the tests for species of interest and to establish the range of normal values for animals living in their natural environments.

Clinical biochemistry can be divided into several subdisciplines, each with its own techniques and applications. These are: clinical enzymology, products of metabolism, hematology, clinical endocrinology, and diagnostic immunology. Many of the approaches used in all of these disciplines are noninvasive and all of them are conducted on live animals. Measurements may be made on samples taken from blood, tissues (e.g., biopsies), urine, feces, cerebral spinal fluid, synovial fluid, ocular fluids, or salivary secretions. This chapter will briefly review the procedures involved in each subdiscipline and discuss their applications as they relate to free-ranging animals, with a particular emphasis on toxicological problems. For more detailed descriptions of the various tests and their physiological bases, the reader is referred to the excellent text on clinical biochemistry of domestic animals edited by Kaneko[2] and several of the other chapters of this book.

II. CLINICAL ENZYMOLOGY

In general, enzymes in blood that are important in clinical diagnosis fall into two groups: those that are secreted into the blood for a certain purpose (e.g., coagulation enzymes) or those that have a specific function only within an organ and are released into the blood incidentally or as a result of organ damage. The latter group is the most important diagnostically. Clinical enzymology has been utilized since 1927 as a marker of cell integrity.[3] Increases in tissue enzymes within the blood serum or plasma are indicative of cell damage such as in cardiac infarction or acute hepatitis or of functional impairment of cells such as occurs during hypoxia.

Cellular necrosis results in leakage into the bloodstream of the enzymes that are normally produced by and contained within the damaged organ. These processes frequently are manifested by concurrent changes in the histological structure of the organ and are demonstrable during a biopsy or necropsy examination. However, leakage of enzymes through the cell membranes may also be caused by dysfunctional membrane pumps, pores, or other processes. These changes would not be observable histologically and could only be demonstrated by measurements of the blood enzyme values. Increased amounts of tissue enzymes in the blood serum can also be caused by increased tissue mass during neoplastic growth or impaired excretion of enzymes due to obstruction of excretory ducts resulting in secondary damage to the organ and leakage of enzymes into the bloodstream.[4]

Decreased tissue enzyme concentrations or activity in the blood also have diagnostic value and result from tissue atrophy, poor nutritional condition, or toxicological insult. For enzymes that are normally secreted by tissues, pathology more frequently results in a decrease in activity than an increase. Cholinesterase is actively secreted within the synaptic cleft of neurons in the sympathetic nervous system, and a decrease in activity is indicative of exposure to cholinesterase-inhibiting compounds such as organophosphorus insecticides (see Chapter 2). Lead specifically inhibits the activity of δ-aminolevulinic acid dehydratase (δ-ALAD) in red blood cells, a necessary enzyme in the heme synthesis pathway.

Interpretation of the results of measurement of enzyme activity in blood or fecal material necessitates an understanding of the site of normal production and function of the enzymes. Most enzymes such as the transaminases and lactic dehydrogenase (LDH) are produced by several tissues (liver, kidney, and muscle) with one tissue generally predominating. The pattern of changes in the various serum constituents will be indicative of malfunction or injury to a particular organ system. The predominant tissue of production for a particular enzyme differs among species, frequently confounding our ability to generalize results to a particular animal of interest.[5,6] Fortunately, there has developed over the past 10–15 years a body of data on clinical biochemistry values for captive zoo animals and caged birds that can also be applied to free-ranging wildlife.[7-9] However, it is extremely important to verify which tissues are the primary producers of each enzyme as this varies considerably among species. For example, in the dog, cat, or rat, elevation of serum activity of γ-glutamyl transferase(γ-GT) and alanine aminotransferase (ALT) is indicative of liver damage. In other animals (such as the pig, sheep, horse, cow) ALT activity is too low to be of diagnostic value,[3] while in many species of birds ALT is produced by kidney and muscle tissue as well and therefore is not specific to liver pathology.[6,10]

Many enzymes exist as several closely related forms, known as isoenzymes. Electrophoretic measurements of the isozyme pattern can be used to accurately diagnosis tissue pathology.[11] The most highly developed application of this system is with the LDH isozyme group that can be utilized to differentiate cellular necrosis due to myocardial infarction, muscle disease, or liver disease from other dysfunctions such as congestive heart failure or impaired renal clearance.[12,13] The LDH enzyme occurs in two isomeric forms in mammals: the M subunit which is indicative of skeletal (striated) muscle-derived enzyme and the H subunit which derives from heart muscle.[14] These are further subdivided into a total of five isoenzymes.[15] In fish, a third subunit — type C — is also found and is characteristic of the liver only.[16,17]

Alkaline phosphatase is actually a group of enzymes produced by nearly every tissue and can be separated into tissue-specific isoenzymes by electrophoresis.[18] The enzyme differs among species, as does the primary tissue of origin; electrophoretic parameters need to be adjusted for the different species in order to achieve proper separation of the isoenzymes. Amylase and lipase are two digestive enzymes, predominantly of pancreatic origin, that are actively secreted into the gut lumen but can be measured in blood, urine, and feces. Increased concentrations of these enzymes are indicative of gastrointestinal disorders such as pancreatitis and intestinal ulceration, but can also be elevated in renal or hepatic diseases due to impaired clearance from the blood, bile duct obstruction, or primary renal disease.[4,19] Electrophoretic separation of the amylase isoenzymes will reveal a pancreatic-specific isoamylase in some species (dog, human) that can be used to aid in a differential diagnosis.[19] Additionally, the change in blood enzyme activity over time can indicate the magnitude of tissue damage within an organ. Delivery of creatinine phosphokinase (CPK) to the bloodstream in dogs is linearly related to the size of a cardiac infarct. Serial determination of activity of CPK over the period of several days will provide a clinical measurement of the severity of the lesion.[4]

While clinical enzymology generally measures enzyme activity in blood serum or plasma, enzyme activity in fecal material or saliva may also prove useful diagnostically. For example, if a pancreatic insufficiency is suspected, fecal proteolytic activity can be measured very simply by immersing a piece of unexposed X-ray film into a fecal suspension, under appropriate temperature and pH conditions. Failure to digest the gelatin on the film indicates possible pancreatic protease deficiency.[19] Saliva can be analyzed for amylase activity in rodents but not in dogs, cats, horses, cattle, and sheep.[20]

Clinical biochemistry alterations are indicative of organ dysfunction but do not discriminate between potential etiologies. Thus, while they may indicate the effect of the toxicant, serum enzyme measurements alone cannot provide information about whether the insult to the organ was a toxin and, if so, which one. However, when combined with changes in other biochemical constituents in the blood, a discrimination often can be made between parasitic, microbiological, or toxicological insults. By identifying which tissue was damaged by the exposure, a list of toxicants that are known to affect that organ can be generated. This then reduces the amount of analytical chemistry that needs to be done to identify the offending chemical. Clinical enzymology alone is most useful for diagnosing toxicant-induced liver or cardiac necrosis or pancreatic disorders because specific enzymes or isoenzymes for these organs can be measured in most species. Renal function is impaired by several toxicants without causing kidney necrosis. Heavy metals, such as mercuric chloride, and aniline dyes may be an exception.[21,22] However, there are no kidney-specific tissue enzymes in routine use. Renal dysfunction generally is diagnosed on the basis of impairment of metabolic processes and is discussed further in Section III.

A. Methods

Historically, enzyme activity instead of concentration has been measured for diagnostic purposes. It generally is assumed that activity is linearly related to concentration. More recently, immunoassays have been developed for a few enzymes whereby the concentration of the enzyme can be measured. Combining the two values of activity and concentration into an index of specific activity would be extremely valuable for determining whether a measured decreased activity was due to a change in the functional capacity of the enzyme or truly related to the amount of enzyme present.

Enzymes catalyze reactions by complexing with the substrate and lowering the energy needed for the reaction to occur. The reaction will occur without the presence of the enzyme but at a much slower rate. Most enzymes are not substrate specific but have markedly different affinities for various substrates, even closely related forms. Substrate affinity is expressed as the Michaelis constant (K_m) which is defined as the concentration of the substrate at one half the maximum velocity of the reaction.[3] It is determined by changing the substrate concentration while holding the enzyme concentration constant and measuring the change in activity. The choice of substrate, therefore, is of critical importance. Interspecies variation in substrate affinity is likely to occur and must be accounted for when adapting an enzyme assay to a different species. This is best exemplified by the difference in binding affinity of acetylcholinesterase to acetylthiocholine or butyrylthiocholine. Although a measure of enzyme activity can occur for either substrate, a significantly higher and more precise measurement can be obtained when using acetylthiocholine. The reverse is true when measuring butyrylcholinesterase activity. As species of wild animals differ in the ratios of butyrylcholinesterase:acetylcholinesterase in the serum, the most accurate measurement of cholinesterase activity is dependent on the correct choice of substrate. The rate at which the reaction occurs in the presence of the enzyme is also dependent on other factors such as pH and temperature and is reviewed in detail by Fairbrother et al.[23] and Kramer.[3]

In clinical enzymology, most biochemical reactions that convert substrate to product in the presence of catalytic enzymes are measured spectrophotometrically, either as simple one substrate-one product reactions or as a coupling of two or three substrate-product reactions. Measurements are made of the amount of time needed to produce a predetermined amount of product ("fixed endpoint" assay), the amount of product formed in a predetermined length of time ("fixed-time" assay), or the rate of the reaction (amount of substrate converted to product per unit time). For all assays, substrate is present in excess and physical parameters (pH, temperature, etc.) are established to maximize the reaction rate. This results in a zero-order kinetic reaction so that the rate is determined solely by the amount of enzyme present.

Coupled reactions are used when the equilibrium constant for the reaction catalyzed by the enzyme of concern is relatively low. A low K_{eq} means that the reaction reaches equilibrium before all of the substrate is converted to product and, frequently, that the product may then be converted back to substrate. This can be avoided by converting the product to an irreversible by-product as soon as it is formed (a process called "trapping"). It may be necessary to convert the by-product to yet another form through a third coupled reaction in order to produce a colored product that can be measured spectrophotometrically. Sometimes, a cofactor such as reduced nicotinamide adenine dinucleotide (NADH) or reduced NAD phosphate (NADPH) can be measured instead of formation of product. NADH, for example, is a colored product with a maximum absorption of 340 nm. Its rate of appearance or disappearance is stoichiometrically related to substrate consumption and, therefore, can be used as a surrogate measure of enzyme activity in oxidation-reduction reactions. It is extremely important that the substrates, cofactors, and secondary enzymes be present in excess so that the rate-limiting step is the amount of primary enzyme present. The following is an example of a coupled reaction utilizing measurement of the formation of a cofactor:

$$\text{Creatine phosphate} + \text{ADP} \xrightarrow{\text{CPK}} \text{creatine} + \text{ATP}$$

$$\text{ATP} + \text{glucose} \xrightarrow{\text{hexokinase}} \text{glucose-6-phosphate} + \text{ADP}$$

$$\text{Glucose-6-phosphate} + \text{NADP} \xrightarrow{\text{G-6-PD}} \text{6-phosphogluconate} + \text{NADPH} + \text{H}^+$$

Because of the widespread use of clinical enzymology in human medicine, commercial kits are available that contain all the necessary substrates and cofactors for generating a measurable reaction in the presence of the enzyme of interest. It is tempting to use these kits for any animal species of interest with any available spectrophotometer. However, it is extremely important to realize that this likely will result in misleading information. Differences in substrate specificity may exist among species, as do optimum temperature or pH (e.g., human mean body temperature is 37°C while avian body temperature generally averages 41°C). Therefore, each assay must be revalidated for each species. Additionally, not all spectrophotometers are equivalent (e.g., differences exist in cuvette type and size, light path length, thermal regulation, etc.) and commercial assay kits are generally produced with a specific system in mind. Once again, an assay kit must be revalidated for each piece of equipment used. Any other changes that are made in the procedure (e.g., reduction in sample size) also necessitate a reexamination of the entire assay system. A more detailed discussion of this process and concomitant quality assurance/quality control procedures is available in Fairbrother et al.[23]

B. Toxicology

Some enzymes are specifically induced by the presence of exogenous chemicals in response to the need to detoxify or eliminate the compounds. Liver cytochrome P-450 enzymes are induced by dioxins, polycyclic aromatic hydrocarbons, and other organochlorines and are discussed in detail in Chapter 1. Enzymes hydrolyzing organophosphorus compounds (such as A esterases, paraoxonase, and organophosphatase) are also specifically induced by a particular class of chemicals.[24] These enzymes, while present to some degree in the bloodstream, are primarily intracellular in nature (primarily in the liver) and have not been utilized routinely as diagnostic indicators of exposure. Inhibition of cholinesterase activity is also a function of exposure to organophosphorus pesticides and has proved to be a more useful noninvasive diagnostic tool (see Chapter 2 for a detailed discussion of blood esterases).

One enzyme that has been shown to be directly inhibited by exposure to a toxicant is δ-aminolevulinic acid dehydrase (δ-ALAD), one of the enzymes necessary for heme synthesis. This enzyme is very sensitive to the effects of increased blood lead levels. Erythrocyte δ-ALAD is inhibited by lead at levels as low as 10–15 μg/dL and is frequently used as a nondestructive screening procedure in suspected lead toxicosis for both birds and mammals.[2,25-30]

Toxicant-induced liver damage, such as cirrhosis resulting from exposure to carbon tetrachloride, has been shown to alter liver-specific enzymes in a predictable manner in rats.[31,32] In marmosets *(Callithrix jacchus),* lactate dehydrogenase (LDH), ALT, and aspartate aminotransferase (AST) could not differentiate between carbon tetrachloride-induced liver damage and chlorpromazine-induced muscle damage. Only by measuring different isozyme patterns of isocitrate dehydrogenase (ICDH) could these two be distinguished.[32] Other compounds cause acute liver necrosis and likely would show similar changes in serum levels of γ-GT, ALT, or ICDH. These include the following: azaserine, beryllium, bromobenzine, bromotrichloromethane, chloroform, dimethylnitrosamine, galactosamine, phosphorus, tannic acid, tetrachloroethane, thioacetamide, trichloroethylene, and urethane.[33,34] Natural toxins also cause liver necrosis. Aflatoxins are of particular concern due to the sensitivity of chickens and other birds to their toxicological effects. Pyrrolizidine alkaloids and the toxins from the mushrooms *Amanita phalloides* are also potent hepatotoxins.

Most organochlorine, organophosphorus, and carbamate compounds are classified as neurotoxins and effects on the systemic organs generally are considered negligible. However, several studies recently have documented chemical-induced changes in serum enzymes indicative of organ damage following exposure to sublethal concentrations of these compounds. Recent studies by Gupta et al.[15] have demonstrated measurable increases in total serum LDH activity and changes in the isozyme patterns in rats following a single exposure to carbofuran. Embryonic chickens inoculated *in ovo* with methyl parathion, phosmethylan, carbendazim, or 2,4-dichlorophenoxyacetic acid (2,4-D) experienced significant increases in plasma AST and LDH activities at day 19 of incubation.[35] Similarly, lindane (hexachlorocyclohexane), a widely used organochlorine insecticide, significantly increased serum and liver ALT and AST activity in rats fed contaminated feed for 2 weeks.[36] Methyl mercury, dichlorodiphenyldichloroethylene (DDE), and Aroclor 1254 increased LDH, creatine kinase, and AST activities in starlings *(Sturnus vulgaris)* and Japanese quail *(Coturnix coturnix japonica)* two- to fourfold while malathion increased starling LDH by 50%.[37,38] Plasma LDH was also elevated in kestrels *(Falco sparverius)* by exposure to paraquat, and alkaline phosphatase was lowered.[39] The administration of carbophenothion to Japanese quail, pigeons *(Columba livia);* chickens *(Gallus domesticus);* grayleg, pink-footed, and Greenland white-fronted geese *(Anser* spp.); and Canada geese *(Branta canadensis)* caused an increase in glutamate oxaloacetate transaminase in all species except pigeon and chickens, an increase in glutamate dehydrogenase in quail, and an increase in sorbitol dehydrogenase in the geese.[40,41] These enzymes also were variously increased in Japanese quail exposed to methiocarb, oxamyl, primicarb, thiofax, chlorfenvinphos, dimethoate, and primiphos ethyl.[41,42] Carp *(Cyprinus carpio)* exposed to methidathion had significantly increased LDH activity for the following 2 weeks while paraquat exposure resulted in a significant decrease of enzyme activity.[43] These examples illustrate that a broad panel of clinical biochemistry assays can detect previously unsuspected subtle effects of exposure to environmental contaminants.

III. PRODUCTS OF METABOLISM

End products or intermediates of metabolism also can be measured in blood and excreta. Changes in the amounts of these compounds may reflect malfunctions of various systems and frequently are not as organ specific as are tissue enzymes. For example, elevated blood urea nitrogen (BUN) values may indicate kidney malfunction (decreased excretion rate), an accelerated rate of protein catabolism (e.g., during starvation), or a high protein diet. Plasma bilirubin concentration is directly proportional to the amount produced due to heme turnover and inversely related to hepatic clearance. Bile pigments can be measured in urine as well as in blood. Other metabolites that are measured routinely in both blood and urine include creatinine (a measure of kidney glomerular function), glucose and ketones (indicative of pancreatic function and status of digestion of complex carbohydrates or proteins), and proteins (both alpha and gamma globulins as well as nitrogenous compounds). Fat content can be measured in fecal material as well as in blood serum through the measurement of degree of lipemia and by cholesterol and triglyceride concentrations, and has been used as an indicator of general body condition of wild animals.

Serum bilirubin is most useful for differentiating jaundice due to intrahepatic congestion such as occurs with toxic hepatitis or cirrhosis from extrahepatic obstructive jaundice due to bile duct obstructions by stones, neoplasms, or stricture. Hemolytic jaundice caused by acute or chronic hemolytic anemia also can be differentiated. This can be accomplished by measuring total bilirubin and direct (or conjugated) bilirubin, with indirect bilirubin calculated by difference. An increased level of indirect (or unconjugated) bilirubin is caused by prehepatic icterus due to excessive hemolysis. Elevated direct (conjugated) bilirubin is associated with intrahepatic congestion. Bile duct obstructions result in the highest levels of direct bilirubin with minor amounts of indirect bilirubin also present.[44] Birds, amphibians, reptiles, sloths, and nutrias lack sufficient biliverdin reductase activity to convert biliverdin to bilirubin.[44,45] Frequently, bilirubin values are reported for these species but they are relatively meaningless given the metabolic inability to convert significant amounts of biliverdin to bilirubin. No standard clinical methods exist for measurement of biliverdin, although spectrophotometric or high-performance liquid chromatography (HPLC) analyses are used in research laboratories.

Interspecies differences in metabolism are particularly important when attempting to diagnose renal dysfunction. In contrast to mammals, creatinine and blood urea nitrogen are of little diagnostic value for most birds. Birds convert little of their creatine to creatinine and what little is produced normally is reabsorbed by the renal tubule. Tubular necrosis could result in a measurable creatinine value due to reduced absorption; however, the precision of the assay at these low levels is questionable. Blood urea nitrogen is also present in very low concentrations in birds. Uric acid measurements are a more reliable indicator of kidney function in birds than are creatinine or blood urea nitrogen. See Fairbrother et al.[23] for a more in-depth review of differences in clinical biochemistry profiles between birds and mammals.

Minerals and electrolytes are necessary for maintaining homeostasis. Measurements of elements such as magnesium, phosphorus, calcium, sodium, potassium, and chloride can provide a means of assessing this balance. Ablation or immune-mediated atrophy of the adrenal cortex results in imbalanced mineral metabolism as well as decreased glucocorticoid secretion (Addison's disease), a relatively common disease in the domestic dog.[46] Other than the therapeutic agent, o,p-DDD (Lysodren™),* there are no known chemicals that directly reduce the size or functioning of the adrenal glands. Dioxin (2,3,7,8-tetrachlorodibenzo-p-dioxin [TCDD]) may have an indirect effect through its increase of glucocorticoid secretions and the resulting negative feedback onto the adrenals. In addition, electrolyte imbalances also result from excessive vomiting or diarrhea or other malfunctions of the digestive system (e.g., malabsorption). Therefore, measurements of these blood constituents help to provide an assessment of the overall health of the animal and clues as to adverse symptoms it might be experiencing, thereby contributing toward the ultimate diagnosis of the etiology of suspected disease processes.

A. Methods

Total bilirubin is measured by differential spectrophotometry at 454 nm. Direct bilirubin is determined by adding diazotized sulfanilic acid to the serum sample

---

* Registered trademark of Bristol-Meyers, New York, NY.

and measuring the absorbance of the resulting purple-colored azobilirubin at 540 nm. These assays can be conducted on a conventional spectrophotometer or on a dedicated bilirubinometer.[18] Measurement of urinary direct bilirubin follows the same principles but is generally conducted visually by mixing the sample with a commercially available diazo tablet or placing a drop on a dye-impregnated paper on the end of a stick (Dip-stick™).* Serum and urinary glucose are also measured colorimetrically (either with a spectrophotometer or visually) via a coupled reaction resulting in the reduction of nicotinamide adenine dinucleotide phosphate (NADP). If accurate measures of serum glucose are required, the serum must be separated from the red blood cells within 30 min to minimize leaching of glucose from the red blood cells into the serum and consequent elevation of the glucose levels.[18] Total plasma or serum proteins may be estimated rapidly through the use of a hand-held refractometer. For more accurate determinations, the biuret spectrophotometric method is preferable. In this method, copper ions are reacted with the protein in an alkaline solution to form a colored chelate. Serum protein fractions (alpha, beta, and gamma globulins) are determined by electrophoresis.[18] Triglycerides and blood urea nitrogen are also determined spectrophotometrically with coupled reactions resulting in the oxidation of NADH. In birds, reptiles, and other animals that produce uric acid as the major waste product of protein catabolism, the preferred method for determination of uric acid levels is the production of allantoin by uricase with a series of coupled reactions that produce a yellow-colored dye.[18]

Plasma electrolytes include sodium, potassium, calcium, magnesium, chloride, phosphate, and bicarbonate. Sodium, potassium, and calcium can be measured in a variety of ways, the most common being through the use of flame photometry although ion-specific electrodes are now receiving wider use.

Determinations performed by flame photometry tend to be about 7% lower than those obtained using ion-specific electrodes.[47] Flame photometry operates on the principle that, when heated in a flame, the sodium or potassium ion emits spectra with sharp lines at set wavelengths. Light emitted from the thermally excited ions is passed through appropriate filters and measured by a photodetector. The intensity of the light is proportional to the concentration of the ion in the sample. Calcium, magnesium, chloride, and phosphate generally are measured by coupled reactions producing a colored product whose optical density is measured spectrophotometrically and compared to a standard curve to determine the relationship between optical density and concentration. Bicarbonate is somewhat more difficult to measure since the carbon dioxide gas tends to escape from the sample. Either an enzyme-catalyzed reaction or a continuous flow titration apparatus is used to produce a colored product for spectrophotometric measurement. Details of these procedures can be found in Reference 48.

B. Toxicology

Kidney damage as a result of heavy metal exposure (e.g., mercuric chloride) or exposure to chlorinated anilines can be diagnosed through observations of

---

* Registered trademark of Miles, Inc., Elkhart, IN.

polyuria, glucosuria, proteinuria, and elevated blood urea nitrogen and creatinine in conjunction with increased serum enzymes such as ALT and AST. Elevated urine glutamic-oxaloacetic transaminase activity has also been measured in the rat poisoned with mercuric chloride.[49] These changes result from necrosis of the proximal tubular epithelium.[21,22,50]

Cadmium is another potent nephrotoxin while other metals such as chromium, arsenic, gold, lead, iron, antimony, uranium, and thallium cause lesser, albeit significant, renal damage[21] with similar clinical biochemical changes. Chloroform and other halogenated hydrocarbons and bromobenzine are hepatotoxic in most animals and also nephrotoxic in some species.[21] Hexachlorobutadiene and petroleum hydrocarbon vapors are nephrotoxic in rats, mice, and other mammals. Paraquat primarily affects the lungs, but in high doses also causes kidney damage which, in turn, reduces excretion and leads to greater lung damage.

Exposure of embryonated chicken eggs to high doses of methyl parathion or 2,4-D caused a significant decrease in serum glucose and cholesterol levels measured at day 19 of incubation. 2,4-D caused a decrease in total plasma protein while methyl parathion, phosmethylan, and carbendazim increased total plasma protein.[35] Toxaphene, an organochlorine insecticide, increased serum cholesterol levels of rats.[51] Carp exposed to copper sulfate or paraquat experienced increased plasma glucose levels.[43]

IV. HEMATOLOGY

The subdiscipline of clinical hematology is devoted to assessing the functioning of the blood itself, primarily the blood cells and their progenitors in the bone marrow. Hematologic parameters can provide evidence of pathology including anemias, dehydration, infectious processes, excessive parasitism, or toxicoses.

A large number of immature red cells in the circulating blood is an indication of loss of blood, an increased rate of destruction of mature erythrocytes, or an increased rate of release from the bone marrow. An increase in immature granulocytes or lymphocytes is indicative of a response to an infectious or inflammatory process. Because the packed cell volume is a function of both the number and size of the red blood cells, it can be used as an estimate of the presence of an anemia. Allowance must be made, however, if the animal is dehydrated because this reduces the plasma volume and artificially increases the packed cell volume percent. The size of red blood cells (mean corpuscular volume [MCV]) can be calculated by dividing the packed cell volume by the erythrocyte count. A high MCV (i.e., a smaller number of cells per unit volume) is caused by a larger than normal number of large, immature red cells and often is indicative of regenerative anemia such as caused by infectious agents, internal blood loss, autoimmune disease, or deficiencies of vitamin B_{12} or folic acid. The amount of hemoglobin per erythrocyte (mean corpuscular hemoglobin [MCH]) is calculated by dividing the amount of hemoglobin per milliliter by the number of red blood cells per milliliter. Hemolytic anemias and iron deficiencies lead to increased MCH. The mean corpuscular hemoglobin concentration (MCHC) is the ratio of the weight of the hemoglobin to the volume of the erythrocyte. Iron deficiency, immature erythrocytes (reticulocytosis), and methemoglobinemia will result in decreased MCHC values. Increased MCHC (hyperchromasia) cannot exist because normal erythrocytes already contain the maximum amount of hemoglobin. Normal hematologic values for domestic animals, caged birds, and some wildlife species may be found in Schalm et al.[52] and Kirk;[9] values for zoo animals may be found in Fowler;[7] values for wild waterfowl may be found in Driver,[53] Fairbrother and O'Loughlin,[54] and Perry et al.;[55] and values for several other species of birds may be found in Harrison and Harrison.[8]

A. Methods

The number of red blood cells and leukocytes per milliliter of blood can be determined through the use of automated particle counters (e.g., Coulter®* counters) for animals with nonnucleated red blood cells (most mammals) or by appropriate staining and counting utilizing a hemocytometer and a microscope. White blood cells may be further differentiated into subclasses (granulocytes, lymphocytes, monocytes, and thrombocytes) by microscopic examination of a thin-film smear on a glass slide followed by a Wright-Geimsa stain.[52] Blood smears also can be examined for the ratio of mature cells to immature cells. The amount of hemoglobin in the blood can be measured spectrophotometrically by converting the hemoglobin to oxyhemoglobin with ammonium hydroxide and measuring the resultant absorbance at 545 nm. The packed cell volume, or hematocrit, is the percent of the total blood volume comprised of red blood cells. It is readily measured through centrifugation of a blood sample at ≥ 2300 g for 15 min.

Hemostasis is a complex process by which bleeding is stopped. It depends on the sequential activation of a cascade of events involving production of fibrin and formation of a physical plug (thrombus) at the site of the lesion followed by activation of platelets and biochemical reactions of the clotting factors. Several quick screening tests are available to diagnose coagulopathies.[52,56] Clotting time can be measured by withdrawal of a blood sample by venipuncture and placing 1 mL of whole blood into a plastic or siliconized glass tube (or a smaller volume taken into a siliconized glass capillary tube). The tubes are held in a constant temperature warm water bath, and the time until the tube can be inverted without the blood flowing out is recorded as the clotting time. The one-stage prothrombin time measures the functioning of five of the clotting factors. Venous blood is mixed with sodium citrate, and the plasma is removed after centrifugation. Commercially available thromboplastin solution is forcibly blown into the tube of plasma, and the time required for visible clot formation is recorded. It is important to note that the commercial solution is made from a crude extract of rabbit brain and may not be suitable for all animal species. The activated partial thromboplastin time is an indicator of the intrinsic clotting cascade. It is useful in human or veterinary medicine, but may have lesser application as a biomarker because it has wide interspecific variability; and it is dependent on the source of partial thromboplastin, with the exact method used and with the type of instrument used. The

---

* Registered mark of Coulter Electronics, Inc., Hialeah, FL.

thrombin clotting time is a measure of the rate of conversion of fibrinogen to fibrin, the first step in hemostasis. This is simply conducted by mixing equal volumes of plasma with commercially available thrombin diluted with phenobarbital buffer in sodium chloride and calcium chloride. These tests can be followed by specialized assays designed to detect defects in any of the clotting factors or platelet functions.[52,56] Care should be used in collecting blood samples from animals with suspected coagulopathies.

B. Toxicology

Toxic chemicals can alter the functioning of the hematologic system through interference with cell production in the bone marrow or with heme synthesis, by direct cytotoxicity to the circulating cells, or by lesions in other tissues resulting in loss of blood cells through sequestration in organs or external to the body. Other chapters in this book cover in detail the development of porphyrins in blood and excreta due to impacts on heme synthesis (Chapter 4) and direct cytological effects (Chapters 6 and 8). Here, we will investigate what is known about the effect of toxins on cell production in the bone marrow and on functioning of the red blood cells as oxygen carriers.

Respiratory exposure to carbon monoxide results in carboxyhemoglobin formation in preference to oxyhemoglobin formation. This is a fully reversible reaction as soon as exposure ceases. Carboxyhemoglobin can be detected spectrophotometrically because it is a bright cherry red in color. Environmental exposure to carbon monoxide arises through inhalation of cigarette smoke or combustion products of fossil fuels (e.g., automobile exhaust). Methemoglobin, in contrast, is formed when hemoglobin is oxidized and is a dark greenish-brown to black in color. Sodium nitrite and hydroxylamine hydrochloride, aminophenols, amyl nitrite and other aliphatic esters of nitrous acid, phenylhydroxylamine, aniline, or nitrobenzene also cause methemoglobin formation *in vivo*. Methemoglobin formation can be measured in blood or in urine.

Heinz bodies are visible on microscopic examination of thin-film blood smears. They are dark-staining, dense, refractile granules on the interior surface of the red cell membrane. They consist of denatured hemoglobin and may sufficiently alter the cell size and structure so that the cell is prematurely phagocytosed by the spleen, or cause impairment of ion transport resulting in intravascular hemolysis of the cell. Heinz bodies may be formed due to congenital malformations or through prolonged exposure to many of the same oxidant chemicals that cause methemoglobin formation. Indeed, Heinz body formation frequently follows methemoglobinemia. In addition to the chemicals listed above, Heinz body formation has been shown to occur following exposure to phenols, propylene glycol, ascorbic acid, sulfite dichromate, arsine, and stibine. Heinz body hemolytic anemia has been shown to be the primary toxic effect of ingestion of crude oil by seabirds,[57] and dimethyl disulfide produces Heinz bodies in chickens.[58] Hemolytic anemia due to Heinz body formation in cattle, sheep, horses, cats, and dogs can also be caused by ingestion of naturally occurring compounds in onions and similarly by ingestion of *Brassica* species by ruminants.[59]

Coagulopathies generally are caused by exposure to anticoagulant rodenticides, either directly or by ingestion of rodents containing residual poison in their gastrointestinal tracts. Warfarin, 3-(α-acetonylbenzyl)-4-hydroxycoumarin, is a vitamin K antagonist and interferes with the synthesis of several clotting factors. Brodifacoum, bromadiolone, coumachlor, and diphencoumarin are second generation anticoagulants that have a similar mode of action to warfarin but reduced genetic resistance by rodents. The indanediones (including diphacinone, chlorophacinone, and pindone) are more water soluble and have a similar mode of action.[60] Exposure to all compounds in all three groups can be diagnosed by an extended clotting time and confirmed or differentiated from hereditary coagulopathies with more specific assays for individual clotting factors or through successful vitamin K therapy.

V. CLINICAL ENDOCRINOLOGY

Following the development of the radioimmunoassay (RIA) and the more recent technique of enzyme-linked immunosorbent assays (ELISA), measurement of serum hormone concentrations was greatly simplified and became applicable to a clinical laboratory situation. Prior to the development of these assays, hormone measurements were performed utilizing high-performance liquid chromatography (HPLC) methods. It is now possible to measure hormones routinely in blood, feces, or urine samples. Of particular importance in environmental toxicology and animal health programs are the steroid hormones (including both the corticosteroids and reproductive hormones) and thyroid hormones.

Measurement of hormones in urine and fecal samples was pioneered in zoos about a decade ago in order to follow the estrus cycle of animals in response to the need for higher success rates in captive breeding programs. This method has two advantages over measurements of blood values: (1) there is no need to handle the animal in order to collect a sample (truly "noninvasive") and (2) the hormone levels reflect an average value over the period of time between voidance of the sample (approximately daily for fecal samples from most animals). In order to relate the concentration of a hormone or its metabolite in urine to the concentration of the parent hormone in the blood, indexing of hormone levels with creatinine measurements should be done since creatinine is cleared from the blood at a uniform rate.[61] The technique has successfully been applied to cheetahs,[62] monomorphic birds,[63] blackfooted ferrets,[64] and feral horses[65] in captive situations but has not yet been applied to free-ranging animals.

When interpreting the results of hormone assays, it is of the utmost importance to recognize that hormones fluctuate on a circadian and/or circannual basis, have sex and age biases, and are influenced by stress of handling and blood collection (particularly corticosteroid levels). Therefore, concurrent samples from control populations of animals should be taken, if possible, or baseline values for the species of interest established through laboratory studies.

A. Methods

The underlying principles of the RIA and ELISA tests are the same. An antibody is produced to the substance to be measured and is labeled either with a radioisotope (RIA — usually ^{125}I) or an enzyme capable of producing a colored compound (ELISA — usually horseradish peroxidase). Unlabeled antibody is then bound to the inside of a plastic tube or onto a 96-well microtitration plate. The sample is added and allowed to bind for a specified time period after which the tube or plate is washed several times to remove the unbound sample. Labeled antibody is then added and allowed to bind. Following additional washing steps, the amount of radiation or intensity of the color reaction is measured and compared against a standard curve to determine the amount of antigen present in the test sample. The RIA and ELISA methods have equal sensitivity, but the ELISA has the advantage of not requiring radiation safety and control procedures. Many RIA and ELISA kits have been developed and are available commercially. As discussed above for enzyme assays, each kit needs to be validated for every new species to be certain that it is measuring within the correct range (e.g., normal values for the mallard *[Anas platyrhynchos]* thyroxine are lower than for the dog or cat [unpublished data]) and that the antigenic structure of the hormone molecule is conserved across species sufficiently so that the antibody has the same relative binding affinity (e.g., corticosterone is well-conserved across species as are the thyroid hormones, but prolactin is not).

B. Toxicology

Polychlorinated biphenyl (PCB) exposure causes thymic involution in mice[66] and elevated corticosteroid levels in rats and mice.[67,68] Other hormones modulated by acute PCB exposure include prolactin and the sex steroids.[68-70] Polybrominated biphenyls (PBBs) — another group of halogenated aromatic hydrocarbons — decrease serum corticosteroid levels following chronic (5–7 months) exposure,[71] decrease thyroxine (T4) and triiodothyronine (T3) levels, and increase serum cholesterol and γ-glutamyl transpeptidase.[72,73] Both PCBs and PBBs decrease serum glucose as part of the wasting syndrome that likely is modulated through chronically elevated glucocorticoid levels. Compounds that increase plasma cortisol levels include benzyl alcohol, ethanol, and ether.[74] Ingestion of polycyclic aromatic hydrocarbons such as those found in petroleum crude oil, on the other hand, severely diminishes levels of plasma corticosterone in experimentally dosed mallards which may contribute to reproductive failures in oil-exposed marine birds.[75]

Many chemicals, including PCBs and other halogenated industrial chemicals, carbon tetrachloride, and pesticides such as dichlorodiphenyltrichloroethane (DDT), endosulfan, malathion, and carbaryl have been shown to alter the T3:T4 ratio by direct action on the thyroid gland, through interference with the transport protein, or by alterations in liver function inhibiting the conversion of T3 to T4.[75-78] Some chemicals, such as polybrominated diphenyl ethers, bear structural similarity to the thyroid hormones and may cause a decrease in normal production of these hormones through a negative feedback mechanism. Studies of herring gulls from colonies on the Great Lakes that were suffering reproductive failures revealed thyroid follicular hypoplasia and decreased T3, T4, and free thyroid hormone concentrations (reviewed by Rattner et al.[75]).

It has been known for some time that the chlorinated hydrocarbon o,p'-DDT produces an estrogenic response in both mammals and birds, including feminization of embryos in ovo.[75,79,80] Reijnders[81] demonstrated that the harbor seal *(Phoca vitulina)* fed fish contaminated with PCBs and p,p'-DDE had significantly lowered estradiol concentrations resulting in reduced implantation of embryos and the observed reduction in reproductive rate of seal populations in the Dutch Waddenzee. More recently, it has been demonstrated that the chlorinated hydrocarbons p,p'-DDE, o,p'-DDE, chlordane, and Aroclor 1242 also interact with progesterone receptors of domestic fowl, ducks, and rabbits.[82] These compounds as well as p,p'-DDT, (p,p'-DDD), toxaphene, dieldrin, and mercury have now been shown to interfere with the binding of progesterone, dexamethasone, and testosterone to their receptors in the cytoplasmic fraction of the shell gland mucosa in domestic fowl and ducks.[83] While it has not been demonstrated how these effects are reflected in serum concentrations of the various hormones, it is now evident that persistent chlorinated hydrocarbons do have physiological effects on the reproductive system beyond eggshell thinning.

The organophosphorus pesticides also have been shown to affect reproductive hormones. Subacute exposure of nesting mallard *(Anas platyrhynchos)* hens to methyl parathion caused them to abandon their nests.[84] This behavioral alteration may have been a result of a sharp decrease in prolactin levels caused by the chemical exposure.[85] Additionally, it is likely that chemically induced tumors of the reproductive organs (e.g., ovarian tumors, endotheliomas, testicular tumors, etc.) cause measurable imbalances in the reproductive hormones.

The imbalances in hormonal regulation of the reproductive system, including corticosterone and thyroid hormones, results in reduced reproductive output of the individual and, potentially, a reduction in annual recruitment to the population as a whole. While hormone analyses from free-ranging animals are rare, several elegant laboratory studies referred to above have provided evidence that endocrine dysfunction is a consequence of exposure to environmental contaminants and a likely arbiter of observed reductions in reproduction. Thus, it is apparent that inclusion of hormone assays in a panel of nondestructive biomarkers is becoming more imperative and with the advent of radioimmunoassays and ELISA kits, more tractable.

VI. DIAGNOSTIC IMMUNOLOGY

Immunotoxicology is a newly emergent subdiscipline of ecotoxicology and is defined as the study of the effects of poisons on the immune system in humans or animals. During the past 15 years, a broad spectrum of xenobiotics has been shown to alter immune function (see reviews in References 86–88). Additionally, the close linkage between the endocrine and immune systems necessitates evaluation of both systems following exposure to such compounds as the halogenated aromatic hydrocarbons. Luster et al.[89] have developed a tiered approach for testing the immunotoxicity of compounds to white mice in a laboratory setting. This panel of assays has been conducted by several laboratories in an effort to validate its use. Although significant interlaboratory differences exist and the assays are still not standardized, it was found that the splenic plaque-forming cell assay was uniformly the most sensitive method for detection of immunotoxic effects. This assay measures the number of active antibody producing cells in the spleen. Unfortunately, it is an invasive (and generally fatal) procedure and not recommended for clinical applications.

The immune system is a complex interaction of many different cell types (e.g., macrophages, granulocytes, lymphocytes) and soluble circulating proteins (e.g., gamma globulins, lymphokines) that requires the interaction of the endocrine system (e.g., prostaglandins) for proper functioning. Different parts of the system are activated in response to bacteria, viruses, parasites, neoplastic cells, or other protein antigens. Environmental chemicals interact with various parts of this complex system resulting in either suppression or enhancement (hypersensitivity) of immune activity and surveillance. Due to feedback loops and built-in redundancies in the system, an alteration in one part may not result in an overall decrease in immune status at the organismal level. Therefore, a proper evaluation of the immune system must include methods for testing each of the various components. In laboratory studies, a definitive assessment of immune function requires a challenge of the animal with one or more antigenic stimulants such as a pathogenic bacteria or virus.[89] Results of an immunotoxicology test panel can be confounded by other immunosuppressive physiological functions, such as malnutrition or prolonged elevation of corticosteriods in response to a chronic stress. As with most of the other clinical biochemistry subcategories, the immune system should not be evaluated in isolation; instead it should be included in a panel of biomarkers designed to systematically evaluate the overall health and potential exposure of animals in relation to environmental toxicants.

A. Methods

The generally recognized panel of clinical immune function assays includes the complete and differential white blood cell count described previously, electrophoretic measurements of serum immunoglobulins, macrophage function, and lymphocyte blastogenesis. In my laboratory, we have had success with the natural killer cell assay in many diverse bird species and have added a hemagglutination assay to our standard panel of tests (either as a primary response to injected sheep red blood cells or a passive production of antibodies to rabbit red blood cells). This group of assays will provide information about all three legs of the immune system: humoral immunity, cellular immunity, and nonspecific immunity.

Total leukocyte counts and macrophage functions have received the widest application in ecotoxicology and have been used to assess immune competence of fish,[90-96] earthworms,[97] chickens,[98] bobwhite quail *(Colinus virginianus)*,[99] mallards[100,101] and deer mice[102] as well as a plethora of studies with laboratory rats and mice. Macrophage phagocytosis studies can be conducted in a variety of ways. If the cells to be assayed are harvested from a blood sample (as opposed to peritoneal or lung lavage), they should be allowed to adhere to a surface and mature for 24–48 hr prior to conducting the assay. Phagocytic ability can then be measured by conjugating heat-killed yeast cells, bacteria, or microscopic plastic beads to a fluorochrome; opsonizing the particles with appropriate serum, incubating them with the macrophages up to 1 hr; and examining the washed preparation either by using a microscope with ultraviolet light[101] or in a flow cytometer in order to quantify the number of cells containing particles and the number of particles per cell, or by lysing the cells and examining the supernatant spectrophotometrically.

The hemagglutination assay is relatively simple and can be conducted to measure "resting" antibodies such as occur in many bird species against rabbit red blood cells or, as is more commonly done, a primary or secondary response to a prior challenge with sheep red blood cells. Test serum is added to wells in a 96-well microtitration plate containing either sheep or rabbit red blood cells. Following incubation either at room temperature or at 37°C, the wells are examined for agglutination. A positive response is indicated by diffuse red blood cells across the entire well bottom whereas a negative response will form a "button" of cells at the bottom of the well. Classes of antibodies can be differentiated by the addition of 2-mercaptoethanol which selectively inhibits IgM activity. IgM activity can then be calculated as the difference between total antibody titer and the 2-mercaptoethanol-inhibited (IgG) titer. Lymphocyte blastogenesis and natural killer cell assays measure the amount of uptake of tritiated thymidine by the dividing cells exposed to specific mitogens such as concanavalin A (blastogenesis) or the amount of lysis of the target cells following uptake of a radioisotope (natural killer cell utilizing ^{51}Cr). Detailed methodologies can be found in Fowles et al.[101] and Baecher-Steppan et al.[103] for birds and in Luster et al.[89] for mammals.

B. Toxicology

There now exists a large body of data on immunotoxic properties of chemicals as measured in rabbits, rodents, chickens, mallard, and various fish under laboratory conditions. These studies include environmental contaminants such as heavy metals, benzene, pentachlorophenol, nonpesticide organochlorines, polycyclic aromatic hydrocarbons, and dioxins as well as all classes of pesticides. These laboratory studies have been reviewed by Dean et al.,[86] Exon et al.,[104] Fairbrother,[105] Sharma,[87] and Vos.[88] To date, there have been no published reports of immunotoxicological tests on free-ranging wildlife, although several are in progress. Two studies utilized mallards maintained in seminatural conditions of outdoor enclosures to study the effect of lead shot on antibody production to sheep red blood cells[106] and the effect of aqueous selenium in streams on antibody production, phagocytosis, white cell counts, and juvenile survival.[107] Additional immunotoxicity studies have been conducted with penned mallards.[100,108] Wild-caught cotton rats *(Sigmodon hispidus)* were exposed to benzene intraperitoneally, with lymphocyte proliferation, splenic plaque-forming cell assays, and cutaneous delayed-type hypersensitivity tests measured as endpoints.[109]

A few field studies have been done with fish that substantiate the immune suppression observed in laboratory tests. Spot *(Leiostomus xanthurus)* collected from the Chesapeake Bay have been shown to have reduced phagocytic, chemi-luminescent, and chemotactic responses of kidney macrophages.[93,110] Wild-caught perch *(Perca fluviatitus)* from a cadmium-contaminated river exhibited increased numbers of lymphocytes.[111] Earthworms *(Lumbricus terrestris* and *Eisenia foetida)* are frequently used in bioassays of hazardous waste site material. In addition to standard morbidity and mortality measures, the effect of chemical contaminants on the earthworm immune system can be measured through observation of rosette formation by coelomocytes (functionally analogous to mammalian antibody) and through measurement of phagocytic activity of the coelomocytes of laboratory-exposed worms or those retrieved from contaminated soils.[97]

VII. SUMMARY

An assessment of the health status of free-living fish and wildlife can be achieved through the use of nondestructive biomarkers. A single blood sample can be analyzed for characteristics and functioning of both the red and white cells and for levels of enzyme activity, products of metabolic pathways, hormones, and antibodies. The appropriate combination of these assays for predictive or diagnostic purposes depends on several factors, including the species of concern, suspected exposures and/or disease state, size of sample, and availability of appropriate expertise and equipment. In general, a screening panel should include at least one indicator of the functioning of the major organs and physiological processes and should be limited primarily by the amount of available blood or serum. A thin-film blood smear requires a single drop of blood and should always be done. With appropriate stains, a blood smear can provide general information about whether the animal is anemic, whether it has recently been exposed to an infectious or parasitic agent, whether it is experiencing a nonspecific stress, and whether certain lymphocyte subpopulations (e.g., T helper cells, cytotoxic T cells) have been induced or suppressed. Hematocrits and a refractometer analysis of total proteins should also be run routinely on all samples because they are simple to do; require small volumes of blood; and provide information about hydration, anemia, nutri-tional state, and infection. At least one enzyme specific for liver function should be included in a general screening panel (this will vary by species but examples include γ-GT, AST, and sorbitol dehydrogenase [SDH]) as should kidney function analytes such as BUN, or uric acid, and creatinine. Enzymes such as CPK or LDH should receive next priority to provide information about muscle damage (e.g., cardiac infarcts). Cholinesterase activity should be given a high priority only if pesticide exposure is suspected, but should be included on all panels if sufficient serum quantity is available. Triglyceride or lipoprotein analysis should be consid-ered for inclusion next to provide information about the nutritional state of the animal; glucose should not be analyzed unless the serum can be separated from the red blood cells within 10 min of collection. The assays listed here constitute a general screening panel and can be done with 1 mL of blood, given the proper equipment.

With the information provided by the general screening panel — combined with knowledge of the physiological effects of toxicants, pathogens, and parasites — it is possible to make a short list of potential causes of disease and then to take a rational course of differential diagnosis that eventually leads to a determination of the etiology of the condition.

Sometimes this can be done quickly, such as when the activity of a toxicant-specific enzyme is altered. This is exemplified by the inhibition in δ-ALAD activity induced by exposure to lead. Other times, a nonspecific response may be observed — such as an increase in liver enzyme activity (e.g., γ-GT) — and additional tests or analytical chemistry must be done to confirm a diagnosis. However, clinical biochemistry provides an extremely useful, relatively inexpensive approach for monitoring and assessing the health of any free-living species from soil invertebrates such as earthworms to fish, mammals, and birds. Although applications of clinical biochemistry technology to free-ranging wild animals have been limited, the laboratory and field studies reviewed in this chapter provide the knowledge base for utilizing these diagnostic tools in field-oriented research efforts.

VIII. LIST OF ABBREVIATIONS

γ-GT	γ-glutamyl transferase
δ-ALAD	δ-aminolevulinic acid dehydratase
ALT	alanine aminotransferase
AST	aspartate aminotransferase
BUN	blood urea nitrogen
CPK	creatine phosphokinase
2,4-D	2,4dichlorophenoxyacetic acid
DDE	dichlorodiphenyldichloroethylene
DDT	dichlorodiphenyltrichloroethane
ELISA	enzyme-linked immunosorbant assay
HPLC	high-performance liquid chromatography
ICDH	isocitrate dehydrogenase
LDH	lactic dehydrogenase
MCH	mean corpuscular hemoglobin
MCHC	mean corpuscular hemoglobin concentration
MCV	mean corpuscular volume
PBB	polybrominated biphenyl
PCB	polychlorinated biphenyl
RIA	radioimmunoassay
SDH	sorbitol dehydrogenase

T3   triiodothyronine
T4   thyroxine
TCDD   2,3,7,8-tetrachlorodibenzo-p-dioxin

REFERENCES

1. Buttner, J., Ed., *History of Clinical Chemistry,* Walter de Gruyter, Berlin, 1983, p. 83.
2. Kaneko, J. J., Ed., *Clinical Biochemistry of Domestic Animals,* 4th ed., Academic Press, San Diego, 1989, p. 932.
3. Kramer, J. W., Clinical enzymology, in *Clinical Biochemistry of Domestic Animals,* 4th ed., Kaneko, J. J., Ed., Academic Press, San Diego, 1989, p. 338.
4. Kaldor, G., Clinical enzymology, *Methods in Laboratory Medicine,* Vol. 3, Praeger Publishers, New York, 1983, p. 222.
5. Clampitt, R. B. and Hart, R. J., The tissue activities of some diagnostic enzymes in ten mammalian species, *J. Comp. Pathol.,* 88, 607, 1978.
6. Fairbrother, A., Craig, M. A., Walker, K., and O'Loughlin, D., Changes in mallard *(Anas platyrhynchos)* serum chemistry due to age, sex, and reproductive condition, *J. Wildl. Dis.,* 26, 67, 1990.
7. Fowler, M., *Zoo and Wild Animal Medicine,* 2nd ed., W.B. Saunders, Philadelphia, 1986, p. 1127.
8. Harrison, G. J. and Harrison, L. R., Eds., *Clinical Avian Medicine and Surgery,* W.B. Saunders, Philadelphia, 1986, p. 717.
9. Kirk, R. W., Ed., *Current Veterinary Therapy: Small Animal Practice,* Vols. 1–10, W.B. Saunders, Philadelphia, 1991.
10. Lewandowski, A. H., Campbell, T. W., and Harrison, G. J., Clinical chemistry, in *Clinical Avian Medicine and Surgery,* Harrison, G. J. and Harrison, L. R., Eds., W.B. Saunders, Philadelphia, 1986, p. 192.
11. Boyd, J. W., The mechanisms relating to increases in plasma enzymes and isoenzymes in diseases of animals, *Vet. Clin. Pathol.,* 12, 9, 1983.
12. Bruns, D. E., Emerson, J. C., Inteman, S., Bertholf, R., Hill, K. E., and Savory, J., Lactate dehydrogenase isoenzyme-1: changes during the first day after acute myocardial infarction, *Clin. Chem.,* 27, 1821, 1981.
13. Cornish, H., Barth, M. L., and Dodson, V. N., Isozyme profiles and protein patterns in specific organ damage, *Toxicol. Appl. Pharmacol.,* 16, 411, 1970.
14. Appella, E. and Markert, C. L., Dissociation of lactate dehydrogenase into subunits with guanidine hydrochloride, *Biochem. Biophys. Res. Commun.,* 6, 171, 1961.
15. Gupta, R. C., Goad, J. T., and Kadel, W. L., *In vivo* alterations in lactate dehydrogenase (LDH) and LDH isoenzymes patterns by acute carbofuran intoxication, *Arch. Environ. Contam. Toxicol.,* 21, 263, 1991.
16. Markert, C. L. and Faulhaber, I., Lactate dehydrogenase isozyme patterns of fish, *J. Exp. Zool.,* 159, 319, 1965.
17. Shaklee, J. B., Kepes, K. L., and Whitt, G. S., Specialized lactate dehydrogenase isozymes: the molecular and genetic basis for the unique eye and liver LDHs of teleost fishes, *J. Exp. Zool.,* 185, 217, 1973.
18. Suber, R. L., Clinical pathology for toxicologists, in *Principles and Methods of Toxicology,* 2nd ed., Hayes, A. W., Ed., Raven Press, New York, 1989, p. 485.
19. Brobst, D. F., Pancreatic function, in *Clinical Biochemistry of Domestic Animals,* 4th ed., Kaneko, J. J., Ed., Academic Press, San Diego, 1989, p. 398.

20. Tennant, B. C. and Hornbuckle, W. E., Gastrointestinal function, in *Clinical Biochemistry of Domestic Animals,* 4th ed., Kaneko, J. J., Ed., Academic Press, San Diego, 1989, p. 417.

21. Hewitt, W. R., Goldstein, R. S., and Hook, J. B., Toxic responses of the kidney, in *Casarett and Doull's Toxicology, The Basic Science of Poisons,* 4th ed., Amdur, M. O., Doull, J. and Klaassen, C. D., Eds., Pergamon Press, New York, 1991, p. 354.

22. Lo, H., Brown, P. I., and Rankin, G. O., Acute nephrotoxicity induced by isomeric dichloroanilines in Fischer 344 rats, *Toxicology,* 63, 215, 1990.

23. Fairbrother, A., Marden, B. T., Bennett, J. K., and Hooper, M. J., Methods used in determination of cholinesterase activity, in *Cholinesterase-inhibiting Insecticides,* Mineau, P., Ed., Elsevier Science, Amsterdam, 1991, p. 35.

24. Reiner, E., Aldridge, W. N., and Hoskin, F. C. G., *Enzymes Hydrolysing Organophosphorus Compounds,* Ellis Horwood, Chichester, England, 1989, p. 266.

25. Hoffman, D. J., Franson, J. C., Pattee, O. H., Bunck, C. M., and Murray, H. C., Biochemical and hematological effects of lead ingestion in nestling American kestrels *(Falco sparverius), Comp. Biochem. Physiol.,* 80C, 431, 1985.

26. Pain, D. J., Haematological parameters as predictors of blood lead and indicators of lead poisoning in the black duck *(Anas rubripes), Environ. Pollut.,* 60, 67, 1989.

27. Redig, P. T., Lawler, E. M., Schwartz, S., Dunnett, J. L., Stephenson, B., and Ducke, G. E., Effects of chronic exposure to sublethal concentrations of lead acetate on heme synthesis and immune function in redtailed hawks, *Arch. Environ. Contam. Toxicol.,* 21, 72, 1991.

28. Scheuhammer, A. M., Erythrocyte δ-aminolevulinic acid dehydratase in birds, II. The effects of lead exposure in vivo, *Toxicology,* 45, 165, 1987.

29. Scheuhammer, A. M., Monitoring wild bird populations for lead exposure, *J. Wildl. Manage.,* 53, 759, 1989.

30. Scheuhammer, A. M. and Wilson, L. K., Effects of lead and pesticides on δ-aminolevulinic acid dehydratase of ring doves *(Streptopelia risoria), Environ. Toxicol. Chem.,* 9, 1379, 1990.

31. Blair, P. C., Thompson, M. B., Wilson, R. E., Esber, H. H., and Maronpot, R. R., Correlation of changes in serum analytes and hepatic histopathology in rats exposed to carbon tetrachloride, *Toxicol. Lett.,* 55, 149, 1991.

32. Davy, C. W., Fulleylove, M., Edmunds, J. G., Eichler, D. A., Rushton, B., Tudor, R. J., and Walker, J. M., The diagnostic usefulness of isocitrate dehydrogenase (ICDH) in the marmoset *(Callithrix jacchus), J. Appl. Toxicol.,* 9, 33, 1989.

33. Plaa, G. L., Toxic responses of the liver, in *Casarett and Doull's Toxicology, The Basic Science of Poisons,* 4th ed., Amdur, M. O., Doull, J., and Klaassen, C. D., Eds., Pergamon Press, New York, 1991, p. 334.

34. Zimmerman, H. J., *Hepatotoxicity: The Adverse Effects of Drugs and Other Chemicals on the Liver,* Appleton-Century-Crofts, New York, 1978, p. 595.

35. Somlyay, I. M. and Varnagy, L. E., Changes in blood plasma biochemistry of chicken embryos exposed to various pesticide formulations, *Acta Vet. Hung.,* 37, 179, 1989.

36. Srinivasan, K. and Radhakrishnamurty, R., Biochemical changes produced by β- and γ-hexachlorocyclohexane isomers in albino rats, *J. Environ. Sci. Health,* B23, 367, 1988.

37. Dieter, M. P., Plasma enzyme activities in coturnix quail fed graded doses of DDE, polychlorinated biphenyl, malathion, and mercuric chloride, *Toxicol. Appl. Pharmacol.,* 27, 86, 1974.

38. Dieter, M. P., Further studies on the use of enzyme profiles to monitor residue accumulation in wildlife: plasma enzymes in starlings fed graded concentrations of morsodren, DDE, Aroclor 1254, and malathion, *Arch. Environ. Contam. Toxicol.,* 3, 142, 1975.
39. Hoffman, D. J., Franson, J. C., Pattee, O. H., Bunck, C. M., and Murray, H. C., Toxicity of paraquat in nestling birds: effects on plasma and tissue biochemistry in American kestrels, *Arch. Environ. Contam. Toxicol.,* 16, 177, 1987.
40. Westlake, G. E., Bunyan, P. J., and Stanley, P. I., Variation in the response of plasma enzyme activities in avian species dosed with carbophenothion, *Ecotoxicol. Environ. Saf.,* 2, 151, 1978.
41. Westlake, G. E., Bunyan, P. J., Martin, A. D., Stanley, P. I., and Steed, L. C., Organophosphate poisoning: effects of selected organophosphate pesticides on plasma enzymes and brain esterases of Japanese quail *(Coturnix coturnix japonica), J. Agric. Food Chem.,* 29, 772, 1981.
42. Westlake, G. E., Bunyan, P. J., Martin, A. D., Stanley, P. I., and Steed, L. C., Carbamate poisoning: effects of selected carbamate pesticides on plasma enzymes and brain esterases of Japanese quail *(Coturnix coturnix japonica), J. Agric. Food Chem.,* 29, 779, 1981.
43. Asztalos, B., Nemcsok, J., Benedeczky, I., Gabriel, R., Szabo, A., and Refaie, O. J., The effects of pesticides on some biochemical parameters of carp *(Cyprinus carpio* L.), *Arch. Environ. Contam. Toxicol.,* 19, 275, 1990.
44. Cornelius, C. E., Liver function, in *Clinical Biochemistry of Domestic Animals,* 4th ed., Kaneko, J. J., Ed., Academic Press, San Diego, 1989, p. 364.
45. Hill, K. J., Physiology of the digestive tract, in *Physiology and Biochemistry of the Domestic Fowl,* Vol. 4., Freeman, B. M., Ed., Academic Press, London, 1983, p. 31.
46. Rijnbert, A. and Mol, J. A., Adrenocortical function, in *Clinical Biochemistry of Domestic Animals,* 4th ed., Kaneko, J. J., Ed., Academic Press, San Diego, 1989, p. 610.
47. Carlson, G. P., Fluid, electrolyte, and acidbase balance, in *Clinical Biochemistry of Domestic Animals,* 4th ed., Kaneko, J. J., Ed., Academic Press, San Diego, 1989, p. 543.
48. Tietz, N. W., *Textbook of Clinical Chemistry,* W.B. Saunders, 1986, p. 1919.
49. Prescott, L. F. and Ansari, S., The effects of repeated administration of mercuric chloride on exfoliation of renal tubular cells and urinary glutamicoxaloacetic transaminase activity in the rat, *Toxicol. Appl. Pharmacol.,* 14, 97, 1969.
50. Davies, D. J. and Kennedy, A., Course of the renal excretion of cells after necrosis of the proximal convoluted tubule by mercuric chloride, *Toxicol. Appl. Pharmacol.,* 10, 62, 1967.
51. Chu, I., Secours, V., Villeneuve, D. C., Valli, V. E., Nakamura, A., Colin, D., Clegg, D. J., and Arnold, E. P., Reproduction study of toxaphene in the rat, *J. Environ. Sci. Health.,* B23, 101, 1988.
52. Schalm, O. W., Jain, J. C., and Carroll, E. J., *Veterinary Hematology,* Lea & Febiger, Philadelphia, 1975, p. 806.
53. Driver, E. A., Hematological and blood chemical values of mallard, *Anas platyrhynchos,* drakes before, during and after remige moult, *J. Wildl. Dis.,* 17, 413, 1981.
54. Fairbrother, A. and O'Loughlin, D., Differential white blood cell values of the mallard *(Anas platyrhynchos)* across different ages and reproductive states, *J. Wildl. Dis.,* 26, 78, 1990.

55. Perry, M. C., Obrecht, H. H., Williams, B. K., and Kuenzel, W. J., Blood chemistry and hematocrit of captive and wild canvasbacks, *J. Wildl. Manage.*, 50, 435, 1986.
56. Dodds, W. J., Hemostasis, in *Clinical Biochemistry of Domestic Animals,* 4th ed., Kaneko, J. J., Ed., Academic Press, San Diego, 1989, p. 274.
57. Leighton, F. A., Peakall, D. B., and Butler, R. G., Heinz body hemolytic anemia from the ingestion of crude oil, a primary toxic effect in marine birds, *Science,* 220, 871, 1983.
58. Maxwell, M. H., Production of a heinz body anaemia in the domestic fowl after ingestion of dimethyl disulfide: a haematological and ultrastructural study, *Res. Vet. Sci.,* 30, 233, 1981.
59. Harvey, J. W., Erythrocyte metabolism, in *Clinical Biochemistry of Domestic Animals,* 4th ed., Kaneko, J. J., Ed., Academic Press, San Diego, 1989, p. 185.
60. Mount, M. E., Woody, B. J., and Murphy, M. J., The anticoagulant rodenticides, in *Current Veterinary Therapy IX Small Animal Practice,* Kirk, R. W., Ed., W.B. Saunders, Philadelphia, 1986, p. 156.
61. Lasley, B. L., Endocrine research advances in breeding endangered species, *Int. Zoo Yearb.,* 20, 166, 1981.
62. Gross, T. S., Asa, C. S., Plotka, E. D., Junge, R. E., Bircher, J. S., Noble, G. A., Patton, M., and Tharnish, T., Non-invasive assessment of reproductive function in the cheetah utilizing fecal steroid hormone analyses, *Am. Assoc. Zool. Parks Aquaria Annu. Proc.,* 1991.
63. Czekala, N. M. and Lasley, B. L., A technical note on sex determination in monomorphic birds using fecal steroid analysis, *Int. Zoo Yearb.,* 17, 209, 1977.
64. Gross, T. S., Wieser, C. M., Armstrong, D. L., Bradley, J. E., Pettit, G. J., Cassidy, D. G., and Simmons, L. G., Analysis of the ovarian cycle in black-footed ferrets *(Mustela nigripes)* by vaginal cytology and fecal hormone measurement, *Biol. Reprod.,* 42 (Suppl.), 142, 1990.
65. Kirkvliet, J. F., Shideler, S. E., Lasley, B. L., and Turner, J. W., Pregnancy diagnosis in feral horses by means of fecal steroid conjugates, *Biol. Reprod.,* 42 (Suppl.), 143, 1990.
66. Kerkvliet, N. I. and Baecher-Steppan, L., Suppression of allograft immunity by 3,4,5,3',4',5'-hexachlorobiphenyl. I. Effects of exposure on tumor rejection and cytotoxic T cell activity in vivo, *Immunopharmacology,* 16, 1, 1988.
67. Kerkvliet, N. I., Baecher-Steppan, L., Smith, B. B., Youngberg, J. A., Henderson, M. C., and Buhler, D. R., Role of the Ah locus in suppression of cytotoxic T lymphocyte activity by halogenated aromatic hydrocarbons (PCBs and TCDD): Structure-activity relationships and effects in C57BL/6 mice congenic at the Ah locus, *Fund. Appl. Toxicol.,* 14, 532, 1990.
68. Sanders, O. T., Kirkpatrick, R. L., and Scanlon, P. E., Polychlorinated biphenyls and nutritional restriction: their effects and interactions on endocrine and reproductive characteristics of white male mice, *Toxicol. Appl. Pharmacol.,* 40, 91, 1977.
69. Jones, M. K., Weisenburger, W., Sipes, I. G., and Russell, D. H., Circadian alterations in prolactin, corticosterone, and thyroid hormone levels and down-regulation of prolactin receptor activity by 2,3,7,8-tetrachlorodibenzo-p-dioxin, *Toxicol. Appl. Pharmacol.,* 87, 337, 1987.
70. Umbriet, T. H. and Gallo, M. A., Physiological implications of estrogen receptor modulation by 2,3,7,8-tetrachlorodibenzo-p-dioxin, *Toxicol. Lett.,* 42, 5, 1988.
71. Byrne, J. J., Carbone, J. P., and Pepe, M. G., Suppression of serum adrenal cortex hormones by chronic low-dose polychlorobiphenyl or polybromobiphenyl treatments, *Arch. Environ. Contam. Toxicol.,* 17, 47, 1988.

72. Allen-Rowlands, C. F., Catracane, V. D., Hamilton, M. G., and Seifter, J., Effect of polybrominated biphenyls (PBB) on the pituitary-thyroid axis of the rat, *Proc. Soc. Exp. Biol. Med.,* 166, 506, 1981.
73. Gupta, B. N., McConnell, E. E., Moore, J. A., and Haseman, J. K., Effects of a polybrominated biphenyl mixture in the rat and mouse. II. Lifetime study, *Toxicol. Appl. Pharmacol.,* 68, 19, 1983.
74. Ismail, A. A. A., *Biochemical Investigations in Endocrinology,* Academic Press, New York, 1981, p. 275.
75. Rattner, B. A., Eroshenko, V. P., Fox, G. A., Fry, D. M., and Gorsline, J., Avian endocrine responses to environmental pollutants, *J. Exp. Zool.,* 232, 683, 1984.
76. Itoh, S., Yamagishi, F., and Matsuyama, Y., Relationship between liver microsomal function and serum thyroid hormones in rats treated with carbon tetrachloride, *Res. Commun. Chem. Pathol. Pharmacol.,* 65, 111, 1989.
77. Sinha, N., Lal, B., and Singh, T. P., Pesticide induced changes in circulating thyroid hormones in the freshwater catfish *Clarias batrachus, Comp. Biochem. Physiol.,* 100C, 107, 1991.
78. Van den Berg, K. J., van Raaij, J. A. G. M., Bragt, P. C., and Notten, W. R. F., Interactions of halogenated industrial chemicals with tranthyretin and effects on thyroid hormone levels in vivo, *Arch. Toxicol.,* 65, 15, 1991.
79. Bitman, J., Cecil, H. C., Harris, S. J., and Fries, G. F., Estrogenic activity of *o,p′*-DDT in the mammalian uterus and avian oviduct, *Science,* 162, 371, 1968.
80. Welch, R. M., Levin, W., and Conney, A. H., Estrogenic action of DDT and its analogs, *Toxicol. Appl. Pharmacol.,* 14, 358, 1969.
81. Reijnders, P. J. H., Reproductive failure in common seals feeding on fish from polluted coastal waters, *Nature (London),* 324, 456, 1986.
82. Lundholm, C. E., The effects of DDE, PCB, and chlordane on the binding of progesterone to its cytoplasmic receptor in the eggshell gland mucosa of birds and the endometrium of mammalian uterus, *Comp. Biochem. Physiol.,* 89C, 361, 1988.
83. Lundholm, C. E., Influence of chlorinated hydrocarbons, Hg^{2+} and methyl-Hg^+ on steroid hormone receptors from eggshell gland mucosa of domestic fowls and ducks, *Arch. Toxicol.,* 65, 220, 1991.
84. Bennett, R. S., Williams, B. A., Schmedding, D. W., and Bennett, J. K., Effects of dietary exposure to methyl parathion on egg laying and incubation in mallards, *Environ. Toxicol. Chem.,* 10, 501, 1991.
85. Bennett, R. S., Fairbrother, A., Bennett, J. K., El Halawani, M. E., and Smith, B., Effects of dietary organophosphorus insecticide exposure on incubation behavior, reproductive hormones, and corticosterone in mallards, in preparation.
86. Dean, J. H., Luster, M. I., Munson, A. E., and Amos, H., Eds., *Immunotoxicology and Immunopharmacology,* Raven Press, New York, 1985, p. 511.
87. Sharma, R. P., Ed., *Immunologic Considerations in Toxicology,* Vol. 1, CRC Press, Boca Raton, FL, 1981, p. 159.
88. Vos, J. G., Immune suppression as related to toxicology, *CRC Crit. Rev. Toxicol.,* 5, 67, 1977.
89. Luster, M. I., Munson, A. E., Thomas, P. T., Holsapple, M. P., Fenters, J. D., White, K. L., Lauer, L. D., Germolec, D. R., Rosenthal, G. J., and Dean, J. H., Development of a testing battery to assess chemical-induced immunotoxicity: National Toxicology Program's guidelines for immunotoxicity evaluation in mice, *Fund. Appl. Toxicol.,* 10, 2, 1988.
90. Chung, S. and Secombes, C. J., Activation of rainbow trout macrophages, *J. Fish Biol.,* 31, 51, 1987.

91. Secombes, C. J., Chung, S., and Jeffries, A. H., Superoxide anion production by rainbow trout macrophages detected by the reduction of ferricytochrome C, *J. Dev. Comp. Immunol.,* 12, 201, 1988.

92. Warinner, J. E., Mathews, E. S., and Weeks, B. A., Preliminary investigation of the chemiluminescent response in normal and pollutant-exposed fish, *Mar. Environ. Res.,* 24, 281, 1988.

93. Weeks, B. A. and Warinner, J. E., Effects of toxic chemicals on macrophage phago-cytosis in two estuarine fishes, *Mar. Environ. Res.,* 14, 327, 1984.

94. Weeks, B. A., Warinner, J. E., Mason, P. L., and McGinnis, D. S., Influence of toxic chemicals on the chemotactic response of fish macrophages, *J. Fish Biol.,* 28, 653, 1986.

95. Weeks, B. A., Keisler, A. S., Myrvik, Q. N., and Warinner, J. E., Differential uptake of neutral red by macrophages from three species of estuarine fish, *Dev. Comp. Immunol.,* 11, 117, 1987.

96. Weeks, B. A., Keisler, A. S., Warinner, J. E., and Mathews, E. S., Preliminary evaluation of macrophage pinocytosis as a fish health monitor, *Mar. Environ. Res.,* 22, 205, 1987.

97. Goven, A. J., Venables, B. J., Fitzpatrick, L. C., and Cooper, E. L., An invertebrate model for analyzing effects of environmental xenobiotics on immunity, *Clin. Ecol.,* 4, 150, 1988.

98. Prescott, C. A., Wilkie, B. N., Hunter, B., and Julian, R. J., Influence of a purified grade of pentachlorophenol on the immune response of chickens, *Am. J. Vet. Res.,* 43, 481, 1982.

99. Zeakes, S. J., Hansen, M. F., and Robel, R. J., Increased susceptibility of bobwhites *(Colinus virginianus)* to *Histomonas meleagridis* after exposure to sevin insecticide, *Avian Dis.,* 25, 981, 1981.

100. Fairbrother, A. and Fowles, J., Subchronic effects of sodium selenite and selenomethionine on several immune functions in mallards, *Arch. Environ. Contam. Toxicol.,* 19, 836, 1990.

101. Fowles, J. R., Fairbrother, A., Fix, M., Schiller, S., and Kerkvliet, N. I., Glucocor-ticoid effects on natural and humoral immunity in mallards, *Dev. Comp. Immunol.,* 17, 165, 1993.

102. Fairbrother, A., Yuill, T. M., and Olson, L. J., Effects of three plant growth regulators on the immune response of young and aged deer mice *(Peromyscus maniculatus),* *Arch. Environ. Contam. Toxicol.,* 15, 265, 1986.

103. Baecher-Steppan, L., Nakaue, H. S., Matsumoto, M., Gainer, J. H., and Kerkvliet, N. I., The broiler chicken as a model for immunotoxicity assessment. I. Standardiza-tion of *in vitro* immunological assays, *Fund. Appl. Toxicol.,* 12, 773, 1989.

104. Exon, J. H., Kerkvliet, N. I., and Talcott, P. A., Immunotoxicity of carcinogenic pesticides and related chemicals, *Environ. Carcino. Revs. (J. Environ. Sci. Health),* C5, 73, 1987.

105. Fairbrother, A., Immunotoxicology of captive and wild birds, in *Wildlife Toxicology and Population Modeling: Integrated Studies of Agroecosystems,* Kendall, R. and Lacher, T. E., Eds., Lewis Publishers, Boca Raton, FL, 1993.

106. Rocke, T. E. and Samuel, M. D., Effects of lead shot ingestion on selected cells of the mallard immune system, *J. Wildl. Dis.,* 27, 1, 1991.

107. Whiteley, P. and Yuill, T. M., Effects of selenium on mallard duck reproduction and immune function, U.S. Environmental Protection Agency, Environmental Research Laboratory, Corvallis, EPA/600/3–89/078, 1989, PB90 120 692/AS.

108. Trust, K. A., Miller, M. W., Ringelman, J. K., and Orme, I. M., Effects of ingested lead on antibody production in mallards *(Anas platyrhynchos), J. Wildl. Dis.,* 26, 316, 1990.

109. McMurry, S. T., Lochmiller, R. L., Vestey, M. R., Qualiis, C. W., and Elangbam, C.S., Acute effects of benzene and cyclophosphamide exposure on cellular and humoral immunity of cotton rats, *Sigmodon hispidus, Bull. Environ. Contam. Toxicol.,* 46, 937, 1991.

110. Weeks, B. A. and Warinner, J. E., Functional evaluation of macrophages on fish from a polluted estuary, *Vet. Immunol. Immunopathol. (Neth.),* 12, 313, 1986.

111. Sjobeck, M. L., Haux, C., Larsson, A., and Lithner, G., Biochemical and hematological studies on perch, *Perca fluviatilis,* from the cadmium-contaminated river, *Ecotoxicol. Environ. Saf.,* 8, 303, 1984.

SECTION THREE

Metabolic Products as Biomarkers

---

4. Porphyrins as "Nondestructive" Indicators
of Exposure to Environmental Pollutants

CHAPTER 4

Porphyrins as "Nondestructive" Indicators of Exposure to Environmental Pollutants

Francesco De Matteis and Chang K. Lim

TABLE OF CONTENTS

0-87371-648-5/94/$0.00+$.50
© 1994 by Lewis Publishers

93

I. INTRODUCTION

In this chapter the effects of chemicals on the heme biosynthetic pathway will be briefly reviewed, and the use of porphyrins as "nondestructive" indicators of exposure of wildlife to environmental pollutants will be advocated. Although most of our knowledge about the effect of chemicals on porphyrin metabolism relates to disorders described in experimental animals or man, nevertheless, from the limited number of studies available in the field it appears very probable that similar alterations also occur in wildlife. As discussed below, the pathway of porphyrin biosynthesis is very sensitive to a number of foreign chemicals, and the methodology is now available for detecting porphyrins in biological samples at very low concentrations. We hope that this chapter will encourage environmental biologists to investigate porphyrins as biomarkers of exposure more extensively than they have done in the past.

Porphyrins are tetrapyrrolic pigments widely distributed in nature. They possess a characteristic absorption spectrum with an intense band of absorbance at about 400 nm (the Soret band) and a diagnostic fluorescence spectrum (see below) which make their detection and estimation both sensitive and specific. They occur as pigmented deposition in certain animal tissues, for example, the eggshell and feathers of some birds where they play no clear physiological role apart from an ornamental and camouflage function. Their main physiological significance lies in the pathway of heme biosynthesis, of which they can be considered as intermediary metabolites or oxidized by-products. The pathway of heme biosynthesis has been almost completely elucidated, and its regulation has also been largely

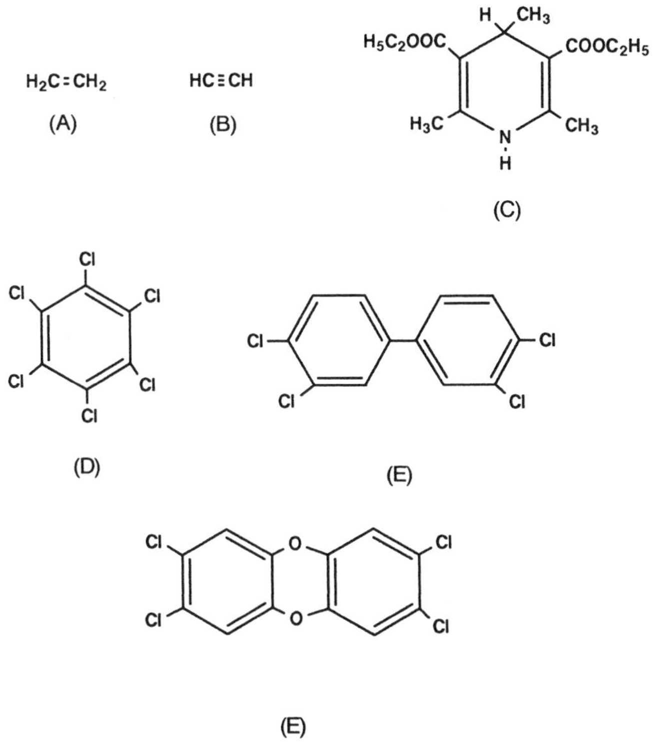

$H_2C{=}CH_2$ $HC{\equiv}CH$

(A)   (B)

(C)

(D)   (E)

(E)

FIGURE 1. Electronic absorption spectrum (A) and fluorescence excitation (B) and emission (C) spectra of uroporphyrin I in 0.5 M HCl. A 3.6 μM solution was used to determine the absorption spectrum, using a 20-fold greater sensitivity (full-scale deflection = 0.1) between 450 and 700 nm. Both fluorescence spectra were measured with a 12 nM solution of uroporphyrin, setting the emission wavelength at 599 nm (in B) and the excitation wavelength at 405 nm (in C).

clarified (see References 1–3 for recent reviews). Before outlining the main steps of porphyrin and heme biosynthesis and some of the modern concepts concerning their regulation, it is important to describe the main absorption and fluorescence characteristics of the porphyrins and also to stress that the pathway is very sensitive to a number of foreign chemicals. These can affect the activity of one or more enzymes of the biosynthetic chain, sometimes very selectively, inducing alterations in the profile of metabolites which accumulate or are excreted. Because of the great sensitivity with which porphyrins can be assayed in biological fluids and because of the varied and often diagnostic pattern of metabolic alterations observed in various intoxications, the porphyrins can be utilized as a sensitive indicator of exposure to foreign chemicals.

Figure 1 shows the electronic absorption spectrum (panel a) and emission and excitation fluorescence spectra (panels b and c, respectively) of dilute solutions of uroporphyrin in 0.5 NHCl. The marked absorption band with the maximum at approximately 400 nm (also called the Soret band) and emission and excitation spectra (similar to those shown in Figure 1 for uroporphyrin) are properties shared by several other porphyrins occurring in biological fluids or tissues and can be used to detect them with a high degree of sensitivity. Concentrations of porphyrins as low as 0.1 to 1.0 μM are easily detected by absorption spectrophotometry and much lower concentrations (in the nanomolar to picomolar range) by spectrophotofluorimetry where a fluorimeter with a red-sensitive photomultiplier tube should be employed. In addition to providing sensitive methods for measuring porphyrins in low concentrations, these spectroscopic techniques — when applied concurrently — can identify an unknown compound as a porphyrin almost conclusively and also provide qualitative information on the particular type of porphyrin (whether uro-, copro-, or protoporphyrin, see below) which predominates in a biological extract. To this purpose use can be made of slight differences between various porphyrins in their absorption maxima[4] or in the excitation and emission maxima of their fluorescence spectra.[5] The same absorption and fluorescence characteristics of the porphyrin pigments can be utilized to detect them with high sensitivity in high-performance liquid chromatography (HPLC) systems (see Section VI) with the added advantage that the retention time of each individual porphyrin can also be used as a criterion for identification. Finally, the characteristic red fluorescence of porphyrins under ultraviolet light (such as that provided by a mercury lamp) is very useful to detect them in biological tissues both by the naked eye and by fluorescence microscopy.

II. BIOSYNTHESIS OF PORPHYRINS AND HEME AND ITS REGULATION

The pathway of heme biosynthesis is given in Figure 2. The first step is the condensation of glycine with a derivative of succinate (succinyl-CoA) catalyzed by the enzyme 5-aminolevulinate (ALA) synthetase in the matrix compartment of the mitochondrion. Two molecules of ALA are then condensed with each other in the soluble part of the cytoplasm to the monopyrrolic precursor, porphobilinogen (PBG), and in the next step four molecules of PBG join together to yield the symmetrical, linear tetrapyrrole, hydroxymethylbilane. This undergoes cyclization under the influence of uroporphyrinogen (uro'gen) III synthetase to the asymmetrical uro'gen III, which is then decarboxylated stepwise by a cytoplasmic decarboxylase to produce coproporphyrinogen (copro'gen) III. The latter is taken up into the mitochondrion, where the remaining steps of heme biosynthesis take place. Copro'gen is first oxidized to protoporphyrinogen by copro'gen oxidase (this step involves oxidative decarboxylation of the two propionic acid side chains to vinyl groups). Protoporphyrinogen is then oxidized to protoporphyrin by another oxidase, and finally heme is produced by insertion of ferrous iron into protoporphyrin, a step catalyzed by the last enzyme of heme biosynthesis, ferrochelatase.

It should be noted that the real intermediates in the pathway are not uroporphyrin and coproporphyrin, but their colorless reduced porphyrinogens (hexahydroporphyrins) in which the pyrrole rings are joined together by methylene bridges. The corresponding porphyrins, oxidative by-products of the pathway, cannot

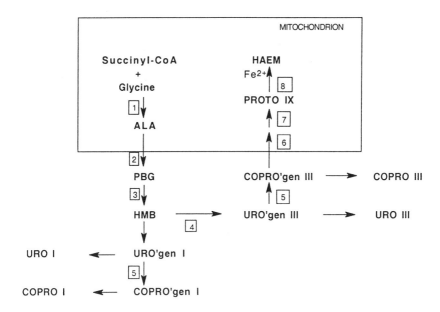

FIGURE 2. Outline of the pathway of heme biosynthesis. ALA: 5-aminolevulinate; PBG: porphobilinogen; HMB: hydroxymethylbilane; URO'gen: uroporphyrinogen; COPRO'gen: coproporphyrinogen; PROTO: protoporphyrin. Reactions are catalyzed by (1) ALA synthetase; (2) ALA dehydratase; (3) PBG deaminase; (4) uroporphyrinogen III synthetase; (5) uroporphyrinogen decarboxylase; (6) coproporphyrinogen oxidase; (7) protoporphyrinogen oxidase; and (8) ferrochelatase. Also shown is oxidation of porphyrinogens to porphyrins, resulting in oxidative escape of some of these intermediates from the pathway.

themselves be metabolized. This has led to the proposal[6] that in certain porphyrias where uroporphyrin accumulates, the mechanism responsible may be at least in part — accelerated oxidation of uro'gen, causing an "oxidative escape" and a loss of the intermediate from the metabolic pathway. This concept will be discussed at some lengths later, when considering the alterations of porphyrin metabolism caused by certain environmental pollutants. It should also be noted that the isomer of uro'gen which is utilized as the precursor of heme is the asymmetrically substituted isomer III. The isomer I of uro'gen — where the β substituents of the four pyrrole rings, acetic acid, and propionic acid side chains are distributed symmetrically along the periphery of the macrocycle — arises from chemical cyclization of the linear tetrapyrrole precursor, hydroxymethylbilane. This is seen when uro'gen III synthetase is either deficient or inhibited and the physiologically preferred, asymmetric, isomer of the III series is formed with difficulty. Under these conditions there may be accumulation of porphyrins of the unphysiological series I isomeric type, not only of uroporphyrin I but also of less carboxylated porphyrins up to the stage of the tetracarboxylated coproporphyrin I. Uro'gen decarboxylase will accept uro'gen I as a substrate and catalyze its decarboxylation. In contrast, copro'gen oxidase is highly specific for the series III isomer, so that once copro'gen I has been formed by stepwise decarboxylation of uro'gen I, it will not be further metabolized.

The first enzyme of the pathway, ALA synthetase, is thought to play a key role in the regulation of heme biosynthesis for two main reasons. First, at least under normal conditions, its activity appears to be rate-limiting in most tissues,[7] in the sense that the rate of heme synthesis is mostly determined by the formation of ALA. This is best illustrated by the observation that an increased supply of ALA will readily increase the formation of porphyrins and heme, while increasing the supply of precursors of ALA will produce little or no increase in synthesis. The second reason why ALA synthetase is of great regulatory importance is that heme, the end product, exercises a negative feedback control on its own synthesis,[7-9] by modulating the amount of this enzyme. The regulatory effect of heme was originally considered to involve repression of the synthesis of the messenger ribonucleic acid (mRNA) for ALA synthetase, to occur, that is, at the transcriptional level.[10-13] However, more recent work suggests that physiological concentrations of heme do not decrease the rate of transcription of ALA-synthetase mRNA, but produce a significant decrease in the stability of this mRNA.[14,15] Thus, in contrast to the inducing effect of lipid-soluble drugs which may involve direct stimulation of ALA synthetase gene transcription (see later), the feedback regulation exercised by heme is thought to be posttranscriptional, resulting in changes of the rate of synthesis of the enzyme through modulation of mRNA stability. Additional effects of heme appear to be at the level of translocation of the ALA synthetase precursor protein into the matrix compartment of the mitochondrion,[16,17] the intracellular site where the enzyme is fully functioning, and also on the activity of this enzyme.[18] Heme has been shown to impair the mitochondrial uptake of the newly formed enzyme and also to inhibit directly its activity, but it is not yet certain in what measure these different mechanisms contribute to the overall feedback control exercised by heme on the activity and amount of ALA synthetase.

III. MAIN MECHANISMS FOR ACCUMULATION OF INTERMEDIARY METABOLITES

Although the property of synthesizing porphyrins and heme is shared by all animal cells, quantitatively the most significant contribution to their formation is afforded in the body by the erythropoietic system, the liver, and — to a lesser extent — the kidney. In the erythroid cells the biosynthesis is mostly geared to production of hemeoglobin; while in the liver most of the newly synthesized tetrapyrroles appear to be destined for cytochrome P-450, a family of hemoproteins, some of which are highly inducible by lipid-soluble drugs and will therefore pose a variable and often considerable demand on heme supply. Under normal conditions, the biosynthetic chain is effectively regulated so that the amount of heme required for the formation of the various hemoproteins is readily made, with little waste of the intermediates. Therefore, normally the various heme precursors, ALA, PBG, and porphyrinogens (and porphyrins) accumulate in various tissues and appear in the excreta only in relatively small amounts (Figure 3A).

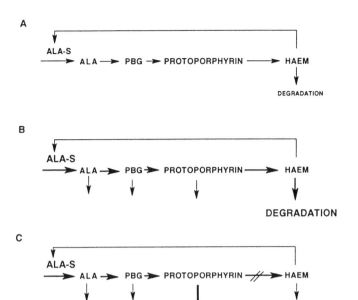

FIGURE 3. Two different mechanisms by which porphyrins and other intermediates of the pathway may accumulate. ALA: 5-aminolevulinate; PBG: porphobilinogen; ALAS: 5-aminolevulinate synthetase. Under normal conditions (A) the biosynthesis of ALA is tightly coupled to formation of heme and the intermediary metabolites do not accumulate in any significant amount. Certain drugs increase the rate of degradation and heme and this leads to a situation of ALAS (B); the increased supply of precursors will then exceed the capacity of the intermediary enzymes of the pathway, even though the activity of these enzymes is still normal. Other chemical agents may inhibit one of the intermediate enzymes of the pathway (for example, ferrochelatase, C). A compensatory stimulation of ALAS will then follow, leading to accumulation of different metabolites, particularly of the substrate of the inhibited enzyme (protoporphyrin).

There are conditions, however, some genetically determined (the hereditary porphyrias) and others induced by exposure to environmental chemicals (the acquired porphyrias and porphyrinurias), where these control mechanisms break down. Far more intermediates of the pathway are made than are turned into heme, so that there is increased accumulation and excretion of one or more of the intermediary metabolites. These abnormalities can in theory apply to the operation of the biosynthetic chain in all cell types; however, in practice for the quantitative considerations alluded to above, the most striking metabolic defects of the pathway originate either in the erythropoietic system or in the liver.

There are two major mechanisms by which the operation of the pathway may become disregulated and wasteful (Table 1). First, the accumulation of the intermediates may be secondary to an increased activity of the first enzyme of the pathway, ALA synthetase. The increased supply of ALA will exceed the metabolic capacity of the subsequent enzymes (even though their activity may still be perfectly normal); one or more of these enzymes will therefore become limiting,

Table 1. Possible Mechanisms by Which Foreign Chemicals May Cause Alterations of Heme Biosynthesis Leading to Accumulation of the Intermediary Metabolites

1. The accumulation of intermediates is secondary to increased activity of ALA synthetase (one or more of the subsequent enzymes becomes limiting):
   a. Failure of heme feedback on ALA synthetase because of accelerated heme degradation
   b. Decreased feedback of heme because of increased utilization of heme for biosynthesis of hemoproteins
   c. Primary activation of transcription of ALA synthetase gene caused by lipid-soluble drugs inducers
2. The accumulation of intermediates is secondary to an acquired partial block in their utilization:
   a. Deficiency of an intermediary enzyme in the pathway, leading to accumulation of the substrate of the inhibited enzyme
   b. Inhibition of uro'gen III synthetase, causing chemical cyclization of hydroxymethylbilane to the nonphysiological uro'gen I, and leading to accumulation of isomer I porphyrins
   c. Increased rate of oxidation of intermediary porphyrinogens, causing their "oxidative escape" from the metabolic pathway
   d. Diversion of a precursor of heme into a nonphysiological pathway of utilization, such as incorporation of a metal other than iron into protopophyrin

causing accumulation of the corresponding substrate(s) (Figure 3B). Foreign chemicals may cause this type of metabolic alteration either directly by activating the transcription of the ALA synthetase gene or indirectly, for example, by accelerating the degradation of heme to abnormal, alkylated products. This will decrease the concentration of free "regulatory" heme in the cell; diminish its negative feedback on ALA synthetase; and lead therefore to increased synthesis of the enzyme, presumably[15] by stabilizing its mRNA and assisting the translocation of the newly formed enzyme into the mitochondrion.

The second major mechanism by which foreign chemicals may lead to accumulation of the intermediates of the pathway is by inducing an acquired defect in an intermediary enzyme (Figure 3C) or some other molecular lesions (see Table 1) leading to impaired conversion of a metabolite. In this case there may also be induction of ALA synthetase, a compensatory response intended to overcome the enzymatic defect; however, the biochemical picture is clearly dominated by the partial block in heme biosynthesis, with a marked accumulation of the intermediate which cannot be readily metabolized. Since different enzymes in the biosynthetic chain are inhibited (or become limiting) after administration of different toxic agents, the profile of metabolites excreted in excess or accumulating in the tissues will also be different, so that distinct metabolic syndromes can be recognized. The great variety of mechanisms by which foreign chemicals may cause accumulation of the intermediates, together with the sensitivity of detection of the porphyrin pigments in biological systems, explains why this pathway is a useful and sensitive indicator of exposure to a number of environmental pollutants.

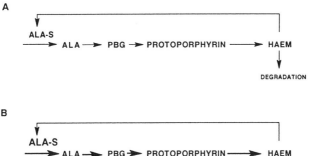

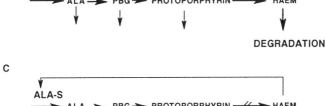

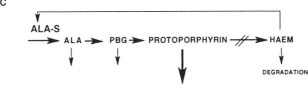

FIGURE 3. Two different mechanisms by which porphyrins and other intermediates of the pathway may accumulate. ALA: 5-aminolevulinate; PBG: porphobilinogen; ALAS: 5-aminolevulinate synthetase. Under normal conditions (A) the biosynthesis of ALA is tightly coupled to formation of heme and the intermediary metabolites do not accumulate in any significant amount. Certain drugs increase the rate of degradation and heme and this leads to a situation of ALAS (B); the increased supply of precursors will then exceed the capacity of the intermediary enzymes of the pathway, even though the activity of these enzymes is still normal. Other chemical agents may inhibit one of the intermediate enzymes of the pathway (for example, ferrochelatase, C). A compensatory stimulation of ALAS will then follow, leading to accumulation of different metabolites, particularly of the substrate of the inhibited enzyme (protoporphyrin).

There are conditions, however, some genetically determined (the hereditary porphyrias) and others induced by exposure to environmental chemicals (the acquired porphyrias and porphyrinurias), where these control mechanisms break down. Far more intermediates of the pathway are made than are turned into heme, so that there is increased accumulation and excretion of one or more of the intermediary metabolites. These abnormalities can in theory apply to the operation of the biosynthetic chain in all cell types; however, in practice for the quantitative considerations alluded to above, the most striking metabolic defects of the pathway originate either in the erythropoietic system or in the liver.

There are two major mechanisms by which the operation of the pathway may become disregulated and wasteful (Table 1). First, the accumulation of the intermediates may be secondary to an increased activity of the first enzyme of the pathway, ALA synthetase. The increased supply of ALA will exceed the metabolic capacity of the subsequent enzymes (even though their activity may still be perfectly normal); one or more of these enzymes will therefore become limiting,

Table 1. Possible Mechanisms by Which Foreign Chemicals May Cause Alterations of Heme Biosynthesis Leading to Accumulation of the Intermediary Metabolites

1. The accumulation of intermediates is secondary to increased activity of ALA synthetase (one or more of the subsequent enzymes becomes limiting):
   a. Failure of heme feedback on ALA synthetase because of accelerated heme degradation
   b. Decreased feedback of heme because of increased utilization of heme for biosynthesis of hemoproteins
   c. Primary activation of transcription of ALA synthetase gene caused by lipid-soluble drugs inducers
2. The accumulation of intermediates is secondary to an acquired partial block in their utilization:
   a. Deficiency of an intermediary enzyme in the pathway, leading to accumulation of the substrate of the inhibited enzyme
   b. Inhibition of uro'gen III synthetase, causing chemical cyclization of hydroxymethylbilane to the nonphysiological uro'gen I, and leading to accumulation of isomer I porphyrins
   c. Increased rate of oxidation of intermediary porphyrinogens, causing their "oxidative escape" from the metabolic pathway
   d. Diversion of a precursor of heme into a nonphysiological pathway of utilization, such as incorporation of a metal other than iron into protopophyrin

causing accumulation of the corresponding substrate(s) (Figure 3B). Foreign chemicals may cause this type of metabolic alteration either directly by activating the transcription of the ALA synthetase gene or indirectly, for example, by accelerating the degradation of heme to abnormal, alkylated products. This will decrease the concentration of free "regulatory" heme in the cell; diminish its negative feedback on ALA synthetase; and lead therefore to increased synthesis of the enzyme, presumably[15] by stabilizing its mRNA and assisting the translocation of the newly formed enzyme into the mitochondrion.

The second major mechanism by which foreign chemicals may lead to accumulation of the intermediates of the pathway is by inducing an acquired defect in an intermediary enzyme (Figure 3C) or some other molecular lesions (see Table 1) leading to impaired conversion of a metabolite. In this case there may also be induction of ALA synthetase, a compensatory response intended to overcome the enzymatic defect; however, the biochemical picture is clearly dominated by the partial block in heme biosynthesis, with a marked accumulation of the intermediate which cannot be readily metabolized. Since different enzymes in the biosynthetic chain are inhibited (or become limiting) after administration of different toxic agents, the profile of metabolites excreted in excess or accumulating in the tissues will also be different, so that distinct metabolic syndromes can be recognized. The great variety of mechanisms by which foreign chemicals may cause accumulation of the intermediates, together with the sensitivity of detection of the porphyrin pigments in biological systems, explains why this pathway is a useful and sensitive indicator of exposure to a number of environmental pollutants.

IV. SOURCE OF PORPHYRINS IN BLOOD AND EXCRETA

Before giving examples of abnormalities of porphyrin metabolism caused by foreign chemicals, it is necessary to discuss briefly the source of porphyrins and early precursors that are found in blood and excreta since the porphyrin analyses — which are intended for nondestructive monitoring of exposure — will usually be conducted on these biological materials.

The distribution within the body of the various metabolites of the pathway depends on which tissue is responsible for their production and also on the characteristic properties of each metabolite (for example, lipid to water partition coefficient and charge) that control binding to carrier proteins; ability to cross biological membranes; and, ultimately, elimination into urine or bile.

ALA and PBG are carried by the circulating blood to the kidney and excreted almost entirely in the urine. Their normal levels in blood plasma are below the limits of detection, but their daily urinary excretion in healthy individuals is approximately 1–2 mg. Correspondingly lower, but still measurable, amounts are excreted by normal rodents. At the normal blood levels they both undergo significant tubular reabsorption, particularly ALA;[19] when their plasma concentrations increase, however, either because of overproduction or underutilization, they are almost quantitatively excreted in the urine.[20,21]

Normal urine also contains small amounts of porphyrins, predominantly coproporphyrin (which is excreted almost entirely as the reduced precursor, copro'gen). Uroporphyrin and traces of porphyrins with seven, six, and five carboxyl groups are also present.[22-25] From studying the fate of exogenously administered porphyrins,[26,27] it is apparent that uroporphyrin and uro'gen, as well as copro'gen are normally excreted in the urine; while coproporphyrin is eliminated almost completely in the bile, unless the bile ducts are occluded or the liver excretory function is damaged. Protoporphyrin is also excreted exclusively by the liver and is normally absent from the urine. However, under conditions where the hepatic excretory mechanism is damaged, protoporphyrin made in excess in the liver may accumulate in the blood;[28] there it becomes bound to plasma proteins and also, to some extent, to the membranes of the red blood cells.[29]

The study of the fecal excretion of porphyrins, as a measure of the amount of porphyrins made in the body, is therefore subject to several potential sources of error. As already mentioned, it can be affected by changes in bile flow and liver function. Also, in addition to the protoporphyrin and coproporphyrin which have reached the intestinal tract with the bile, fecal porphyrins may be derived from chlorophyll and hemoproteins of ingested food or from intestinal hemorrhages. In rodents with a highly developed and active harderian gland,[30] another source of fecal porphyrin may be the ingestion of the protoporphyrin-rich secretion produced by this gland and discharged in the nasal cavity. This explains why normal values for fecal excretion of porphyrins are sometimes very variable and also suggests that moderately raised fecal porphyrin concentrations should not, in themselves, be considered as evidence of disordered porphyrin metabolism, unless these alternative sources of fecal porphyrins have been excluded.

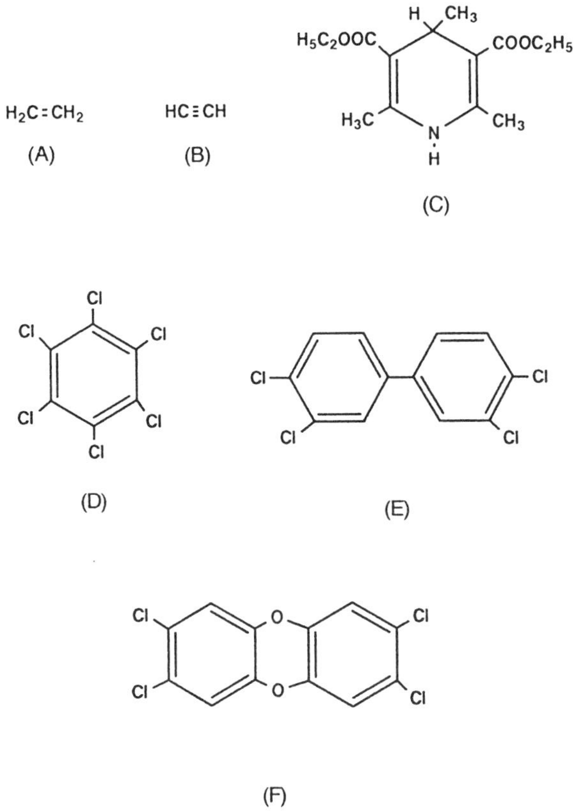

$H_2C \colon CH_2$ $HC \vcentcolon CH$

(A)   (B)

(C)

(D)

(E)

(F)

FIGURE 4. Examples of chemical agents which can induce accumulation and increased excretion of porphyrins. (A) Ethylene; (B) acetylene; (C) 3,5-diethoxycarbonyl-1,4-dihydrocollidine; (D) hexachlorobenzene; (E) 3,3′,4,4′-tetrachlorobiphenyl; (F) 2,3,7,8-tetrachlorodibenzo-p-dioxin (TCDD).

Erythrocytes from humans or mice normally contain approximately 20–60 µg protoporphyrin and 1–2 µg coproporphyrin per 100 mL of packed cells.[31,32] These values may be greatly increased in lead poisoning, when most of the accumulating protoporphyrin is present as the zinc complex.[33] Since the circulating erythrocytes of mammals are devoid of mitochondria, the increased erythrocyte protoporphyrin seen under these conditions must be a reflection of chronic exposure to lead of the erythroblast precursors (present in the bone marrow) which still contain mitochondria and actively synthesize protoporphyrin and heme.[34] The situation may be quite different in other species (for example, in birds) where circulating erythrocytes retain a great deal of the heme biosynthetic activity[35] and changes in red cell protoporphyrin may represent a more acute response.

In conclusion, measurements of ALA and PBG are best performed in urines, where uroporphyrin, other highly carboxylated porphyrins, and copro'gen are also excreted. Coproporphyrin is usually excreted in the bile, but appears in the urine when the liver excretary function is impaired. Protoporphyrin is also excreted by the biliary route; but under conditions when there is increased production by the liver and the liver excretory function is also impaired, it accumulates in blood plasma and in the membranes of the erythrocytes. This "adventitious" accumulation of protoporphyrin can be distinguished from a truly "erythropoietic" increase in red cell protoporphyrin because the former is often accompanied by jaundice and by marked accumulation of protoporphyrin in blood plasma and is usually represented by the porphyrin free base, instead of its zinc complex.

V. ALTERATIONS OF PORPHYRIN METABOLISM CAUSED BY CHEMICALS

Table 2 summarizes various chemically induced abnormalities of heme biosynthesis and gives for each of them the salient feature of the biochemical picture, the main molecular lesion, and also the mechanism that is thought to be responsible for accumulation of porphyrins. A more detailed account of these chemically induced disorders will be found in the following extensive monographs or reviews.[36-39] Most of the information discussed below refers to disorders described in experimental animals or man, and the relevance of these findings to the environmental exposure of wildlife needs further investigation. Nevertheless, as discussed below under "uroporphyria induced by polyhalogenated aromatics" and "lead poisoning," several examples have been described of free-living animals which, on exposure to these chemicals, show the same abnormalities of the pathway that are observed in man and experimental animals. The majority of the chemical agents considered here have the liver as their main target of toxicity; in the case of inorganic lead there is good evidence for the involvement of the erythropoietic system while with several other metals which are discussed at the end of this section, the main site of abnormality either is the kidney or has not yet been clearly defined.

A. Unsaturated Suicide Substrates of Cytochrome P-450

This group of compounds include chemicals containing a terminal unsaturated bond (double or triple bond) between two carbons, as in allyl barbiturates or in ethynyl-substituted steroids, and also the simplest unsaturated compounds of both the alkene and alkyne series, that is, ethylene and acetylene themselves, which are also active. All these compounds are metabolized by cytochrome P-450 in the liver producing a very reactive monooxygenated metabolite (through addition of an oxygen atom to their unsaturated bond), and this metabolite becomes bound to one nitrogen of the heme prosthetic group[40-43] giving rise to N-alkylated protoporphyrins. These functionally inactive derivatives of the heme prosthetic group leave the apoprotein moiety of the cytochrome. Fresh heme is then taken up; and in this way a functionally active cytochrome P-450 is again obtained, only to initiate a new cycle of drug-dependent heme degradation. As a result of these cyclic events[44,45] the concentration of cellular free "regulatory" heme declines, a

Table 2. Summary of the Chemically Induced Disorders of Porphyrin Metabolism Considered Here

Chemical agent	Salient feature of porphyric disturbance	Main molecular lesion	Main organ involved	Mechanism of accumulation on intermediates
Ethylene, acetylene and related unsaturated chemicals	ALA and PBG in liver and urine; also a variety of porphyrins	Alkylation of cytochrome P-450 heme leading to depletion of cellular heme	Liver	Increased supply of precursors exceeds the capacity of intermediary enzymes of the pathway
3,5-Diethyoxycarbonyl-1,4-dihydrocollidine and griseofulvin	Marked accumulations of protoporphyrin in liver, blood and feces	Conversion of cytochrome P-450 heme into N-methyl-protoporphyrin, a targeted inhibitor of ferrochelatase	Liver	Increased synthesis, combined with decreased utilization of protoporphyrin
Hexachlorobenzene and other polyhalogenated aromatic compounds	Marked accumulation of uroporphyrin and other highly carboxylated porphyrins in liver and urine	Still hypothetical; an uncoupled cytochrome P-450 may be responsible for increased oxidation of uroporphyrinogen III	Liver	Oxidative escape of uroporphyrinogen, combined with impaired enzymatic decarboxylation
Inorganic lead	ALA and coproporphyrin in urine; zinc protoporphyrin in red blood cells	Inhibition of ALA dehydratase in erythroid cells and inhibition of iron supply for heme biosynthesis	Bone marrow	Decreased utilization of ALA and diversion of protoporphyrin in a nonphysiological pathway of utilization (zinc protoporphyrin)
Arsenic and mercury	Increased excretion of coproporphyrin and other porphyrins in urine	Still hypothetical; inhibition of coproporphyrinogen oxidase	Kidney	Oxidative escape of coproporphyrinogen, combined with impaired activity of relevant enzyme

secondary stimulation of ALA synthetase follows, and several intermediates of the pathway accumulate as the intermediary enzymes become limiting (mechanism 1.a in Table 1). This type of metabolic alteration has been described in man, experimental rodents, adult chickens;[46] and Japanese quails;[47] and in all these species it involves primarily the liver. There is uncertainty as to whether the kidney may be similarly affected.[48] Although an increase in red fluorescence of the kidney has also been reported in some studies,[47] the kidney accumulation of porphyrins may be mostly secondary to a primary hepatic disorder (through deposition in the kidney of metabolites originating from the liver).

B. Chemically Induced Protoporphyria, a Disorder Associated with Inhibition of Ferrochelatase

3,5-Diethoxycarbonyl-1,4-dihydrocollidine (DDC) and griseofulvin cause a disorder of liver heme metabolism characterized by very marked accumulation of protoporphyrin. As the unsaturated compounds discussed above, these two drugs also promote conversion of the heme prosthetic group of cytochrome P-450 into N-alkylated porphyrins. However, in this case the alkylated porphyrin produced (N-methylprotoporphyrin) has the additional property of inhibiting ferrochelatase,[49,50] thus accounting for the very pronounced accumulation of protoporphyrin (mechanism 2.a in Table 1). A stimulation of ALA synthetase is also seen with these drugs, again reflecting a depletion of the "regulatory" pool; and the increased supply of ALA which follows is an important contributing factor to the accumulation of protoporphyrin. This can be so pronounced that solid casts of protoporphyrin can be demonstrated in the intrahepatic biliary ducts and as pigmented stones in the gall bladder — leading to such complications as jaundice, accumulation of protoporphyrin in blood plasma and erythrocytes, and photosensitivity.

Confirmation of the central role of N-methyl protoporphyrin in hepatic protoporphyria has been obtained in mice and chick embryos[51,52] by showing that the various biochemical effects characteristic of protoporphyria, namely, loss of ferrochelatase activity, compensatory stimulation of ALA synthetase, and accumulation of protoporphyrin can all be produced directly by administration of an N-methylated dicarboxylic porphyrin.

Following the discovery of the role of N-methyl-protoporphyrin in the hepatic protoporphyrias induced by DDC and griseofulvin, similar ferrochelatase-inhibitory porphyrins have been isolated after treatment of animals with other protoporphyria-inducing drugs, namely, sydnones,[53] dihydroquinolines,[54] and more recently[55] 1-[4-(3-acetyl-2,4,6-trimethylphenyl)-2,6-cyclohexane dionyl]-O-ethyl propionaldehyde oxime (ATMP). All these drugs are thought to be metabolized by cytochrome P-450 by oxidative fragmentation, releasing a highly reactive fragment which becomes bound to one nitrogen of the prosthetic heme group. In this way several N-alkylated protoporphyrins are produced — bearing on the pyrrole nitrogen either a methyl, an ethyl, or a vinyl substituent and all capable of inhibiting ferrochelatase. This type of metabolic alteration is therefore more common than originally suspected. The main target organ is invariably the liver, but involvement of the olfactory epithelium has also been reported with a DDC analogue, leading to precipitation of a nasal protoporphyria.[56]

C. Drug-Induced Hepatic Uroporphyria: Role of a Partial Block in Uroporphyrinogen Decarboxylation

Following a widespread outbreak of toxic human porphyria in Turkey over 30 years ago, animal experiments demonstrated conclusively that hexachlorobenzene was the causative agent. A marked inhibition of liver uro'gen decarboxylase was observed in animals treated with this chemical; and this explained the characteristic feature of the metabolic disorder seen in both humans and experimental animals, i.e., the preponderance of uroporphyrin and other highly carboxylated porphyrins in the liver and in the urine.[57] Several other polyhalogenated aromatic compounds (including polychlorinated and polybrominated biphenyls and the very toxic 2,3,7,8-tetrachlorodibenzo-p-dioxin [TCDD]) have since been shown to produce uroporphyria in rodents and chick systems,[37] and there is also some evidence that some of these, for example, TCDD, may be similarly active in humans. Hepatic porphyria with increased liver concentrations of highly carboxylated porphyrins has also been described in the wild in herring gulls (both adults and chicks) from the Great Lakes and in mature pikes from the Rhine River,[57a,57b,57c] a reflection of increased exposure to polyhalogenated aromatic hydrocarbons in both cases. Porphyric pikes were also shown to exhibit markedly decreased activity of hepatic uroporphyrinogen decarboxylase, compared to pike obtained from a relatively unpolluted environment.[57c]

Two main hypotheses have been put forward to account for the mechanism of action of the polyhalogenated chemicals, and both of them involve liver cytochrome P-450. The first proposes that these chemicals may be metabolized by cytochrome P-450 to metabolites which directly inhibit uro'gen decarboxylase.[58] The second hypothesis,[59] which has received more support in recent years,[60,61] suggests, on the other hand, that these chemicals may act indirectly by producing an oxidative stress mechanism of some kind. This may then facilitate oxidation of uro'gen to uroporphyrin (which cannot be metabolized) and also oxidation of uro'gen to some other oxidative derivative which may then act as a targeted inhibitor of uro'gen decarboxylase. The most attractive aspect of this second hypothetical mechanism is that on this basis it is easier to visualize the potentiation of the disorder by iron,[62] since it is known that iron can interact with (and exaggerate) oxidative free-radical reactions. Some support for this oxidative mechanism is also offered by the observation that polyhalogenated chemicals are able to activate cytochrome P-450-dependent oxidative mechanisms in the liver, as measured by oxidation of uroporphyrinogen by microsomes in vitro;[63] and also by oxidation of bilirubin,[64] another easily oxidizable tetrapyrrole. There is still uncertainty as to the oxidative species produced, the precise role of iron in these reactions, and also whether the cytochrome P-450 induced by these chemicals is inherently capable of producing oxidative species or whether the oxidative reaction requires interaction of the inducer itself with the cytochrome. In spite of these uncertainties, however, at the present state of knowledge, the oxidative mechanism outlined above appears to be the most plausible. It will lead both to "oxidative escape" of uro'gen III from the pathway and also to a partial block in heme biosynthesis at the level of the uro'gen decarboxylase (a combination, that is, of mechanisms 2.a and 2.c in Table 1).

There is evidence that the kidneys participate in drug-induced uroporphyria, not only with accumulation of porphyrins but also by showing the characteristic enzymic lesion of uroporphyria. A renal deficiency of uro'gen decarboxylase in fact has been reported not only in rats given hexachlorobenzene[65] but also in birds treated with polychlorinated biphenyls.[66] In this connection it may be relevant to note that inducers such as TCDD not only stimulate microsomal drug metaboliz-ing activities in the kidney but, at least judging from the in vitro degradation of bilirubin,[64] also can activate in this organ a cytochrome P-450-dependent oxida-tive mechanism which may play a role (see above) in the pathogenesis of the metabolic disorder.

D. Lead Poisoning

Lead inhibits at least two steps in the biosynthetic chain, the enzyme ALA dehydratase and the incorporation of iron into protoporphyrin.

ALA dehydratase is a zinc-dependent enzyme:[67] it binds a maximum of one atom of zinc per subunit, but the precise role of zinc — whether contributing to the appropriate steric configuration of the enzyme or playing a more direct role in catalysis — has not yet been clarified. ALA dehydratase is very sensitive to inhibition by lead,[68,69] either because lead interacts with the sulfydryl groups of the enzyme[34] or because it displaces zinc from zinc-binding sites.[67] The inhibition can be reversed by incubating the inhibited enzyme with zinc and dithiothreitol in vitro.

The ALA dehydratase of erythrocytes is probably the most sensitive indicator of exposure to lead. The degree of inhibition has been shown to correlate with the lead concentration in whole blood, the logarithm of ALA dehydratase activity decreasing linearly as the blood lead concentration increases.[70] Particularly impor-tant is the ratio of enzyme activity before and after reactivation with zinc and dithiothreitol, since this ratio shows the best correlation with blood lead levels both in lead-exposed children[71] and free-living birds.[71a] In wild birds the ALA dehydratase activity ratios are better correlated with blood lead levels, than blood protoporphyrin concentrations.[71a] ALA dehydratase activity is also inhibited by lead treatment in liver and kidney both in rats[72] and in urban pigeons,[73] where the degree of inhibition found in erythrocytes, liver, and kidney again correlates with the lead content of these various tissues. Erythrocyte ALA dehydratase also has been found to be a useful short-term indicator of lead exposure in fish.[74] A direct consequence of the enzyme inhibition is the increased urinary excretion of ALA which is seen, however, only after the enzyme activity in the erythrocyte has become depressed by 70–80%.[75]

Although the determination of urinary ALA is preferred by some authors, inhibition of ALA dehydratase activity in blood is a far more sensitive marker of lead exposure and the only assay (relating to metabolism of ALA) which is possible in species like birds where collection of urine is not possible.

Another useful indicator of lead exposure is an increase in erythrocyte proto-porphyrin, which probably reflects (as already discussed) a long-term exposure to lead of the erythroblast precursors in the bone marrow.[34] Most of the accumulating protoporphyrin is in the form of its zinc chelate,[76] which is tightly retained by the circulating hemeglobin (bound as a heme analogue at its heme-binding pocket) and remains bound in this form for the lifetime of the erythrocyte.[77] The increased formation of zinc protoporphyrin in lead poisoning was originally thought to arise from inhibition of ferrochelatase, leading to accumulation of protoporphyrin, and followed in turn by nonenzymic incorporation of zinc into protoporphyrin. It now appears more likely that the action of lead may be mostly through interference with the body iron transfer mechanisms, rather than by direct inhibition of ferrochelatase.[33] Lead would then be responsible for an "iron deficiency" state not unlike that encountered in iron-deficiency anemia, where an increase in erythro-cyte zinc protoporphyrin is also seen. Under both these conditions the preferred metal substrate of ferrochelatase, iron, would become limiting; zinc would then be utilized as an alternative substrate by the enzyme, thus explaining the accumula-tion of zinc protoporphyrin.[33] This effect of lead can therefore be viewed as a diversion of the important precursor of heme, protoporphyrin, into an unphysiological pathway of utilization, effectively resulting in loss of the interme-diate from heme biosynthesis (mechanism 2.d in Table 1).

E. Effect of Arsenic and Mercury

In the alterations of heme biosynthesis considered so far the main site of production of the accumulating metabolites is thought to be the liver or the bone marrow; although the kidney may participate, it does not appear to play a central role. The situation is quite different with both arsenic and mercury where by far the main site of abnormality appears to be the kidney.

Prolonged exposure of rats to methylmercury hydroxide or to sodium arsenate in subtoxic dose levels in their drinking water resulted in progressive coproporphyrinuria accompanied by substantial accumulation of these metals in the kidney and by a significant decrease in renal, but not hepatic, coproporphyrinogen oxidase activity. It was therefore concluded that the disorder in porphyrin metabolism caused by these metals was predominantly renal in origin.[78] In addition to a marked (10.5-fold) increase in coproporphyrin, an 8-fold increase in pentacarboxylate porphyrin was seen after a 5-weeks exposure to 10-ppm methylmercury hydroxide. Also, an unknown porphyrin, eluted from the HPLC column in between coproporphyrin and the pentacarboxylate porphyrin, was markedly increased during exposure to methylmercury.[79] More recent work suggests that in addition to inhibiting coproporphyrinogen oxidase, mercury (in the form of mercuric ion [Hg^{2+}]) may react with reduced glutathione and hydrogen peroxide to form reactive species capable of oxidizing porphyrinogens. Oxidation of porphyrinogens can be demonstrated where there is an interaction of Hg^{2+} with reduced glutathione in the presence of authentic hydrogen peroxide in a purely chemical system,[80] or alternatively when hydrogen peroxide is generated by the mitochondrial electron transport system.[81] The reaction may be mediated by oxidizing free-radical species in both cases and may contribute to the pronounced porphyrinuria observed during prolonged exposure to mercury compounds.

It is not yet clear whether the same mechanism applies to the increased porphyrin excretion reported in arsenic poisoning. However, it may be relevant to note in this respect that oxygen free-radical derivatives have been implicated in the genotoxicity of this metal.[82]

In conclusion, oxidation of porphyrinogens (leading to oxidative escape of these intermediates from the biosynthesis of heme) may be an important mechanism by which toxic metals induce porphyrinuria and may represent a useful indicator of a more general oxidative stress mechanism responsible for a variety of cell-damaging reactions. This is similar to what has been discussed above for the hepatotoxicity of polyhalogenated aromatic compounds, with the difference that in the case of mercury and arsenic the main organ affected is thought to be the kidney instead of the liver.

VI. METHODS FOR THE ANALYSIS OF PORPHYRINS

There are many techniques available for the analysis of porphyrins in biological materials including spectrophotometry, fluorimetry, circular dichroism, countercurrent distribution, electrophoresis, paper chromatography, thin-layer chromatography (TLC), and high-performance liquid chromatography (HPLC). Among these HPLC provides faster separation, better resolution, and higher sensitivity of detection. This section will therefore concentrate on the application of HPLC to the analysis of porphyrins. TLC, which is still widely used, will also be briefly described. Detailed description of other methods can be found elsewhere.[83,84]

A. TLC of Porphyrins

The great majority of TLC separation of porphyrins has been conducted with porphyrin methyl esters on commercially available silica gel plates, although separation of porphyrin free acids has also been described.[85-87] High-performance thin-layer chromatography (HPTLC) which employs a smaller and more uniform particle size (6 μm) coating, considerably improves the efficiency and reduces the time of analysis. Many solvent systems have been used for the development of TLC separations.[85,88-91] Efficient separations of porphyrins with 2-(protoporphyrin) to 8-(uroporphyrin) methyl ester groups have been achieved with benzene:ethyl acetate:methanol (85:13.5:1.5 by vol), carbon tetrachloride:dichloromethane:ethyl acetate (1:1:1 by vol) as the developing solvents. Porphyrins with fewer methyl ester groups run faster than those with higher numbers of ester groups; and the following order of migration is observed: protoporphyrin, coproporphyrin, isocoproporphyrin, pentacarboxylic porphyrin, hexacarboxylic porphyrin, heptacarboxylic porphyrin, and uroporphyrin.

The separated porphyrin methyl esters can be detected with a UV lamp. Quantitation by spectrodensitometric or fluorescence scanning[91,92] has also been described. Alternatively, the separated porphyrins can be recovered from the silica gel by extraction with chloroform:ethanol (4:1 v/v), and the concentration of each porphyrin is then determined spectrophotometrically using the millimolar extinction coefficients of the porphyrin methyl esters in chloroform.[93]

B. Preparation of Porphyrin Methyl Esters in Biological Material for TLC or HPLC

Porphyrins in urine, feces, plasma, red cells, and tissue homogenates can be esterified directly with 5–10% sulfuric acid in methanol. Urine (1 mL) is mixed with 10 mL of the methanol-sulfuric acid reagent and left to stand overnight in the dark at room temperature. The resulting porphyrin methyl esters are extracted into chloroform or dichloromethane. After washing successively with saturated sodium bicarbonate and water, the organic layer is filtered through a filter paper prewetted with chloroform or dichloromethane and then evaporated to dryness under N_2.

For the esterification of feces, the sample (ca. 0.5 g) is homogenized in 5 mL of the esterifying agent and left standing overnight in the dark. The solid materials are removed by filtration or centrifugation and then washed twice with methanol:sulfuric acid (95:5 v/v). The combined filtrate or centrifugate is extracted with chloroform or dichloromethane and then processed as described above for urine samples. Tissue homogenates can be similarly esterified.

For plasma and red cells, the sample (1 mL) is treated with 10 mL of the esterifying agent and left to stand overnight in the dark. The mixture is centrifuged and the supernatant is collected. The residue is washed twice with methanol-sulfuric acid and the supernatants are combined. The porphyrin methyl esters are extracted into chloroform or dichloromethane as described above. After evaporation of the organic solvent, the residue is redissolved in chloroform for spotting or spreading on the TLC plate for separation. A set of porphyrin methyl ester standards is available commercially (Porphyrin Products, Logan, UT) for marking the position of each porphyrin. The esterification procedures described above are also suitable for HPLC applications.

C. HPLC of Porphyrins

The development of HPLC, with its high resolution and speed of separation, has greatly improved the efficiency of porphyrin analysis and has made identification and quantitation of porphyrins much more exact. HPLC is now the technique of choice for the analysis of porphyrins in biological materials. HPLC of porphyrins may be broadly divided into the separation of porphyrin methyl esters and underivatized free porphyrins.

D. HPLC of Porphyrin Methyl Esters

The HPLC separation of porphyrin methyl esters was first reported by Cavaleiro et al.[94] who separated hardero- and isohardero-porphyrin methyl esters on Corasil columns with chloroform-cyclohexane as the mobile phase. The subsequent development of microparticulate silica (3–10 μm particle size) has greatly improved the column efficiency and resolving power. The porphyrin methyl esters

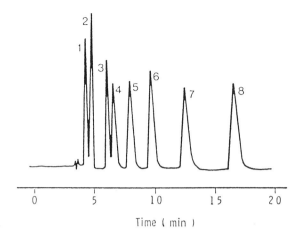

Time (min)

FIGURE 5. HPLC of a standard mixture of porphyrin methyl esters. Column, μ Porasil (300 × 4 mm i.d., 10 μm particle size); eluent, n-heptane:methyl acetate (3:2 v/v); flow rate, 1.5 mL/min. Peaks: 1 = mesoporphyrin; 2 = protoporphyrin; 3 = coproporphyrin; 4 = isocoproporphyrin; 5 = pentacarboxylic porphyrin; 6 = hexacarboxylic porphyrin; 7 = heptacarboxylic porphyrin; 8 = uroporphyrin.

are sequentially eluted with increasing numbers of ester sidechains (in the order of increasing molecular polarity) with a mixture of organic solvents as the eluent. The most commonly employed mobile phases are binary mixtures of ethyl acetate or methyl acetate and a non-polar hydrocarbon such as hexane or n-heptane.[95-98] A typical separation is shown in Figure 5. Ternary[99,100] and quaternary[101] mobile phase systems have also been employed for the separation of porphyrin methyl esters. These systems, however, do not appear to improve on separations achieved with the simpler binary systems.

Silica adsorbents tend to adsorb traces of water present in the organic mobile phase leading to deactivation of the column. This will gradually alter the retention times. All organic solvents must therefore be dried before use in order to achieve the desired reproducibility.

Aminopropyl-bonded silica columns have been suggested as alternatives to bare silica for the normal phase separation of porphyrin methyl esters.[102,103] These columns are less prone to deactivation by water. They are, however, easily poisoned by molecules (solutes or solvents) that can react with the amino function. Aminopropylsilica shows similar selectivity to silica for the separation of porphyrin methyl esters.

Porphyrin methyl esters are best separated by reversed-phase HPLC on ODS-silicas with a mixture of acetonitrile and water as the mobile phase (Figure 6). Reversed-phase HPLC is much more reproducible than normal phase chromatography with the elimination of column deactivation or poisoning encountered on silica or aminopropylsilica. Furthermore, the polar porphyrins (e.g., hydroxylated uroporphyrins) which are difficult to elute from normal phase columns are easily eluted before uroporphyrin from the reversed-phase column because the elution order is the opposite of that encountered in normal phase chromatography.

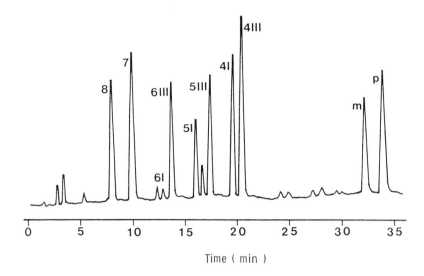

Time (min)

FIGURE 6. Reversed-phase HPLC of porphyrin methyl esters. Column, Hypersil-ODS (250 × 5 mm i.d., 5 μm particle size); eluent, linear gradient elution from 70% (v/v) acetonitrile in water to 100% acetonitrile in 30 min; flow rate, 1 mL/min. Peaks: 4, 5, 6, 7, and 8 refer, respectively, to tetra- (copro), penta-, hexa-, hepta-, and octa- (uro)carboxyl porphyrin; I and III denote type I and type III isomers; m = mesoporphyrin; p = protoporphyrin.

The need for derivatization is an obvious disadvantage of separating porphyrins as methyl esters. Apart from being time-consuming the procedure itself may also lead to undesirable complications. Porphyrin methyl esters, particularly those with a higher number of ester groups, may be partially hydrolyzed during recovery of the esters by solvent extraction following esterification. The presence of a small amount of ethanol in the chloroform extractant has been shown to result in the formation of mixed methyl-ethyl esters,[104] which complicates the identification and quantitation of the porphyrins.

E. HPLC of Porphyrin Free Acids

The problems associated with the derivatization of porphyrins can be overcome by separating the porphyrins as free acids. Early attempts to separate porphyrins by ion-exchange chromatography have met with little success.[95] Porphyrins have been separated on silica with 0.3% water in acetone containing tributylamine at pH 7.6 as the mobile phase.[105] This system probably involves an ion-exchange mechanism since tributylamine may be adsorbed onto the silica and act as ion-exchange sites. Gradient elution systems with acetone-dilute acetic acid[106] and acetonitrile-water containing tetraethylenepentamine[107] have also been described for the separation of porphyrins on silica. The above systems have not been widely adopted because of the development of high efficiency reversed-phase columns which are much more reproducible and easier to equilibrate with solvents.

Adam and Vandemark[108] reported the first separation of porphyrins on a C_{18} column with methanol-0.01% aqueous acetic acid as the eluent. Most of the later

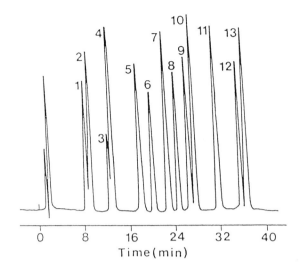

FIGURE 7. Reversed-phase HPLC of porphyrins. Column, Hypersil-SAS (250 × 5 mm i.d., 5 μm particle size); eluent, 10% (v/v) acetonitrile in 1 *M* ammonium acetate buffer, pH 5.16 (solvent A) and 10% acetonitrile in methanol (solvent B); elution, linear gradient from 10 to 65% solvent B in 30 min and then held at 65% B for a further 10 min; flow rate, 1 mL/min. Peaks: 1 = Uroporphyrin I; 2 = uroporphyrin III; 3 = heptacarboxyl porphyrin I; 4 = heptacarboxyl porphyrin III; 5 = hexacarboxyl porphyrin I; 6 = hexacarboxyl porphyrin III; 7 = pentacarboxyl porphyrin I; 8 = pentacarboxyl porphyrin III; 9 = coproporphyrin I; 10 = coproporphyrin III; 11 = isocoproporphyrin; 12 = mesoporphyrin; 13 = protoporphyrin. (From Lim, C. K. and Peter, T. J., *Methods Enzymol.*, 123, 383, 1986. With permission.)

reversed-phase HPLC systems employed gradient elution on ODS-silicas with an increasing proportion of methanol or acetronitrile in phosphate buffer as the eluent.[109-112] Isocratic elution on a phenyl-bonded reversed-phase column has also been described.[113]

Reversed-phase ion-pair chromatography with tetrabutylammonium phosphate as the ion-pairing agent was developed by Bonnett et al.[114] for the separation of porphyrins. The system has been optimized and modified to allow the separation of the type I and type III isomers.[115-117]

Simultaneous separation of porphyrins with from two to eight carboxylic groups, including the resolution of type I and type III isomers, has also been achieved on a C_1- or C_{18}-bonded reversed-phase column by gradient elution with 1 *M* ammonium acetate buffer, pH 5.16, and acetonitrile-methanol as the gradient mixture[118-123] (as shown in Figure 7). This and the ion-pair system above have been successfully applied to the analysis of porphyrins and their isomers in biological and clinical samples.

F. Extraction of Porphyrins from Biological Materials for HPLC Analysis

Urine — Urine (1 mL) is thoroughly mixed with 40 μL of concentrated HCl to dissolve any precipitated material. Precipitates are often seen in stored urine.

They are usually calcium salts which tend to absorb porphyrins. Acid treatment also prevents the formation of metalloporphyrins. The acidified urine may be injected directly (or after a quick centrifugation) into the HPLC. A guard column is used to protect the analytical column from contamination.

Bile — Bile samples can be treated as urine samples.

Feces — The method of Lockwood et al.[124] can be used for the extraction of total fecal porphyrins for HPLC separation. Concentrated HCl (1 mL) is added to a graduated centrifuge tube containing an accurately weighed fecal sample (25–50 mg) and vortex mixed until all particles disintegrate. Diethyl ether (3 mL) is added and thoroughly mixed to give an emulsion, followed by water (3 mL) and further mixing. The mixture is then centrifuged to give an upper ether layer, a pad of insoluble material at the interface, and a lower layer of aqueous acid. The ether layer containing chlorophyll and carotenoid pigments is discarded. The volume of the aqueous acid layer is usually about 4.5 mL, and a sufficient volume is transferred with a Pasteur pipette to a cuvette for spectrophotometry or for HPLC separation.

Plasma — Porphyrins have a strong tendency to adhere to proteins. Simple acidic or alcoholic protein precipitants such as perchloric acid and methanol are ineffective for the quantitative recovery of porphyrins from plasma. The more hydrophobic proto- and coproporphyrin, in particular, tend to coprecipitate with the plasma proteins.

Plasma porphyrins can be extracted by vortex mixing 100–200 μL of plasma with an equal volume of a 1:1 (v/v) mixture of 20% trichloroacetic acid-dimethyl sulfoxide (DMSO). The supernatant after centrifugation is then injected into the HPLC. DMSO, being a good solvent for porphyrins, helps to release them from the proteins.[122,123]

Red blood cells — The porphyrins in red blood cells are mainly protoporphyrin and its Zn complex. Total red blood cell porphyrins are usually estimated by extraction with a mixture of ethyl acetate or diethyl ether and acetic acid followed by back extraction into hydrochloric acid. The porphyrins in the acid extract are then determined spectrophotometrically. Since dilute mineral acids demetallate Zn porphyrins back to the porphyrin free base, this method is not suitable for obtaining a porphyrin extract for the simultaneous separation of porphyrins and Zn porphyrins by HPLC. The ethyl acetate-acetic acid extract, however, has been used for the analysis of porphyrins and metalloporphyrins in red cells.[125,126] The major problem with this approach is that a large amount of hemin is also released from the red cells which may interfere with the separation and detection of porphyrins. Porphyrins and Zn protoporphyrin in red cells are best extracted with neutral organic solvents to minimize contamination by hemin. Acetone:DMSO (9:1 v/v) and methanol:DMSO (4:1 v/v) have been successfully employed.[127] Red cells (100–200 μL) are vortex mixed with 2 vol of methanol:DMSO (4:1 v/v) and centrifuged. The clear supernatant is analyzed by HPLC. The method is also suitable for the extraction of porphyrins in plasma or whole blood.

Tissues — Although analyses on tissues are not common in non-destructive monitoring of exposure, studies on biopsy material may be required in special cases. Porphyrins in tissues may be extracted with the methanol-DMSO mixture used for blood. Tissue (200–300 mg) is homogenized in 2 mL of a homogenizing medium consisting of 8 parts methanol:DMSO (4:1 v/v) and 1 part water. The homogenate is centrifuged at 2600 g for 10 min, 400 μL of the supernatant is mixed with 200 μL of water, and 100–200 μL of the solution is injected. The correct proportion of water in the homogenizing medium and in the final solution is important for satisfactory chromatography. Tissues contain much less water than blood. Injection of tissue extract obtained with methanol:DMSO (4:1 v/v) as the extractant in the absence of water may cause peak distortion. This method is suitable for all tissues, including liver, spleen, kidney, brain, muscle, and skin.

VII. PORPHYRIN EXCRETION PATTERNS IN HUMANS AND ANIMALS GENETICALLY PORPHYRIC OR UNDER THE INFLUENCE OF TOXIC CHEMICALS

As discussed already in the previous sections, enzyme defects in the heme biosynthetic pathway cause genetic disorders called the porphyrias which are characterized by the excessive production and excretion of porphyrins and/or porphyrin precursors. Porphyrias may also be caused by environmental chemicals and toxins which affect one enzyme of the heme biosynthetic pathway. Porphyrias are found in man as well as in animals.[128]

The diagnosis of the porphyrias requires analysis of porphyrins in blood, urine, and feces. Since the porphyrin excretion pattern is characteristic for each of the porphyrias, HPLC porphyrin profiles can be used for their differential diagnosis.[96,120] The patterns illustrated in the following figures are those typical of several inherited or acquired human disorders, compared to the pattern observed in normal subjects (Figure 8). The chemically induced disorder most closely resembled by the human condition is given in each case, since the pattern of porphyrin excretion observed is very similar in both. They also illustrate the main point that since the main metabolite which accumulates is the substrate of the deficient enzyme in all cases, the various patterns observed are diagnostic of the enzymic defect.

A. Congenital Erythropoietic Porphyria (CEP) Porphyrin Excretion Profiles

This rare condition — due to uroporphyrinogen III synthase defect — typically causes the overproduction of type I porphyrin isomers, especially uro- and coproporphyrin I in the urine (Figure 9a) and coproporphyrin I in the feces (Figure 9b). The condition is also found in cattle[128] with similar porphyrin excretion patterns. Increased excretion of the type I isomers is also observed in chemically induced uroporphyria and after exposure to certain toxic metals.

B. Erythropoietic Protoporphyria (EPP) Porphyrin Excretion Profiles

This condition, due to ferrochelatase defect, is characterized by the excessive excretion of fecal protoporphyrin (Figure 10a) and the presence of a high concentration

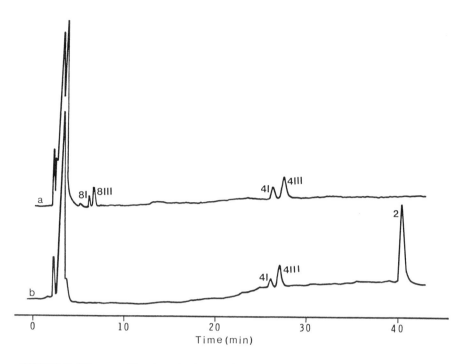

FIGURE 8. Urinary and fecal porphyrin excretion patterns in normal subjects. (a) Urine and (b) feces. HPLC conditions: column, Hypersil-ODS (250 × 5 mm i.d., 5 μm particle size); eluent, 10% (v/v) acetonitrile in 1 *M* ammonium acetate, pH 5.16 (solvent A) and 10% acetonitrile in methanol (solvent B); elution, linear gradient from 10 to 90% B in 30 min followed by isocratic elution at 90% B for a further 10 min., flow rate, 1 mL/min. Peaks: 8I and 8III, uroporphyrin I and III; 4I and 4III, coproporphyrin I and III; 2, protoporphyrin. (From Lim, C. K., in *Liquid Chromatography in Biomedical Analysis,* Hanai, T., Ed., Elsevier, Amsterdam, 1991, p. 209. With permission.)

of protoporphyrin in the blood, particularly in the red cells (Figure 10b). The urinary porphyrin excretion is normal.

Chemicals such as DDC and griseofulvin can also cause protoporphyria by inhibiting ferrochelatase (see Section V.A).

C. Porphyria Cutanea Tarda (PCT) Porphyrin Excretion Profiles

Hepatic uroporphyrinogen decarboxylase activity is decreased in all patients with PCT and as a result the fecal porphyrin profile showed elevation of every porphyrin intermediate between uro- and coproporphyrin (Figure 11a). Two important features of the profile not found in any other porphyria are excessive heptacarboxylic porphyrin III and isocoproporphyrin. The urinary porphyrin profile is dominated by uro- and heptacarboxylic porphyrins (Figure 11b). The uroporphyrin is about 65% type I and 35% type III whereas the heptacarboxylic porphyrin is predominantly type III, a feature typical of the disease.

Toxic PCT follows exposure to certain polyhalogenated aromatic compounds, including hexachlorobenzene, polychlorinated and polybrominated biphenyls. Possibly TCDD can also be demonstrated in animal models treated with these

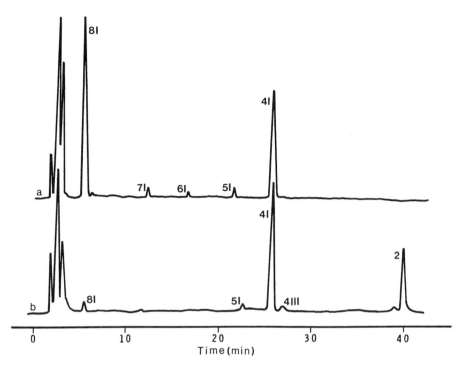

FIGURE 9. Congenital erythropoietic porphyria (CEP) porphyrin excretion profiles. (a) Urine and (b) feces. HPLC conditions as in Figure 6. Peaks: 2, 4, 5, 6, 7, and 8 are di-(proto), tetra-(copro), penta-, hexa-, hepta-, and octa-(uro)carboxyl porphyrin, respectively; I and III denote type I and type III isomers.

chemicals (see Section V.B). Rats and mice poisoned with hexachlorobenzene, for example, have porphyrin excretion patterns similar to human PCT.

D. Porphyrin Accumulation in Lead Poisoning

In lead exposure, excessive Zn protoporphyrin accumulates in the red cells (Figure 12) in contrast to EPP where protoporphyrin accumulates. Urinary ALA and coproporphyrin III are also elevated in lead poisoning.

VIII. DETERMINATION OF ALA DEHYDRATASE ACTIVITY IN ERYTHROCYTES

The determination of ALA dehydratase activity in erythrocytes is important for the diagnosis and confirmation of lead exposure. The enzyme can be assayed by a simple HPLC procedure as follows.

Red cells (30 μL) are thoroughly mixed with 100 μL of 0.2% Triton X-100 in water, 100 μL of 50 mM dithiothreitol in 0.2 M potassium phosphate buffer (pH 6.8), and 100 μL of 20 mM ALA in water in a plastic-stoppered centrifuge tube. The mixture is incubated at 37°C for 1 hr in a shaking water bath. The reaction is terminated by vortex mixing with 250 μL of cold 10% trichloroacetic acid. The

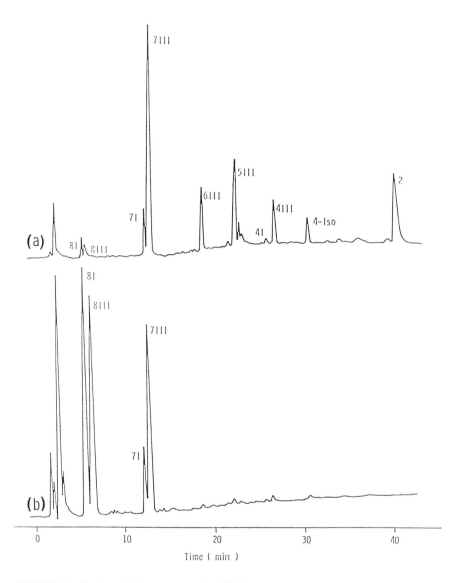

FIGURE 10. Erythropoietic protoporphyria (EPP) porphyrin excretion and accumulation patterns. (a) Fecal porphyrins and (b) red cell porphyrins. HPLC conditions: (a) as in Figure 6; (b) column, Hypersil-SAS; eluent, 80% methanol in 1 M ammonium acetate, pH 5.16. Peaks: ZnPP = Zn protoporphyrin; PP = protoporphyrin; M = mesoporphyrin (internal standard).

supernatant after centrifugation for 5 min at 2600 g is injected (50–100 µL) into the HPLC for separation and quantitation.

The separation is conducted on a Hypersil-SAS column (250 × 5 mm, 5 µm particle size) with 20% methanol in 0.5 M sodium acetate buffer (pH 3.5)

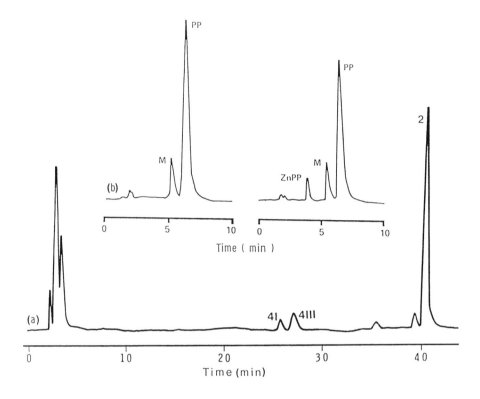

FIGURE 11. Porphyria cutanea tarda (PCT) porphyrin excretion profiles. (a) Feces and (b) urine. HPLC conditions and peak identifications as in Figure 9.

containing 5 m*M* heptanesulfonic acid as the mobile phase at a flow rate of 1 mL/min. PBG is detected with an UV detector set at 240 nm and is quantitated by peak height measurement from a calibration curve constructed by plotting the concentrations of PBG against peak heights. The enzyme activity is expressed as nanomole of PBG formed per millileter of erythrocyte per minute at 37°C.

IX. CONCLUSION

It can be seen in this chapter that heme biosynthesis and porphyrin metabolism are sensitive to many environmental chemicals. Porphyrin levels in blood and excreta may therefore be used as indicators of exposure to such toxic chemicals. Using HPLC with fluorescence detection, porphyrins can be sensitively and accurately measured in relatively small amounts of samples. This provides a potentially useful nondestructive method for monitoring exposure of animals to these environmental pollutants.

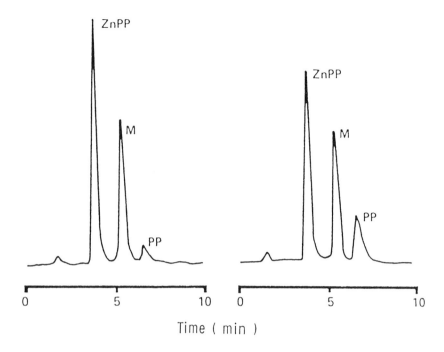

Time (min)

FIGURE 12. Red cell porphyrins in two cases of lead exposure. HPLC conditions and peak
identification as in Figure 10(b).

REFERENCES

1. Jordan, P., Biosynthesis of 5-aminolevulinic acid and its transformation into
    coproporphyrinogen in animals and bacteria, in *Biosynthesis of Heme and Chlorophylls,*
    Daily, H. A., Ed., McGraw-Hill, New York, 1990, p. 55.
2. Dailey, H. A., Conversion of coproporphyrinogen to protoheme in higher eukaryotes
    and bacteria; terminal three enzymes, in *Biosynthesis of Heme and Chlorophylls,*
    Dailey, H. A., Ed., McGraw-Hill, New York, 1990, p. 123.
3. Andrew, T. L., Riley, P. G., and Dailey, H. A., Regulation of heme biosynthesis in
    higher animals, in *Biosynthesis of Heme and Chlorophylls,* Dailey, H. A., Ed.,
    McGraw-Hill, New York, 1990, p. 163.
4. Falk, J. E., *Porphyrins and Metalloporphyrins,* B.B.A. Library, Vol. 2, Elsevier,
    Amsterdam, 1964, p. 236.
5. Grandchamp, B., Deybach, J. C., Greiler, M., De Verneuil, H. and Nordmann, Y.,
    Studies of porphyrin synthesis in fibroblasts of patients with congenital erythropoi-
    etic porphyria and one patient with homozygous coproporphyria, *Biochim. Biophys.
    Acta,* 629, 577, 1980.
6. Heikel, T., Lockwood, W., and Rimington, C., Formation of non-enzymic haem,
    Nature (London), 182, 313, 1958.

7. Granick, S., The induction in vitro of the synthesis of 5-aminolevulinic acid synthetase in chemical porphyria: a response to certain drugs, sex hormones, and foreign chemicals, *J. Biol. Chem.,* 241, 1359, 1966.
8. Burnham, B. F. and Lascelles, J., Control of porphyrin biosynthesis through a negative feedback mechanism, *Biochem. J.,* 87, 462, 1963.
9. Meyer, U. A. and Schmid, R., The porphyrias, in *The Metabolic Basis of Inherited Disease,* 4th ed., Stanburg, J. B., Wyngaarden, J. B., and Frederickson, D. S., Eds., McGraw-Hill, New York, 1977, p. 1166.
10. Yamamoto, M., Hayashi, N., and Kikuchi, G. K., Evidence for the transcriptional inhibition by heme of the synthesis of 5-aminolaevulinate synthetase in rat liver, *Biochem. Biophys. Res. Commun.,* 105, 985, 1982.
11. Ades, I. Z., Stevens, T. M., and Drew, P. D., Biogenesis of embryonic chick liver 5-aminolevulinate synthase; regulation of the level of mRNA by hemin, *Arch. Biochem. Biophys.,* 253, 297, 1987.
12. Maguire, D. J., Day, A. R., Borthwick, I. A., Srivastava, G., Wigley, P. L., May, B., and Elliott, W. H., Nucleotide sequence of chicken 5-aminolevulinate synthase gene, *Nucl. Acids Res.,* 14, 1379, 1986.
13. Srivastava, G., Borthwick, I. B., Maguire, D. J., Elferink, C. J., Bawden, M. J., Mercer, J. F. B., and May, B. K., Regulation of 5-aminolevulinate synthase mRNA in different rat tissues, *J. Biol. Chem.,* 263, 5202, 1988.
14. Drew, P. D. and Ades, I. Z., Regulation of the stability of chicken embryo liver — aminolevulinate synthase mRNA by hemin, *Biochem. Biophys. Res. Commun.,* 162, 102, 1989.
15. Hamilton, J. W., Bement, W. J., Sinclair, P. R., Sinclair, J. F., Alcedo, J. A., and Wetterhahn, K. E., Heme regulates hepatic 5-aminolevulinate synthase mRNA expression by decreasing mRNA half-life and not by altering its rate of transcription, *Arch. Biochem. Biophys.,* 289, 387, 1991.
16. Yamauchi, K., Hayashi, N., and Kikuchi, G., Translocation of 5-aminolevulinate synthase from the cytosol to the mitochondria and its regulation by hemin in the rat liver, *J. Biol. Chem.,* 255, 1746, 1980.
17. Hamilton, J. W., Bement, W. J., Sinclair, P. R., Sinclair, J. F., and Wetterhahn, K. E., Expression of 5-aminolaevulinate synthase and cytochrome P450 mRNAs in chicken embryo hepatocytes in vivo and in culture. Effect of porphyrogenic drugs and haem, *Biochem. J.,* 255, 267, 1988.
18. Scholnick, P. L., Hammaker, L. E., and Marver, H. S., Soluble 5-aminolevulinic acid synthetase of rat liver. II. Studies related to the mechanism of enzyme action and hemin inhibition, *J. Biol. Chem.,* 247, 4132, 1972.
19. Druyan, R., Haeger-Aronsen, B., Von Studnitz, W., and Waldenström, J., Renal mechanism for excretion of porphyrin precursors in patients with acute intermittent porphria and chronic lead poisoning, *Blood,* 26, 181, 1965.
20. Berlin, N. I., Neuberger, A., and Scott, J. J., The metabolism of 5-aminolaevulinic acid. II. Normal pathways, studied with the aid of ^{14}C, *Biochem. J.,* 64, 90, 1956.
21. Goldberg, A., Fate of porphobilinogen, administered enterally or parenterally, in the rat, *Biochem. J.,* 59, 37, 1955.
22. Schwartz, S., Berg, M. H., Bossenmaier, I., and Dinsmore, H., Determination of porphyrins in biological material, in *Methods of Biochemical Analysis,* Vol. 8, Glick, D., Ed., Interscience, New York, 1960.

23. Confort, A., Moore, H., and Weatherall, M., Normal human urinary porphyrins, *Biochem. J.,* 58, 177, 1954.
24. Lockwood, W. H. and Bloomfield, B., Uroporphyrins. III. Crystalline uroporphyrin from normal human urine, *Aust. J. Exp. Biol. Med. Sci.,* 32, 733, 1954.
25. Nicholas, R. E. H. and Rimington, C., Qualitative analysis of the porphyrins by partition chromatography, *Scand. J. Clin. Lab. Invest.,* 1, 12, 1949.
26. Sano, S. and Rimington, C., Excretion of various porphyrins and their porphyrinogens by rabbits after intravenous injection, *Biochem. J.,* 86, 203, 1963.
27. Rimington, C., Patterns of porphyrin excretion and their interpretation, *S. Afr. J. Lab. Clin. Med.,* 9, 255, 1963.
28. Nakao, K., Wada, O., Takaku, F., Sassa, S., Yano, Y., and Urata, G., The origin of the increased protoporphyrin in erythrocytes of mice with experimentally induced porphyria, *J. Lab. Clin. Med.,* 70, 923, 1967.
29. Poh-Fitzpatrick, M. B., Lamola, A. A., Zalar, G. L., Weistein, M., Doleiden, F., and Freeman, M., Comparative study of protoporphyrins in erythropoietic protoporphyria and griseofulvin-induced murine protoporphyria, *J. Clin. Invest.,* 60, 380, 1977.
30. Johnston, H. S., McGadey, J., Thompson, G. G., Moore, M. R., Breed, W. G., and Payne, A. P., The harderian gland, its secretory duct and porphyrin content in the plains mouse *(Pseudomys australis), J. Anat.,* 140, 337, 1985.
31. Marver, H. S. and Schmid, R., The porphyrias, in *The Metabolic Basis of Inherited Disease,* 3rd Ed., Wyngaarden, J. B. and Fredrickson, D. S., Eds., McGraw-Hill, New York, 1972, p. 1087.
32. De Matteis, F. and Rimington, C., Disturbance of porphyrin metabolism caused by griseofulvin in mice, *Br. J. Dermatol.,* 75, 91, 1963.
33. Labbe, R. F., Rettmer, R. L., Shah, A. G., and Turnlund, J. R., *Zinc protoporphyrin. Past, present and future, Ann. N.Y. Acad. Sci.,* 514, 7, 1987.
34. Sassa, S., Toxic effects of lead, with particular reference to porphyrin and heme metabolism, in *Heme and Hemoproteins,* De Matteis, F. and Aldridge, W. N., Eds., *Handbook of Experimental Pharmacology,* Vol. 44, Springer-Verlag, Berlin, 1978.
35. Dresel, E. I. B. and Falk, J. E., Studies on the biosynthesis of blood pigments. I. Haem synthesis in haemolysed erythrocytes of chicken blood, *Biochem. J.,* 56, 156, 1954.
36. De Matteis, F. and Aldridge, W. N., Eds., *Heme and Hemoproteins, Handbook of Exp. Pharmacol.,* Vol. 44, Springer-Verlag, Berlin, 1978.
37. Marks, G. S., Exposure to toxic agents: the heme biosynthetic pathway and hemoproteins as indicator, *CRC Crit. Rev. Toxicol.,* 15, 151, 1985.
38. Silbergeld, E. K. and Fowler, B. A., Eds., *Mechanisms of Chemical-Induced Porphyrinopathies,* Vol. 514, *N.Y. Academy of Sciences,* New York, 1987.
39. De Matteis, F., Drug-induced abnormalities of lvier heme biosynthesis, in *Hepatotoxicology,* Meeks, R. G., Harrison, S. D., and Bull, R. J., Eds., CRC Press, Boca Raton, FL, 1991, p. 437.
40. De Matteis, F. and Cantoni, L., Alteration of the porphyrin nucleus of cytochrome P-450 caused in the liver by treatment with allyl-containing drugs. Is the modified porphyrin *N*-substituted?, *Biochem. J.,* 183, 99, 1979.
41. De Matteis, F., Gibbs, A. H., Jackson, A. H., and Weerasinghe, S., Conversion of liver haem into *N*-substituted porphyrins or green pigments. Nature of the substituent at the pyrrole nitrogen atom, *FEBS Lett.,* 119, 109, 1980.

42. Ortiz de Montellano, P. R. and Correia, M. A., Suicidal destruction of cytochrome P450 during oxidative drug metabolism, *Annu. Rev. Pharmacol. Toxicol.,* 23, 481, 1983.

43. White, I., Suicidal destruction of cytochrome P450 by ethynyl substituted compounds, *Pharmaceut. Res.,* 141, 1984.

44. Unseld, A. and De Matteis, F., Destruction of endogenous and exogenous haem by 2-allyl-2-isopropylacetamide: role of the liver cytochrome P450 which is inducible by phenobarbitone, *Int. J. Biochem.,* 9, 865, 1978.

45. Correia, M. A., Farrell, G. C., Schmid, R., Ortiz de Montellano, P. R., Yost, G. S., and Mico, B. A., Incorporation of exogenous heme into hepatic cytochrome P450 in vivo, *J. Biol. Chem.,* 254, 15, 1979.

46. Goldberg, A. and Rimington, C., Experiemntally produced porphyria in animals, *Proc. R. Soc. B,* 143, 257, 1955.

47. Strik, J. J. T. W. A., Chemical porphyria in Japanese quail *(Coturnix* c. *Japonica),* *Enzyme,* 16, 211, 1973.

48. Yoda, B., Schacter, B. A., and Israels, L. G., Induction of δ-aminolevulinic acid synthetase in the kidney of chicks treated with porphyrinogenic drugs, *Biochim. Biophys. Acta,* 372, 478, 1974.

49. Tephly, T. R., Gibbs, A. H., and De Matteis, F., Studies on the mechanism of experimental porphyria produced by 3,5-diethoxycarbonyl-1,4-dihydrocollidine: role of porphyrin-like inhibitor of protohaem ferro-lyase, *Biochem. J.,* 180, 241, 1979.

50. Holley, A. E., Frater, Y., Gibbs, A. H., De Matteis, F., Lamb, J. H., Farmer, P. B., and Naylor, S., Isolation of two N-monosubstituted protoporphyrins, bearing either the whole drug or a methyl group on the pyrrole ntirogen atom, from liver of mice given griseofulvin, *Biochem. J.,* 274, 843, 1991.

51. De Matteis, F., Gibbs, A. H., and Smith, A. G., Inhibition of protohaem ferro-lyase by N-substituted porphyrins, *Biochem. J.,* 189, 645, 1980.

52. De Matteis, F. and Marks, G. S., The effect of N-methyl protoporphyrin and succinylacetone on the regulation of heme biosynthesis in chicken hepatocytes in culture, *FEBS Lett.,* 159, 127, 1983.

53. Ortiz de Montellano, P. R., Costa, A. K., Grab, A., Sutherland, E. P., and Marks, G. S., Cytochrome P450 destruction and ferrochelatase inhibition, in *Porphyrins and Porphyrias,* Nordmann, Y., Ed., Colloque INSERM 134, 1986, p. 109.

54. Pawels, J. E., Marks, G. S., Ortiz de Montellano, P. R., and Lukton, D., The involvement of cytochrome P450 in the porphyrinogenicity of 1,2-dihydro-2,4,6-trimethyl quinoline (DTMQ) and 2,4-diethyl-1,2-dihydro-2-methyl-quinoline (DDMQ), *Can. Fed. Biol. Soc. Proc.,* 31, 163, 1988.

55. Frater, Y., Mechanism of Action of the Porphyrogenic Agent ATMP and the Identification of Griseofulvin Induced Green Pigments in the Mouse, Ph.D. thesis, Council for National Academic Awards, U.K., 1991.

56. Reed, C. J., Van den Broeke, L. T., and De Matteis, F., Drug-induced protoporphyria in the olfactory mucosa of the hamster, *J. Biochem. Toxicol.,* 4, 161, 1989.

57. Elder, G. H., Porphyria caused by hexachlorobenzene and other polyhalogenated aromatic hydrocarbons, in *Heme and Hemoproteins, Handbook of Experimental Pharmacology,* Vol. 44, De Matteis, F. and Aldridge, W. N., Eds., Springer-Verlag, Berlin, 1978, p. 157.

57a. Fox, G. A., Kennedy, S. W., Norstrom, R. J., and Wigfield, D. C., Porphyria in herring gulls; a biochemical response to chemical contamination of Great Lakes food chains, *Environ. Toxicol. Chem.,* 7, 831, 1988.

57b. Kennedy, S. W. and Fox, G. A., Highly carboxylated porphyrins as a biomarker of polyhalogenated aromatic hydrocarbon exposure in wildlife: confirmation of their presence in Great Lakes Herring Gull chicks in the early 1970s and important methodological details, *Chemosphere,* 21, 407, 1990.

57c. Koss, G., Schüler, E., Arndt, B., Seitel, J., Seubert, S., and Seubert, A., A comparative toxicological study on Pike (*Esox Lucius* L.) from the river Rhine and river Lahn, *Aquat. Toxicol.,* 8, 1, 1986.

58. Van Ommen, B., Adang, A. E. P., Brader, L., Posthumus, M. A., Müler, F., and Van Bladeren, P. J., The microsomal metabolism of hexachlorobenzene. Origin of the covalent binding to protein, *Biochem. Pharmacol.,* 35, 3233, 1986.

59. De Matteis, F. and Stonard, M., Experimental porphyrias as models for human hepatic porphyrias, *Semin. Hematol.,* 14, 187, 1977.

60. Bonkowsky, H. L., Sinclair, P. R., Bement, W. J., Lambrecht, R. W., and Sinclair, J. F., Role of cytochrome P-450 in porphyria caused by halogenated aromatic compounds, *N.Y. Acad. Sci.,* 514, 96, 1987.

61. Smith, A. G. and De Matteis, F., Oxidative injury mediated by the hepatic cytochrome P450 system in conjunction with cellular iron. Effects on the pathway of haem biosynthesis, *Xenobiotica,* 20, 865, 1990.

62. Smith, A. G., Francis, J. E., Kay, S. J. E., Greig, J. B., and Stewart, F. P., Mechanistic studies of the inhibition of hepatic uroporphyrinogen decarboxylase in C57BL/10 mice by iron-hexachlorobenzene synergism, *Biochem. J.,* 238, 871, 1986.

63. Jacobs, J. M., Sinclair, P. R., Bement, W. J., Lambrecht, R. W., Sinclair, J., and Goldstein, J. A., Oxidation of uroporphyrinogen by methylcholanthrene-induced cytochrome P450. Essential role of cytochrome P450d, *Biochem. J.,* 258, 247, 1989.

64. De Matteis, F., Dawson, S. J., Boobis, A. R., and Comoglio, A., Inducible bilirubin-degrading system of rat liver microsomes: role of cytochrome P450IA1, *Mol. Pharmacol.,* 40, 686, 1991.

65. Blekkenhorst, G. H., Day, R. S., and Eales, L., The effect of bleeding and iron administration on the development of hexachlorobenzene-induced rat porphyria, *Int. J. Biochem.,* 12, 1013, 1980.

66. Miranda, C. L., Henderson, M. C., Wang, J. K., Nakaue, H. S., and Buhler, D. R., Induction of acute renal porphyria in Japanese quail by Aroclor 1254, *Biochem. Pharmacol.,* 35, 3637, 1986.

67. Tsukamoto, I., Yoshinaga, T., and Sano, S. P., The role of zinc with special references to the essential thiol gorups of δ-aminolevulinic acid dehydratase of bovine liver, *Biochim. Biophys. Acta,* 570, 167, 1979.

68. Bernard, A. and Lauwerys, R., Metal-induced alterations of δ-aminolevulinic acid dehydratase, *Ann. N.Y. Acad. Sci.,* 514, 41, 1987.

69. Astrin, K. H., Bishop, D. F., Wetmur, J. G., Kaul, B., Davidow, B., and Desnick, R. J., δ-Aminolevulinic acid dehydratase isozymes and lead toxicity, *Ann. N.Y. Acad. Sci.,* 514, 23, 1987.

70. Bonsignore, D., L'Attivita' ALA deidratasica eritrocitaria quale test diagnostico del saturnismo professionale, *Med. Lav.,* 57, 647, 1966.

71. Chisholm, J. J., Jr., Thomas, D. J., and Hamill, T. J., Erythrocyte porphobilinogen synthase activity as an indicator of lead exposure in children, *Clin. Chem.*, 31, 601, 1985.

71a. Scheuhammer, A. M., Monitoring wild bird populations for lead exposure, *J. Wildl. Manage.*, 53, 759, 1989.

72. Meredith, P. A., Moore, M. R., and Goldberg, A., Effects of aluminium, lead and zinc on δ-aminolaevulinic acid dehydratase, *Enzyme*, 22, 22, 1977.

73. Hutton, M., The effects of environmental lead exposure and in vitro zinc on tissue δ-aminolaevulinic acid dehydratase in urban pigeons, *Comp. Biochem. Physiol.*, 74C, 441, 1983.

74. Hodson, P. V., Blunt, B. R., Spry, D. J., and Austen, K., Evaluation of erythrocyte δ-aminolaevulinic acid dehydratase activity as a short-term indicator in fish of a harmful exposure to lead, *J. Fish. Res. Board Can.*, 34, 501, 1977.

75. Lauwerys, R., Buchet, J. P., Roels, H., and Materne, D., Relationship between urinary δ-aminolevulinic acid excretion and the inhibition of red cell δ-aminolevulinate dehydratase by lead, *Clin. Toxicol.*, 7, 383, 1974.

76. Lamola, A. A. and Yamane, T., Zinc protoporphyrin in the erythrocytes of patients with lead intoxication and iron deficiency anemia, *Science*, 186, 936, 1974.

77. Valciukas, J. A., Lilis, R., Fischbein, A., and Selikoff, I. J., Central nervous system disfunction due to lead exposure, *Science*, 201, 465, 1978.

78. Woods, J. S. and Southerne, M. R., Studies on the etiology of trace metal-induced porphyria: effects of porphyrinogenic metals on coproporphyrinogen oxidase in rat liver and kidney, *Toxicol. Appl. Pharmacol.*, 97, 183, 1989.

79. Woods, J. S., Bowers, M. A., and Davis, H. A., Urinary porphyrin profiles as biomarkers of trace metal exposure and toxicity: studies on urinary porphyrin excretion patterns in rats during prolonged exposure to methyl mercury, *Toxicol. Appl. Pharmacol.*, 110, 464, 1991.

80. Woods, J. S., Calas, C. A., Aicher, L. D., Robinson, B. H., and Mailer, C., Stimulation of porphyrinogen oxidation by mercuric ion. I. Evidence of free radical formation in the presence of thiols and hydrogen peroxide, *Mol. Pharmacol.*, 38, 253, 1990.

81. Woods, J. S., Calas, C. A., and Aicher, L. D., Stimulation of porphyrinogen oxidation by mercuric ion. II. Promotion of oxidation from the interaction of mercuric ion, glutathione, and mitochondria-generated hydrogen peroxide, *Mol. Pharmacol.*, 38, 261, 1990.

82. Nordenson, I. and Beckman, L., Is the genotoxic effect of arsenic mediated by oxygen free radicals? *Human Hered.*, 41, 71, 1991.

83. Fuhrhop, J.-H. and Smith, K. M., Laboratory Methods, in *Porphyrins and Metalloporphyrins*, Smith, K. M., Ed., Elsevier, Amsterdam, 1975, chap. 19, p. 839.

84. Jackson, A. H., Modern spectroscopic and chromatographic techniques for the analysis of porphyrin on a microscale, *Semin. Hematol.*, 14, 193, 1977.

85. Smith, S. G., The use of thin layer chromatography in the separation of free porphyrins and porphyrin methyl esters, *Br. J. Dermatol.*, 93, 291, 1975.

86. Friedmann, H. C. and Baldwin, E. T., Reverse-phase purification and silica gel thin-layer chromatography of porphyrin carboxylic acids, *Anal. Biochem.*, 137, 473, 1984.

87. Henderson, M. J., Thin-layer chromatography of free porphyrin for diagnosis of porphyria, *Clin. Chem.*, 35, 1043, 1989.

88. Elder, G. H., Separation of porphyrin methyl esters by two-dimensional thin-layer chromatography, *J. Chromatogr.*, 59, 234, 1971.

89. Doss, M., Analytical and preparative thin layer chromatograophy of porphyrin methyl esters, *Z. Klin. Chem. Klin. Biochem.*, 8, 197, 1970.

90. With, T. K., A simplified system of clinical porphyrin analysis of urine and feces based on thin-layer chromatography, in *Porphyrins in Human Diseases,* Doss, M., Ed., Karger, Basel, 1976, p. 492.

91. Petryka, Z. J. and Watson, C. J., Separation and spectrodensitometric quantitation of porphyrin esters on thin-layer chromatograms, *J. Chromator.*, 179, 143, 1979.

92. Day, R. S., De Salamanca, R. E., and Eales, L., Quantitation of red cell porphyrins by fluorescence scanning after thin-layer chromatography, *Clin. Chim. Acta*, 89, 25, 1978.

93. Smith, K. M., Electronic absorption spectra, in *Porphyrins and Metalloporphyrins,* Smith, K. M., Ed., Elsevier, Amsterdam, 1975, p. 872.

94. Cavaleiro, J. A. S., Kenner, G. W., and Smith, K. M., Porphyrins and related compounds. XXXII. Biosynthesis of protoporphyrin IX from coproporphyrinogen III, *J. Chem. Soc. Perkin Trans.* 1, 1188, 1974.

95. Evans, N., Jackson, A. H., Matlin, S. A., and Towill, R., HPLC analysis of porphyrins in clinical materials, *J. Chromatogr.*, 125, 345, 1976.

96. Gray, C. H., Lim, C. K., and Nicholson, D. C., The differentiation of the porphyrias by high pressure liquid chromatography, *Clin. Chim. Acta,* 77, 167, 1977.

97. Rossi, E. and Curnow, D. H., Porphyrins, in *HPLC of Small Molecules,* Lim, C. K., Ed., IRL Press, Oxford, 1986, chap. 10, p. 261.

98. Straka, J. G., High-performance liquid chromatograophy of porphyrin methyl esters, *Methods Enzymol.*, 123, 352, 1986.

99. Petryka, Z. J. and Watson, C. J., A new method for isolation of naturally occurring porphyrins and their quantitation after high performance liquid chromatography, *Anal. Biochem.*, 84, 173, 1978.

100. Zelt, D. T., Owen, J. A., and Marks, G. S., Second derivative-high performance liquid chromatographic fluorometric detection of porphyrins in chick embryo liver cell culture medium, *J. Chromatogr.*, 189, 209, 1980.

101. Seubert, A. and Seubert, S., High-performance liquid chromatographic analysis of porphyrins and their isomers with radial compression columns, *Methods Enzymol.*, 123, 346, 1986.

102. Sanitrak, J., Krijt, J., Coupek, J., Janousek, V., and Magnus, I., Determination of porphyrins in tissue: preadsorption followed by high-performance liquid chromatography, *J. Chromatogr.*, 415, 129, 1987.

103. Kotal, P., Jirsa, M., Martasek, P., and Kordac, V., Solid phase extraction and isocratic separation of urinary porphyrins by HPLC, *Biomed. Chromatogr.*, 1, 159, 1986.

104. Straka, J. G., Kushner, J. P., and Burnham, B. F., High-performance liquid chromatography of porphyrin esters. Identification of mixed esters generated in sample preparation, *Anal. Biochem.*, 111, 269, 1981.

105. Slavin, W., Williams, A. T. R., and Adams, R. F., A fluorescence detector for high-pressure liquid chromtography, *J. Chromatogr.*, 134, 121, 1977.

106. Longas, M. O. and Poh-Fitzpatrick, M. B., High-pressure liquid chromatography of plasma free acid porphyrins, *Anal. Biochem.*, 104, 268, 1980.

107. Lim, C. K. and Chan, J. Y. Y., Normal-phase high-performance liquid chromatography of porphyrin free acids on silica modified with tetraethylenepentamine, *J. Chromatogr.*, 228, 305, 1982.

108. Adams, R. F. and Vandemark, F. L., The determination of porphyrins in urine by use of high pressure liquid chromatography, *Clin. Chem.,* 22, 1180, 1976.

109. Englert, E., Jr., Wayne, A. W., Wales, E. E., Jr., and Straight, R. C., A rapid, new and direct method for isolation and measurement of porphyrins in biological samples by high performance liquid chromatography, *J. High Resolut. Chromtatogr. Chromatogr. Commun.,* 2, 570, 1979.

110. Chiba, M. and Sassa, S., Analysis of porphyrin carboxylic acids in biological fluids by high performance liquid chromatography, *Anal. Biochem.,* 124, 279, 1982.

111. Ford, R. E., Ou, C. N., and Ellefson, R. D., Liquid chromatographic analysis for urinary porphyrins, *Clin. Chem.,* 27, 397, 1981.

112. Kennedy, S. W., Wigfield, D. C., and Fox, G. A., Tissue porphyrin pattern determination by high-speed high-performance liquid chromatography, *Anal. Biochem.,* 157, 1, 1986.

113. Hill, R. H., Jr., Bailey, S. L., and Needham, L. L., Development and utilization of a procedure for measuring urinary porphyrins by high-performance liquid chromatography, *J. Chromatogr.,* 232, 251, 1982.

114. Bonnett, R., Charalambides, A. A., Jones, K., Magnus, I. A., and Ridge, R. J., The direct determination of porphyrin carboxylic acids. High-pressure liquid chromatography with solvent systems containing phase-transfer agents, *Biochem. J.,* 173, 693, 1978.

115. Meyer, H. D., Vogt, W., and Jacob, K., Improved separation and detection of free porphyrins by high-performance liquid chromatography, *J. Chromatogr.,* 290, 207, 1984.

116. Jacob, K., Sommer, W., Meyer, H. D., and Vogt, W., Ion-pair high-performance liqudi chromatographic separation of porphyrin isomers, *J. Chromatogr.,* 349, 283, 1985.

117. Jacobs, K. and Luppa, P., Applicaton of iron pair high performance liquid chromatography to the analysis of porphyrins in clinical samples, *Biomed. Chromatogr.,* 5, 122, 1991.

118. Lim, C. K., Rideout, J. M., and Wright, D. J., Separation of porphyrin isomers by high-performance liquid chromatography, *Biochem. J.,* 211, 435, 1983.

119. Lim, C. K., Rideout, J. M., and Wright, D. J., High-performance liquid chromatography of naturally occurring 8-, 7-, 6-, 5- and 4-carboxylic porphyrin isomers, *J. Chromatogr.,* 282, 629, 1983.

120. Lim, C. K. and Peters, T. J., Urine and faecal porphyrin profiles by reversed-phase high-performance liquid chromatography in the porphyrias, *Clin. Chim. Acta,* 139, 55, 1984.

121. Lim, C. K. and Peters, T. J., High-performance liquid chromatography of uroporphyrin and coproporphyrin isomers, *Methods Enzymol.,* 123, 383, 1986.

122. Lim, C. K., Li, F., and Peters, T. J., High-performance liquid chromatograophy of porphyrins, *J. Chromatogr.,* 429, 123, 1988.

123. Lim, C. K., Porphyrins, in *Liquid Chromatography in Biomedical Analysis,* J. Chromatogr. Library 50, Hanai, T., Ed., Elsevier, Amsterdam, 1991, chap. 9, p. 209.

124. Lockwood, W. H., Poulos, V., Rossi, E., and Curnow, D. H., Rapid procedure for fecal porphyrin assay, *Clin. Chem.,* 31, 1163, 1985.

125. Smith, R. M., Doran, D., Mazur, M., and Bush, B., High-performance liquid chromatographic determination of protoporphyrin and zinc protoporphyrin in blood, *J. Chromatogr.,* 181, 319, 1980.

126. Scoble, H. A., McKeag, M., Brown, P. R., and Kavarnos, G. J., The rapid determination of erythrocyte porphyrins using reversed-phase high performance liquid chromatography, *Clin. Chim. Acta*, 113, 253, 1981.
127. Rossi, E. and Garcia-Webb, P., Red cell zinc protoporphyrin and protoporphyrin by HPLC with fluorescence detection, *Biomed. Chromatogr.*, 1, 163, 1986.
128. With, T. R., Porphyrias in animals, *Clin. Haematol.*, 9, 345, 1980.

SECTION FOUR

Genotoxic Responses

CHAPTER 5

Genotoxic Responses in Blood*

Lee Shugart

TABLE OF CONTENTS

---

* The submitted manuscript has been authored by a contractor of the U.S. Government under contract No. DE-AC05-84OR21400. Accordingly, the U.S. Government retains a nonexclusive, royalty-free license to publish or reproduce the published form of this contribution, or allow others to do so, for U.S. Government purposes.

I. INTRODUCTION

A. Environmental Toxicity Assessment

With enhanced worldwide industrialization and concomitant demand for chemicals, society is faced with complex ecological and toxicological problems resulting from the release of toxic contaminants into the environment. In response to these expanding stresses on the environment, public awareness and government understanding have grown. Accordingly, toxicity testing for environmental protection is gaining acceptance.

The term "toxicity" refers to a wide range of adverse responses in a living organism, all of which represent impairment to the normal function of the individual, population, or community and which result from a critical exposure to chemicals or changes these chemicals bring about in the physical environment. Contaminant discharges into the environment typically occur as complex mixtures where the presence of one chemical can increase or decrease the toxic effect of another. Therefore, knowledge about the toxicant concentration and the period of exposure alone is insufficient to precisely determine toxicity because the interactive effects of chemicals can rarely be quantified.

Toxicity testing of the physical environment (e.g., water, sediments, soils, air, etc.) may be determined in a variety of laboratory tests using standard test organisms and protocols. A common evaluation of acute toxicity is the determination of the LC_{50}, which is the concentration of substance that causes lethality in 50% of the exposed test organisms. Endpoints may also incorporate chronic effects such as impairment of growth, reproduction, or physiological function. Such testing is invaluable for identifying environments that are contaminated, but provides very little definitive information about the potential impact on the health of the organisms present in that environment.

B. Biological Markers

Biological monitoring is an effective approach to assessing the health of a contaminated environment and has received considerable attention recently by ecologists and toxicologists. Specifically, biological monitoring refers to the use of indigenous organisms, including humans, to determine the status of the environment. The underlying principle is that selected biological responses (biomarkers) measured in the organism and at various levels of biological complexity can provide sensitive indices of exposure and stress.[1-6] Biomarkers are any of a series of biochemical or molecular responses to compounds that have entered an organism, reached sites of toxic action, and are exerting an effect on the organism. In this context, the organism functions as an integrator of exposure, accounting for abiotic and physiological factors that modulate the dose of toxicant taken from the environment. These biological markers can be used to quantify exposure to harmful agents and to detect the response to environmental insults. Often these indices correlate with subsequent deleterious consequences from such exposure.

C. Environmental Genotoxicology

Perhaps the single greatest societal concern related to the release of anthropogenic pollutants into the environment is the potential for exposure to carcinogenic and mutagenic chemicals. However, recent advances in toxicology, clinical medicine, and molecular genetics are fostering a better understanding of the biological, chemical, and physical processes that are involved. More importantly, a stronger scientific basis for assessment of the risk of exposure to these deleterious chemicals is becoming available.

Genetic toxicology is an area of science in which the interaction of deoxyribonucleic acid (DNA)-damaging agents with genetic material is studied in relation to subsequent effect(s) on the health of the organism. Environmental genotoxicology is an approach that applies the principles and techniques of genetic toxicology to assess the potential effect of pollution, in the form of genotoxic agents, on the health of the biota. The success of this approach is based on the predictive and anticipatory capabilities of the techniques and methods employed. In this chapter the *in situ* approach is characterized and methodologies are described for the examination of indigenous species present in a contaminated environment for the effect of genotoxins.[1,7,8] This is as opposed to the approach in which genotoxicity is studied in a laboratory setting using bioassays that range, for example, from the microbial mutation test of Ames[9] to the eukaryotic cell test for anaphase aberration.[10] This is not to imply that the latter approach should not be used to obtain scientifically relevant data that supplement and complement the former approach, or vice versa.

D. Approaches and Methodologies

Early attempts at assessing the effects of genotoxins on species present in the environment most often involved direct observation such as the visual occurrence of neoplasms and chromosomal aberrations in various plants, wild terrestrial mammals, and aquatic vertebrates.[11] Although the development of neoplasia, tumors, or other pathological responses associated with carcinogenesis is a slow process (months to years) that is dependent on numerous unknown and ill-defined factors, the exposure of an organism to a genotoxic agent can result in the induction of a cascade of cellular events in a relatively short time period of days to weeks (Table 1).

Structural alterations to the DNA molecule (e.g., adducts, strand breaks, and changes in minor nucleoside content) usually occur shortly after exposure to a genotoxic agent. However, these types of damage are frequently corrected by the DNA repair machinery of the cells. Biological damage that is not corrected, or is improperly processed, may potentiate irreversible events with the appearance of

Table 1. Biomarkers of Genotoxicity

Biomarker	Biological response	Level of biological organization	Temporal occurrence[a]	Restriction[b] Detection level	Limitations
DNA damage	Adducts	Molecular	Early	Low	Repair/ analysis
DNA damage	Strand breaks	Molecular	Early	Moderate	Repair
DNA damage	Repair	Molecular	Early	Low	Analysis
Mixed function oxidase	Enzyme induction	Molecular	Early	Moderate	Species variability
DNA damage	Minor nu- cleoside	Biochemical	Early/middle	High	Analysis
Chromosomal aberrations	Abnormal DNA	Subcellular	Middle/late	High	

[a] Temporal occurrence subsequent to exposure: early — hours to days; middle — days to weeks/months; late — weeks/months to years.
[b] Constraints on biomarker application: (1) detection level — anticipated probability of detecting biomarker and (2) limitation — factor(s) contributing to detection or affecting use of biomarker.

abnormally processed DNA (e.g., chromosomal aberrations, micronuclei, etc.). It should be emphasized that the events listed are points on a continuum that are subject to change and redefinition as our knowledge in this area increases. Nevertheless, these early biological responses at the molecular level involve changes to the integrity of the genetic material that are amenable to detection and therefore suitable biomarkers to test for genotoxicity.[3,5,7] The rationale for the selection of a given change in DNA integrity to assess exposure to and/or effect of a genotoxin is dictated and limited by two important considerations: (1) the present state of our knowledge concerning those cellular mechanisms involved in the genotoxic response; and (2) the analytical technology available to define the response. The success of current *in situ* efforts in the field of environmental genotoxicology is due mainly to the use of new analytical techniques that are extremely selective and/or sensitive for the detection of changes in DNA integrity. Because the focus of this chapter is on blood as a vehicle for nondestructive sampling, methods applicable to this medium will be described. Furthermore, methods that measure structural damage to DNA will be discussed in relation to their capability to detect and/or describe genotoxic insult by environmental pollutants in living organisms.

II. DNA ALTERATIONS

A. Introduction

The integrity of an organism's DNA is of utmost importance to its survival. To ensure that DNA integrity is maintained, an elaborate monitoring/repair system[12] has evolved that recognizes numerous types of structural damage to this molecule and corrects these deficiencies. Shugart et al.[13] have detailed the various types of structural changes that may occur to DNA under normal cellular conditions as well as after exposure to chemical and physical agents. Structural modifications to the DNA are the best understood genotoxic event for which analytical techniques with the appropriate selectivity and sensitivity are available. If this type of DNA damage persists, then it becomes possible to detect changes in the integrity of the DNA that are indicative of exposure to genotoxins.

The remaining portion of this section on DNA alterations will consider three common types of structural changes that genotoxic agents can cause to DNA (i.e., adducts, strand breakage, and modification to bases). Additional information will elaborate analytical techniques available to detect and measure these alterations as well as examples of the application and utilization of these methodologies *in situ* (i.e., environmentally relevant situations). With regard to the last point, the reader should not anticipate an exhaustive review of the scientific literature, but instead examples that document application of techniques. In most instances the environmental applications cited describe sampling performed in a destructive mode, i.e., the animal is sacrificed to obtain suitable target tissue DNA. However, the techniques used are amenable to nondestructive sampling as well.

B. Adducts

The majority of chemicals that are genotoxic exert their effects only after metabolic conversion to chemically reactive forms which bind covalently to cellular macromolecules, including nucleic acids and proteins, to form addition products (adducts). Currently, methods of varying sensitivity exist to measure DNA adducts and include ^{32}P postlabeling, high-performance liquid chromatography (HPLC)/fluorescence spectrophotometry, and immunoassays using adduct-specific antibodies.

1. ^{32}P Postlabeling

Many lines of evidence implicate DNA adducts as a key element in the initiation of chemical carcinogenesis.[14,15] As a result the ^{32}P-postlabeling technique was originally developed as a general method for studying the binding of authentic carcinogens to DNA.[16] The technique has been well documented for the measurement of extremely low levels of DNA adducts in laboratory animals and human tissue.[17] Currently this methodology is finding application[18-22,63] for the detection of adducted chemicals to DNA in aquatic species suspected of exposure to environmental pollution (Table 2).

The methodology used for the ^{32}P-postlabeling adduct assay consists of the follows steps. DNA isolated from a suitable sample is enzymatically hydrolyzed with micrococcal nuclease and spleen phosphodiesterase. This results in the formation of 3'-monophosphates of both normal and adducted nucleosides. The products are then ^{32}P labeled by T_4 polynucleotide kinase-catalyzed phosphorylation with $[\gamma\text{-}^{32}P]ATP$, leading to 5'-$^{32}P$-labeled 3',5'bisphosphate nucleosides.

^{32}P-labeled adducts are resolved from normal nucleotides, ^{32}Pi, unused $[\gamma\text{-}^{32}P]ATP$, and unknown contaminants by anion exchange thin-layer chromatography (TLC)

**Table 2. Detection of Adducts by [32]P Postlabeling in Environmental
Species**

Environment	Species	Contaminant
Freshwater stream/U.S.	Catfish	Sediment-bound PAHs[18]
Freshwater stream/Europe	Various fish species	Complex industrial waste[19]
Marine/Adriatic	Mussel	Complex industrial waste[20]
Marine harbors/U.S.	English sole and winter flounder	Complex industrial waste[21]
Marine/Arctic and St. Lawrence	Beluga whale	Organics and PAHs[22]
Estuerine river/U.S.	Muskrats	Sediment-bound PAHs Complex industrial waste[63]

on polyethyleneimine-cellulose using a multidirectional developmental scheme.[23,24] Dried chromatograms are placed in contact with X-ray film and exposed at −80°C. Adducts are detected by autoradiography and quantitated by scintillation counting. Carcinogen-related adduct patterns are obtained from carcinogen-exposed DNA.

The sensitivity of the assay for adducts can be enhanced by several orders of magnitude, and is most often accomplished via the preenrichment of aromatic/ hydrophobic nucleosides prior to labeling. Extracting in 1-butanol in the presence of a phase-transfer agent is one method.[25] An alternative procedure[26] involves selective enzyme hydrolysis of the 3′nucleotides with nuclease P1, or combined with reversed-phase HPLC.[27]

Detection of adduct levels as low as 1 in 10^{10} normal nucleotides is possible by this technique; however, ^{32}P-labeled spots of undetermined origin also occur and may interfere with the interpretation of adduct patterns.[28] Nevertheless, with the exception of methylating agents and mycotoxins, numerous known carcinogens have been shown to produce carcinogen-specific fingerprints by this method.[24]

2. HPLC/Fluorescence

An alternative method to the ^{32}P-postlabeling technique is detection of the binding of fluorescent xenobiotic chemicals to DNA.[29] This technique involves removal of the adduct from DNA (usually by acid hydrolysis) and separation by HPLC coupled with fluorescence analysis. Application of this technique is limited by the presence and release of an intact fluorescent moiety. Shugart et al.[30] have shown that adduct formation at the femtomole level between the ubiquitous chemical carcinogen, benzo[a]pyrene, and DNA can be detected and quantitated by this method in both mice and fish exposed under laboratory conditions.[31,32] The technique has also been used in environmental studies to detect benzo[a]pyrene adducts in the DNA of beluga whales from the St. Lawrence River in Canada.[33]

3. Immunological Techniques

A major limitation of this approach is the need to develop specific antibodies and establish the specificity of the antibody for each DNA adduct or class of adducts of interest. This is an important consideration when exposure is to a complex mixture of environmental chemicals. In addition, sufficient quantities of either the modified DNA or individual adduct must be available for immunization and antibody characterization. These limitations may explain why this technique has not been applied in the laboratory or field on a large scale to detect DNA damage in aquatic or terrestrial species. However, in certain cases broad specificity antibodies have been used to detect DNA adducts of a class of carcinogens such as benzo[a]pyrene, benz[a]anthracene, and chrysene with application in human health studies.[17,34]

C. Strand Breaks

DNA strand breakage is not an uncommon occurrence in a cell. Heat energy causes thousands of abasic sites per cell per day which, however, are rapidly repaired.[35] This is an example of an insult to DNA that indirectly results in strand breakage (i.e., the initial damage is a loss of a base from the DNA chain, the repair of this damage results in a temporary gap in the DNA molecule). Ionizing radiation can cause strand breakage directly, whereas other physical agents such as UV light and chemical agents that are genotoxic potentiate alterations to the DNA molecule that are candidates for repair (e.g., photoproducts, adducts, etc.) and thus for the occurrence of strand breaks.[13]

The reader is referred to the scientific literature for detailed explanations of the methodology for performing a particular DNA strand break assay; however, most assays currently in use are based on the general principle that under *in vitro* conditions the rate at which single-stranded DNA is released from the duplex DNA at high pH is proportional to the number of strand breaks in the DNA molecule.[36] It should be noted that under these assay conditions alkaline labile sites in the DNA molecule will also be detected because they are chemically converted to single-strand breaks.

Since the various assays incorporate the same basic principle for determining strand breakage in DNA, their uniqueness has to do with the manipulation of the DNA preparation before release of the single-stranded DNA from the DNA duplex, and/or the method for detection of double-stranded and single-stranded DNA. For example, in both the "alkaline elution" assay of Kohn et al.[37] and the "DNA precipitation" assay of Olive,[38] denatured single-stranded DNA is released from a physical matrix: fibers constituting the filter of the alkaline elution assay or cellular proteins in the DNA precipitation assay. Because these assays allow for the physical separation of single-stranded DNA from double-stranded DNA during the denaturation process, the quantitation of the two DNA species at the end of the assay is easily accomplished.

In the alkaline unwinding assay,[39] on the other hand, some degree of DNA purification must be attained before the denaturation step. Furthermore, in this assay both double- and single-stranded DNA species will be present at the end of the denaturation process; therefore, subsequent steps must be incorporated to accommodate this situation. This can be accomplished by the physical separation of the two species using hydroxyapatite chromatography[40,41] or differential extraction with phenol containing 1 *N* NaCl.[42] A facile procedure that does not require prior separation is based on the difference in fluorescence that occurs when bisbenzimidazole (Hoechst dye no. 33258) binds to the two species of DNA.[39,43]

Gel electrophoresis represents another analytical technique that may be used to assay for strand breaks in the DNA of environmental species.[44] Under alkaline conditions, electrophoresis of DNA on agarose gels results in migration within this matrix that is size dependent. Detection is easily accomplished with ethidium bromide staining. With gel electrophoresis it is possible to detect both single-strand and double-strand breaks. The latter is an extremely important consideration since it can be a lethal event for most cells. A recent interesting application of this assay is the evaluation of DNA damage in individual cells.[45,46] Using this method, subsets of cells varying in DNA damage may be observed.

A summary of the assay methods currently used to detect strand breakage in DNA is found in Table 3. Again it should be noted that the applications cited are for documentation purposes only.

The interpretation of strand breakage observed in the DNA of an organism is not a straightforward matter. As mentioned previously, heat energy within a cell, as well as normal events associated with cellular metabolism, can result in breaks along the backbone of the DNA molecule. In addition, genotoxic agents can cause DNA strand breakage. Thus the problem becomes one of distinguishing between strand breakage that occurs under normal conditions from those that result from abnormal situations. This distinction can often be accommodated through the selection of a suitable control or reference population for comparison. Ideally, the reference population would not be subject to genotoxic stress. However, this population should experience the same natural environmental stresses as the population under study. Any significant increase in strand breakage in the sampled population over the baseline data observed in the reference population would be indicative of the potential for genotoxic insult to the sampled population. Admittedly this explanation oversimplifies the interpretative process, and the reader is referred to recently published articles for more information.[8,13,39,44] For example, many chemicals in the environment may not be genotoxic per se, but can cause necrosis and cell death;[47] that is, exposure to a nongenotoxic agent initiates a pathological event leading ultimately, but indirectly, to the degradation of DNA in the dead cell.

In summary, strand breakage in DNA can be a genotoxic response for which several facile analytical techniques are available. This approach is readily amenable to field sampling and is currently being applied to genotoxicological problems in the environment (Table 3).

D. Modified Bases

1. Minor Nucleoside Content

In eukaryotic DNA, 5-methyl deoxycytidine is generally the only methylated deoxyribonucleoside present, and its level is enzymatically maintained.[51-53] Chemical carcinogens have been shown to produce hypomethylation of DNA as a result of their effect on these enzymes.[54-56] The base composition of DNA is easily

**Table 3. Applications of Methods for
DNA Strand Breaks**

Method	Application
Alkaline elution[37]	Marine mussel[48]
DNA precipitation[38]	Animal cells[38]
Alkaline unwinding[39]	Fish[8,39]
	Rodents[42]
	Turtles[49]
	Marine mussel[50]
Electrophoresis	
Agarose gel[44]	Fish[44]
Single-cell gel[45,46]	Lymphocytes[45]
	Animal cells[46]

determined by ion-exchange chromatography.[57,58] As with the detection of strand breakage, this method measures a loss of DNA integrity but does not identify the chemical responsible.

Hypomethylation of DNA, as measured by the loss of 5methyl deoxycytidine, was demonstrated in fish exposed to benzo[a]pyrene.[58] The onset and persistence of this phenomenon were found to be correlated with other types of DNA-damaging events, such as strand breaks and adduct formation. These observations suggest that hypomethylation may be a specific biological response to genotoxic agents.

2. Free-Radical Damage to DNA

The microsomal metabolism of some hydrocarbons produce free radicals, including oxyradicals, that are thought to play important roles in chemical carcinogenesis. The biochemical mechanisms whereby fluxes of these radicals are produced in aquatic animals has been reviewed.[59] Both the superoxide radical and the hydroxyl radical are known to damage DNA directly through strand scission and oxidation of the bases of DNA, in particular, guanine.

Free-radical damage to DNA can result in the hydroxylation of guanosine to form 8-hydroxydeoxyguanosine (8-OH-dGuo) which is released from intact DNA when hydrolyzed enzymatically to the deoxynucleoside level. The 8-OH-dGuo is separated from other deoxynucleoside by reverse-phase HPLC, and analysis is with an electrochemical detector system.[60] The level of sensitivity is in the order of one 8-OH-dGuo residue per 10^5 normal dGuo residues. Free-radical damage to the guanine base may also result in the formation of 2,6-diamino-4-hydroxy-5-formamidopyrimidine moiety (FapyGua) in which ring opening of the heterocyclic base occurs. The FapyGua lesion is removed from DNA by acid hydrolysis, trimethylsilyated, and subjected to gas chromatography-mass spectrometry with single-ion monitoring. Detection of 0.01 nmol of this compound in 1 mg of DNA has been reported.[61]

Results from field studies with English sole *(Parophrys vetulus)* taken from Puget Sound, U.S. suggested a strong correlation between the levels of chemical contaminants — particularly polycyclic aromatic hydrocarbons (PAHs) and polychlorinated biphenyls (PCBs) — in the sediments and the prevalence of liver neoplasms.[62] Recent studies by Malins et al.[61] have shown that FapyGua DNA lesions also occur in the hepatic DNA of English sole from the same environment. The occurrence of this damage is indicative of insult by a free-radical mechanism of genotoxicity, but is unusual in that the lesion is intimately associated with neoplastic tissue and not with normal tissue.

E. Surrogate DNA Adducts

Since evidence of DNA alterations in most tissues — including that of humans — is sometimes difficult to obtain, damage to ancillary molecules may serve as a surrogate (protein adducts, for example). Blood proteins such as hemoglobin and serum albumin have proved useful in this context.

Because it meets a number of essential requirements, hemoglobin has been proposed as a surrogate for DNA for estimating the *in vivo* dose of chemicals subsequent to exposure. First, it has reactive nucleophilic sites and the reaction products with electrophilic agents are stable. Over 60 compounds to date have been shown to yield covalent reaction products with hemoglobin in animal experiments.[15,17] These compounds include representatives of most of the important classes of genotoxic chemicals currently known. No mutagenic or cancer-initiating compound has failed to produce covalent reaction products with hemoglobin. Second, hemoglobin has a well-established lifespan and is readily available in large quantities in humans and animals; and its concentration is not subject to large variation. Third, modification of hemoglobin has been shown to give an indirect measure of the dose to the DNA in cells which are potential targets for genotoxic agents.

The detection and measurement of adducts to hemoglobin has been seriously studied in humans exposed to hazardous occupational chemicals for some time and a considerable scientific literature exists on methodologies and approaches in this area. Only recently have protein adducts (particularly hemoglobin) received attention as biomarkers to assess environmental contamination (see Chapter 7).

III. CONCLUSIONS

Xenobiotic chemicals, in the forms and concentrations found in the environment, often do not by themselves constitute a hazard to indigenous organisms. However, once exposure has occurred and substances are bioavailable, a sequence of biological responses may progress. Whether the wellbeing of the organism is eventually affected will depend on many factors, some intrinsic (e.g., age, sex, health, and nutritional status of the organism) and others extrinsic (e.g., dose, duration, route of exposure to the contaminant, and the presence of other chemicals). These factors represent obstacles to the assessment of exposure and subsequent risk from that exposure. However, biological markers (responses) can help circumvent these problems to a large extent by focusing on relevant molecular events that occur after exposure and metabolism. Reliable and sensitive analytical methods are being used to clarify the relationship between exposure to xenobiotic

**Table 4. Assays for Detection of DNA
Structural Alterations**

Assay	DNA[a] (µg)	Blood[b] (mL)
DNA adducts		
[32]P postlabeling	0.1	0.02
HPLC/fluorescence	100	20
DNA strand breaks		
Alkaline unwinding	5	1
Gel electrophoresis	1	0.2
DNA modified bases		
8-OH guanine	100	20
[5]m-dCyd	1	0.2

[a] Amount reported is per assay and may vary depending on degree of purification, intrinsic sensitivity of assay, and number of assays to be performed.

[b] Volume will vary depending on species selected. Value reported based on 10^4 white blood cells per cubic millimeter blood. Use of nonmammalian species which contain nucleated red blood cell (10^6 cells per cubic millimeter blood) will require less blood.

compounds and their effects. This is particularly evident in the field of environmental genotoxicology where alterations to DNA serve as biomarkers.

Background levels of DNA adducts, strand breaks, mutations, and other DNA alterations occur as a result of natural phenomena, such as ionizing radiation and dietary components. These levels can vary among species and among tissues within a single species. The ability to measure contamination-induced DNA alterations is directly dependent on an accurate measurement of the background levels of such alterations. These DNA alterations may potentiate irreversible changes to the DNA molecule and result in expression of other cellular responses, such as chromosomal aberrations and oncogene activation. The ability to predict a toxic cellular response from a single biomarker will be limited by the complex interactions that characterize biological systems. It is likely that the prediction of a toxic effect such as tumorigenesis will be enhanced through the use of a battery of biomarkers that span a range of cellular processes. Although not all DNA alterations will produce harmful effects, they have the potential for increasing the likelihood for long-term, deleterious effects, not only for the organism but also for future generations. Therefore, there is the obvious need to correlate DNA damage with genotoxic responses and endpoints other than cancer.

A concluding comment should be made about the availability of DNA in blood for genotoxcity testing. An approximation of the amount of blood/DNA needed to assess DNA alterations by the various techniques discussed is listed in Table 4. DNA will be restricted to a nucleated cell whose DNA content, on the average, is about 5 picograms (pg) per nucleus. In mammalian vertebrates, the availability of a sufficient number of nucleated cells may preclude the use of a specific methodology or technique for nondestructive sampling due to the inherent sensitivity of the test. For this reason nonmammalian vertebrates such as fishes, amphibians, reptiles, and birds are more suitable candidates because they possess nucleated red blood cells. Thus in these species a considerable amount of high molecular weight DNA could be obtained from a very small sample of blood without having to sacrifice the animal.

ACKNOWLEDGMENTS

The Oak Ridge National Laboratory is managed by Martin Marietta Energy Systems Inc., under contract DE-AC05-84OR21400 with the U.S. Department of Energy, Environmental Sciences Division Publication No. 4111.

REFERENCES

1. Shugart, L. R., Adams, S. M., Jimenez, B. D., Talmage, S. S., and McCarthy, J. F., Biological markers to study exposure in animals and bioavailability of environmental contaminants, in *ACS Symposium Series 382, Biological Monitoring for Pesticide Exposure: Measurement, Estimation, and Risk Reduction,* Wang, R. G. M., Franklin, C. A., Honeycutt, R. C., and Reinert, J. C., Eds., American Chemical Society, Washington, DC, 1989, p. 86.
2. McCarthy, J. F. and Shugart, L. R., *Biological Markers of Environmental Contamination,* Lewis Publishers, Boca Raton, FL, 1990.
3. McCarthy, J. F., Halbrook, R. S., and Shugart, L. R., *Conceptual Strategy for Design, Implementation, and Validation of a Biomarker-Based Biomonitoring Capability,* Oak Ridge National Laboratory/TM-11783, Oak Ridge, TN, 1992.
4. Shugart, L. R., McCarthy, J. F., and Halbrook, R. S., Biological markers of environmental and ecological contamination: an overview, *Risk Anal.,* 12, 353, 1992.
5. Huggett, R. J., Kimerle, R. A., Mehrle, P. M., and Bergman, H. L., *Biomarkers: Biochemical, Physiological, and Histological Markers of Anthropogenic Stress,* Lewis Publishers, Boca Raton, FL, 1992.
6. Peakall, D., *Animal Biomarkers as Pollution Indicators,* Ecotoxicology Series 1, Depledge, M. H., Sanders, B., Eds., Chapman & Hall, London, 1992.
7. Shugart, L. R., Biological monitoring: testing for genotoxicity, in *Biological Markers of Environmental Contaminants,* McCarthy, J. F. and Shugart, L. R., Eds., Lewis Publishers, Boca Raton, FL, 1990, p. 205.
8. Shugart, L. R., DNA damage as an indicator of pollutant-induced genotoxicity, in *13th Symp. Aquat. Toxicol. Risk Assessment: Sublethal Indicators of Toxic Stress,* Landis, W. G. and van der Schalie, W. H., Eds., ASTM, Philadelphia, PA, 1990, p. 348.
9. Ames, B. N., Durston, W. E., Yamasaki, E., and Lee, R. D., Carcinogens are mutagens: a simple test system combining liver homogenates for activation and bacteria for detection, *Proc. Natl. Acad. Sci. U.S.A.,* 70, 2281, 1973.

10. Kocan, R. M. and Powell, D. B., Anaphase aberrations: an in vitro test for assessing the genotoxicity of individual compounds and complex mixtures, in *Short-Term Genetic Bioassays in the Analysis of Complex Environmental Mixtures IV*, Water, M. D. S., Sandhu, S. S., Lewtas, J., Claxton, L., Strauss, G., and Nesnow, S., Eds., Plenum Press, New York, 1985, p. 75.

11. Sandhu, S. S. and Lower, W. R., In situ assessment of genotoxic hazards of environmental pollution, *Toxicol. Ind. Health,* 5, 73, 1989.

12. Sancar, A. and Sancar, G. B., DNA repair enzymes, *Annu. Rev. Biochem.,* 57, 29, 1988.

13. Shugart, L. R., Bickham, J., Jackim, G., McMahon, G., Ridley, W., Stein, J., and Steiner, S., DNA alterations, in *Biomarkers: Biochemical, Physiological, and Histological Markers of Anthropogenic Stress,* Huggett, R. J., Kimerle, R. A., Mehrle, P. M., and Bergman, H. L., Eds., Lewis Publishers, Boca Raton, FL, 1992, p. 127.

14. Weinstein, I. B., Current concepts on mechanism of chemical carcinogenesis, *Bull. N.Y. Acad. Med.,* 54, 336, 1978.

15. Wogan, G. N. and Gorelick, N. J., Chemical and biochemical dosimetry to exposure to genotoxic chemicals, *Environ. Health Perspect.,* 62, 5, 1985.

16. Randerath, K., Reddy, M. V., and Gupta, R. C., ^{32}P-Labeling test for DNA damage, *Proc. Natl. Acad. Sci. U.S.A.,* 78, 6126, 1981.

17. Bartsch, H., Hemminki, K., and O'Neill, I. K., *IARC Scientific Publication No. 89, Methods for Detecting DNA Damaging Agents in Humans: Application in Cancer Epidemiology and Prevention,* IRAC, Lyon, France, 1988.

18. Dunn, B., Black, J., and Maccubbin, A., ^{32}P-Postlabeling analysis of aromatic DNA adducts in fish from polluted areas, *Cancer Res.,* 47, 6543, 1987.

19. Kurelec, B., Garg, A., Krca, S., Chacko, M., and Gupta, R. C., Natural environment surpasses polluted environment in inducing DNA damage in fish, *Carcinogenesis,* 7, 1337, 1989.

20. Kurelec, B., Garg, A., Krca, S, and Gupta, R. C., DNA adducts in marine mussel *Mytilus galloprovincialis* living in polluted and unpolluted environments, in *Biomarkers of Environmental Contamination,* McCarthy, J. F. and Shugart, L. R., Eds., Lewis Publishers, Boca Raton, FL, 1990, p. 217.

21. Varanasi, U., Reichert, W. L., and Stein, J. E., ^{32}P-Postlabeling analysis of DNA adducts in liver of wild English Sole *Paraophrys vetulus* and winter flounder *Pseudopleuronectes americanus, Cancer Res.,* 49, 1171, 1989.

22. Ray, S., Dunn, B. P., Payne, J. F., Fancey, L., Helbig, R., and Beland, P., Aromatic DNA-carcinogen adducts in beluga whales from the Canadian Arctic and Gulf of St. Lawrence, *Mar. Environ. Pollut. Bull.,* 22, 392, 1991.

23. Gupta, R. C., Reddy, M. C., and Randerath, K., ^{32}P-Labeling analysis of nonradioactive aromatic carcinogen-DNA adducts, *Carcinogenesis,* 3, 1081, 1982.

24. Gupta, R. C. and Randerath, K., Analysis of DNA adducts by ^{32}P-labeling and thin layer chromatography, in *DNA Repair,* Vol. 3, Friedberg, E. C. and Hanawalt, P. C., Eds., Marcel Dekker, New York, 1988, p. 399.

25. Gupta, R. C., Enhanced sensitivity of ^{32}P-postlabeling analysis of aromatic carcinogen-DNA adducts, *Cancer Res.,* 45, 5656, 1985.

26. Reddy, M. V. and Randerath, K., Nuclease P1-mediated enhancement of sensitivity of ^{32}P-postlabeling test for structurally diverse DNA adducts, *Carcinogenesis.* 7, 1543, 1985.

27. Dunn, B. P. and San, R. H. C., HPLC enrichment of hydrophobic DNA-carcinogen adducts for enhanced sensitivity of ^{32}P-postlabeling analysis, *Carcinogenesis,* 9, 1055, 1988.

28. Randerath, K., Reddy, M. C., and Disher, R. M., Age- and tissue-related DNA modifications in untreated rats: detection by ^{32}P-postlabeling assay and possible significance for spontaneous tumor induction and aging, *Carcinogenesis,* 7, 1615, 1986.

29. Rahn, R., Chang, S., Holland, J. M., and Shugart, L. R., A fluorometric-HPLC assay for quantitating the binding of benzo[a]pyrene metabolites to DNA, *Biochem. Biophys. Res. Commun.,* 109, 262, 1982.

30. Shugart, L. R., Holland, J. M., and Rahn, R., Dosimetry of PAH carcinogenesis: covalent binding of benzo[a]pyrene to mouse epidermal DNA, *Carcinogenesis,* 4, 195, 1983.

31. Shugart, L. R. and Kao, J., Examination of adduct formation in vivo in the mouse between benzo[a]pyrene and DNA of skin and hemoglobin of red blood cells, *Environ. Health Perspect.,* 62, 223, 1985.

32. Shugart, L. R., McCarthy, J. M., Jimenez, B. D., and Daniel, J., Analysis of adduct formation in the Bluegill Sunfish *Lepomis macrochirus* between benzo[a]pyrene and DNA of the liver and hemoglobin of the erythrocyte, *Aquat. Toxicol.,* 9, 319, 1987.

33. Martineau, D., Legace, A., Beland, P., Higgins, R., Armstrong, D., and Shugart, L. R., Pathology of stranded beluga whales *Delphinapterus leucas* from the St. Lawrence estuary, Quebec, Canada, *J. Comp. Pathol.,* 98, 287, 1988.

34. Santella, R. M., Application of new techniques for the detection of carcinogen adducts to human population monitoring, *Mutat. Res.,* 205, 271, 1988.

35. Alberts, B., Bray, D., Lewis, L., Raff, M., Roberts, K., and Watson, J. D., *Molecular Biology of the Cell,* 2nd ed., Garland Publishing, New York, 1989, p. 220.

36. Rydberg, B., The rate of strand separation in alkali of DNA of irradiated mammalian cells, *Radiat. Res.,* 61, 274, 1975.

37. Kohn, K. W., Erickson, L. C., Ewig, A. G., and Friedman, C. A., Fractionation of DNA from mammalian cells by alkaline elution, *Biochemistry,* 15, 4629, 1976.

38. Olive, P. L., DNA precipitation assay: a rapid and simple method for detecting DNA damage in mammalian cells, *Environ. Mol. Mutagen.,* 11, 487, 1988.

39. Shugart, L. R., Quantitation of chemically induced damage to DNA of aquatic organisms by alkaline unwinding assay, *Aquat. Toxicol.,* 13, 43, 1988.

40. Kanter, P. M. and Schwartz, H. S., A hydoxylapatite batch assay for quantitation of cellular DNA damage, *Anal. Biochem.,* 97, 77, 1979.

41. Daniel, F. B., Haas, D. L., and Pyle, S. M., Quantitation of chemically induced DNA strand breaks in human cells via an alkaline unwinding assay, *Anal. Biochem.,* 144, 390, 1985.

42. Morris, S. R. and Shertzer, H. G., Rapid analysis of DNA strand breaks in soft tissues, *Environ. Mutagen.,* 7, 871, 1985.

43. Kanter, P. M. and Schwartz, H. S., A fluorescence enhancement assay for cellular DNA damage, *Mol. Pharmacol.,* 22, 145, 1982.

44. Theodorakis, C. W., D'Surney, S. J., Bickham, J. W., Lyne, T. B., Bradley, B. P., Hawkins, W. E., Farkas, W. L., McCarthy, J. F., and Shugart, L. R., Sequential expression of biomarkers in bluegill sunfish exposed to contaminated sediment, *Ecotoxicology,* 1, 45, 1992.

45. Singh, N. P., McCoy, M. T., Tice, R. R., and Schneider, E. L., A simple technique for quantitation of low levels of DNA damage in individual cells, *Exp. Cell Res.,* 175, 184, 1988.

46. Olive, P. L., Wlodek, D., Durand, R. E., and Banath, J. P., Factors influencing DNA migration from individual cells subjected to gel electrophoresis, *Exp. Cell Res.*, 198, 259, 1992.

47. Corcoran, G. B. and Ray, S. D., The role of the nucleus and other compartments in toxic cell death produced by alkylating hepatotoxicants, *Toxicol. Appl. Pharmacol.*, 113, 167, 1992.

48. Bihari, N., Batel, R., and Zahn, R. K., DNA damage determination by alkaline elution technique in the haemolymph of mussel *Mytilu galloprivincialis* treated with benzo[a]pyrene and 4-nitroquinolin-*N*-oxide, *Aquat. Toxicol.*, 18, 13, 1990.

49. Meyers-Schone, L., Shugart, L. R., Beauchamp, J. J., and Walton, B. T., Comparison of two freshwater turtle species as monitors of chemical contamination: DNA damage and residue analysis, *Environ. Toxicol. Chem.*, 12, 1993.

50. Nacci, D. and Jackim, E., Using the DNA alkaline unwinding assay to detect DNA damage in laboratory and environmentally exposed cells and tissues, *Mar. Environ. Res.*, 28, 333, 1989.

51. Razin, A. and Riggs, A. D., DNA methylation and gene function, *Science*, 210, 604, 1980.

52. Ehrlich, M. and Yang, R. Y. H., 5-Methylcytosine in eukaryotic DNA, *Science*, 212, 1350, 1981.

53. Holliday, R., The inheritance of epigenetic defects, *Science*, 238, 163, 1987.

54. Boehim, T. L. and Drahovsky, D., Alteration of enzymatic methylation of DNA cystosines by chemical carcinogens: a mechanism involved in the initiation of carcinogenesis, *J. Natl. Cancer Inst.*, 71, 429, 1983.

55. Wilson, V. L. and Jones, P. A., Inhibition of DNA methylation by chemical carcinogens in vitro, *Cell*, 32, 229, 1983.

56. Pfeifer, G. P., Grungerger, D., and Drahovsky, D., Impaired enzymatic methylation of BPDE-modified DNA, *Carcinogenesis*, 5, 931, 1984.

57. Uziel, M., Koh, C. K., and Cohn, W. E., Rapid ion-exchange chromatographic microanalysis of ultraviolet-absorbing materials and its application to nucleosides, *Anal. Biochem.*, 25, 77, 1965.

58. Shugart, L. R., 5-Methyl deoxycytidine content of DNA from Bluegill Sunfish *Lepomis macrochirus* exposed to benzo[a]pyrene, *Environ. Toxicol. Chem.*, 9, 205, 1990.

59. DiGiulio, R. T., Washburn, P. C., Wenning, R. J., Winston, G. W., and Jewell, C. C., Biochemical responses in aquatic animals: a review of determinants of oxidative stress, *Environ. Toxicol. Chem.*, 8, 1103, 1989.

60. Floyd, R. A., Watson, J. J., Wong, P. K., Altmiller, D. H., and Rickard, R. C., Hydroxyl free radical adduct of deoxyguanosine: sensitive detection and mechanisms of formation, *Free Radical Res. Commun.*, 1, 163, 1986.

61. Malins, D. C., Ostrander, G. K., Haimanot, R., and Williams, P., A novel DNA lesion in neoplastic livers of feral fish: 2,6,-diamino-4-hydroxy-5-formamidopyrimidine, *Carcinogenesis*, 11, 1045, 1990.

62. Malins, D. C., McCain, B. B., Brown, D. W., Chan, S. L., Myers, M. S., Landahl, J. T., Prohaska, P. G., Friedman, A. J., Rhodes, L. D., Burrows, D. G., Gronlund, W. D., and Hodgins, H. O., Chemical Pollutants in Sediments and Diseases in Bottom-dwelling Ash in Puget Sound, Washington, *Environ. Sci. Technol.*, 18, 705, 1984.

63. Halbrook, R. S., Kirkpartick, R. L., Bevan, D. R., and Dunn, B. P., DNA adducts detected in muskrats by ^{32}P-Postlabeling analysis, *Environ. Toxicol. Chem.*, 11, 1605, 1992.

CHAPTER 6

Genotoxic Responses in Blood Detected by Cytogenetic and Cytometric Assays

TABLE OF CONTENTS

0-87371-648-5/94/$0.00+$.50
© 1994 by Lewis Publishers

I. INTRODUCTION

A. The Biomarker Approach

As environmental degradation resulting from human activities continues, the need for sensitive, accurate, and repeatable assays for use in risk assessment increases. Many traditional tests are useful for the measurement and detection of chemicals in the environment, and often the toxicity of such chemicals has been determined by laboratory experiments using well-accepted toxicity assays.[1] Knowledge of the chemistry and mode of action within living systems of toxic chemicals has resulted in a greater understanding of their potential risks to human health. Unfortunately, the environment places severe constraints on the applicability of traditional physicochemical assays, as well as on the inference of the toxicity of chemicals from laboratory assays, in environmental risk assessment. This is because organisms in polluted environments are usually exposed to complex mixtures of potentially toxic chemicals. The additive, antagonistic, or synergistic effects of these complex mixtures are not well understood. Our ability to predict the effects of environmental pollutants on an ecosystem is further complicated by the differential sensitivity found among species[2,3] as well as the ability of such chemicals to be taken up by resident populations of organisms and passed, or concentrated, through the food chain.

The biomarker approach was proposed as an alternative means of environmental risk assessment.[4-8] This approach uses indigenous populations of animals and attempts to assess the effects of xenobiotic chemicals on the health of individuals. Biomarkers are the biochemical, molecular, physiological, morphological, or behavioral responses that result from toxic action of chemicals to which the organism is exposed. Thus, the biomarker approach goes beyond simply measuring or detecting the presence of toxic chemicals in tissues or the environment. Instead it is a measure of the biological effects of those chemicals on sentinel organisms. There are many possible approaches to designing biomarker studies. In my laboratory, we are interested in the development of cytological, cytometric, and genetic assays to detect the effects of a major class of xenobiotics — environmental mutagens. We follow the concept of Depledge[4] who espouses the use of multiple biomarkers for environmental risk assessment.

B. Environmental Genotoxicology

The exposure of animals to environmental genotoxins can result in damage to the structural integrity of chromosomes.[9] Chromosomal alterations can be expressed as chromatid damage such as gaps; and chromosomal rearrangements such as inversions, translocations, and acentric fragments that can be observed in metaphase cells. Acentric fragments that lag at anaphase lead to the formation of micronuclei that can be observed in interphase cells.[10-12] Subsequent cell divisions that occur after the integrity of chromosomes has been compromised lead to the production of cells and cell populations with altered DNA content. Such cells can be detected by flow cytometry which measures the amount of DNA, as well as other chemical and physical features of cells. The purpose of this chapter is to review the application of chromosome analyses, micronucleus assays, and flow cytometry to studies of environmental contamination and its effects on vertebrate animals. In particular, I emphasize the potential use of these procedures in blood and other tissues that can be obtained by relatively noninvasive biopsy.

II. APPROACHES AND METHODOLOGIES

A. Chromosome Studies

The detection of chromosomal aberrations is done by direct visual examination of metaphase spreads stained by conventional techniques. In vertebrates, such studies can be performed by in vivo or in vitro procedures. The necessity to obtain a high frequency of metaphase cells limits the source of cells to tissues that proliferate rapidly such as bone marrow in mammals and birds, or to cell types that can be induced to proliferate in culture such as white blood cells or fibroblasts.

One of the most widely used procedures for chromosome studies is the short-term leukocyte culture.[13] This procedure is commonly used in medical labs and simply requires a sterile blood sample from which leukocytes are extracted and cultured. The advantages of this technique are that it can be performed from blood and thus is nondestructive to the animal, it can potentially be modified for use with mammalian and nonmammalian blood, and it can result in very high quality metaphase spreads obtained under controlled conditions. Also, it is a relatively short-term test; cells are usually cultured for 3 days. Disadvantages of the procedure are that, although the procedure and culture media are well developed for use with humans, the procedure works less well with lower vertebrates and even with other species of mammals. The difficulty with culturing nonhuman leukocytes can be daunting. Nonetheless, excellent formulations for culture media have been developed in recent years for blood cell culture, and several mitogenic chemicals are available as well. It would seem that the technical difficulties associated with blood cell cultures of animals can be overcome with some patience and experimentation.[13-15]

Another procedure to obtain high-quality chromosome spreads is the fibroblast culture technique.[16] An advantage of this technique is that fibroblasts can be obtained from many tissues including external tissues such as fish fins or mammalian pinnae. Moreover, fibroblasts can be grown for many generations whereas blood cells cannot be maintained. However, as with leukocytes, the culture of fibroblasts from lower vertebrates is not as easily accomplished as in mammals, in general, and humans, in particular. The establishment of fibroblast cultures usually takes a matter of weeks, not days.

In vivo procedures for obtaining metaphase cells have been developed for most vertebrates.[14] These procedures work well and are faster and cheaper than any culture technique. However, they often require the sacrifice of the animal because bone marrow or spleen are two of the most commonly used sources of cells. Moreover, the quality and number of metaphases are not usually as good as with cell cultures. Nonetheless, in vivo karyotype procedures can be adapted for use on

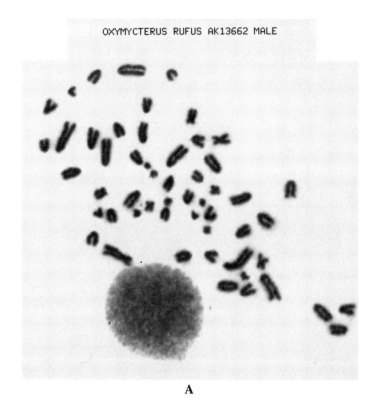

A

leukocytes and potentially are an important source of information regarding genotoxic effects.

B. Micronucleus Assay

The micronucleus assay[10-12] is a simpler and more rapid test for clastogenic activity or aneuploidy than standard metaphase chromosome analysis. Micronuclei are formed in cells as a result of chromosomes or chromosome fragments that lag at anaphase and are thus not included in the main nucleus when it is formed. Micronuclei are observed as small pieces of chromatin outside the nucleus in cells that normally have only a single nucleus or in mature mammalian red blood cells that normally have no nucleus. In theory, almost any tissue could be examined for micronuclei but usually blood cells or bone marrow cells are used.[18-19] Advantages of this procedure are that larger numbers of cells are examined in micronucleus assays (n = 1000) than in chromosome studies (n = 50–100), and the procedure can be performed noninvasively. A disadvantage is that only a subset of types of chromosomal aberrations produce micronuclei.

Both chromosome studies and the micronucleus assay are well accepted toxicological endpoints.[9] Extensive research has shown that both of these endpoints respond positively with increased dosage and exposure to mutagenic agents. Moreover, both are relatively sensitive assays. However, both assays are time-consuming and laborious. The use of imaging systems to analyze data will help

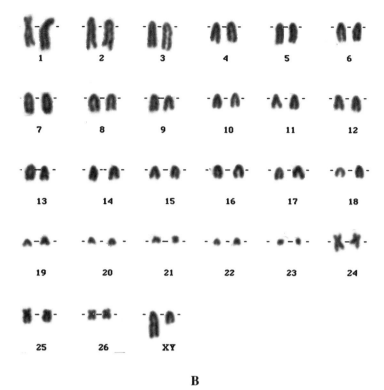

B

FIGURE 1. (A) Metaphase plate and (B) corresponding karyotype produced by use of the Genetiscan automated karyotyping program.

streamline these procedures. However, processing tissues, culturing cells, slide preparation, and visual examination of slides will always be major impediments in these studies. More fully automated systems that are less costly are needed for large-scale studies. One such system is being used by us through the Center for Biosystematics and Biodiversity at Texas A&M University. This system, called the Genetiscan (Perceptive Systems, Inc.), captures and digitizes the image of a metaphase plate. Software allows for the enhancement of the image and for the construction of the karyotype. Prints of the image and karyotype are produced on a laser printer (Figure 1). The use of the Genetiscan negates the need for extensive darkroom work and allows for the storage of images as computer files rather than photographic negatives. We have found this system to greatly enhance our ability to conduct environmental toxicological research with chromosomal or cytological techniques.

C. Flow Cytometry

Flow cytometry (FCM) is a rapid technique for the quantification of biochemical and structural characteristics of cells.[20] Macromolecules, including DNA, RNA, and proteins, that can be labeled with a fluorescent tag can be measured in large cell populations — tens of thousands to hundreds of thousands of cells are typically used. Structural features such as cell size and internal granularity also are measured by light-scatter parameters. Advantages of FCM for mutagen testing are numerous. Because very large numbers of cells are studied, FCM potentially has very high resolution. Subtle effects or rare events that might not be statistically significant in studies utilizing smaller numbers of cells might be observed with the use of flow cytometry. Virtually any tissue can be analyzed by flow cytometry, not just rapidly proliferating ones.[21] Moreover, different cell types within tissues can be differentiated and analyzed separately. Because of these properties, target-specific effects of mutagens can be examined as well as general effects.

Flow cytometry is a rapid procedure; flow rates of up to 1000 cells per second are used which make it possible to examine large numbers of individuals. This is ideal for the initial screening of potential effects in populations because large sample sizes offer the possibility of identification of subsets of the population that might be at risk or highly sensitive. Flow cytometry is also easily adapted to studies of a diversity of organisms.[3] Data can be obtained from lower vertebrates as easily as from mammals. Flow cytometry also is well suited for analyses of blood and other tissues that can be nondestructively sampled. Finally, FCM is highly versatile in that multiple biomarkers can be simultaneously assayed including mutagenic, cytotoxic, and cytostatic effects.[22]

Although flow cytometry is clearly a useful procedure for toxicological studies, it nonetheless has its disadvantages. Flow cytometry lacks the history of validation through controlled laboratory studies that chromosomal and micronucleus assays possess. Thus, the endpoints produced by FCM are not as well accepted and are not as well understood, as are chromosome aberrations and micronuclei in the field of toxicology. This is particularly true in the regulatory arena.

D. Validation

Before toxicity assays should be applied in field studies they should be extensively validated by controlled laboratory studies. Such studies have been performed for many years for the chromosome and micronucleus assays. These assays are already well accepted as valid biomarkers in toxicity studies. Ample evidence now exists in the literature to make a strong case for accepting DNA alterations as measured by FCM as an equally valid endpoint for genetic toxicity.

The studies of Otto and Oldiges[23] and Otto et al.[24] were among the earliest to demonstrate the effects of exposure to mutagenic chemicals on DNA histograms of cell populations in vivo and in vitro. These authors showed that exposure to a variety of chemical mutagens and X-rays resulted in dosage-dependant increases in the coefficients of variation (CVs) of chromosomes and nuclei as measured by flow cytometry. Subsequent laboratory dosage studies of chemical mutagens and ionizing radiation have confirmed the findings of Otto and co-workers.[23,24] We studied the effects of the anticancer drug triethylenemelamine (TEM) using various dosages and times of exposure[21,25] to determine whether our procedures could reproduce effects similar to those observed by Otto and Oldiges[23] and Otto et al.[24] We chose TEM because it is a powerful clastogen that has been shown to cause cancer, point mutations, heritable translocations, DNA adducts, chromosomal aberrations, and spermatogenic effects in male mice.[26-29] Our studies[21,25] demonstrated that exposure of male rats to TEM in vivo causes increased CVs of the G_1 peak of blood cells, spleen, and bone marrow; and of the 1C peak of testis. These results were observed in some tissues at relatively low dosage levels (as low as 0.1 mg/kg) and after as little as 24-hr exposure with some dosages. We also observed that different tissues show different response times and recovery rates, and that the results of the flow cytometry assays paralleled a chromosome assay. We therefore conclude that DNA alterations are detectable by flow cytometry and that this assay should be well suited for environmental toxicity studies as previously proposed.[3,30]

E. Field Studies

The cytological and cytometric assays described above have been applied to studies of natural populations of vertebrates exposed in a variety of situations to environmental mutagens. In this section we review several studies that exemplify the use of chromosome aberrations, micronuclei, and flow cytometry to detect genetic damage in the environment.

One of the first studies to demonstrate that exposure to chemical mutagens in the environment causes chromosomal aberrations utilized native rodents collected at the Firemen's Training School (FS) in College Station, TX.[31] Two species of rodents *(Peromyscus leucopus* and *Sigmodon hispidus)* were collected in the vicinity of retention ponds contaminated with a complex mixture of petrochemical wastes, PCBs, and heavy metals. Mutagenicity and contamination of the site has been well documented by chemical and bacterial assays.[32-34] Examination of metaphase cells revealed a statistically significant increase in the number of chromosomal lesions per cell and in the number of aberrant cells per individual in both species of rodents taken from the FS compared to matched control sites.[31] Moreover, significant increases were observed in all types of chromosomal aberrations scored including chromatid breaks, chromosome breaks, acentric fragments, dicentric chromosomes, ring chromosomes, and translocations at the FS. However, no significant differences in aberrations were observed between rodents taken from two control sites. Although both species showed strong evidence for genetic damage, *Peromyscus leucopus* appeared to be the most sensitive species because it possessed significantly more lesions per cell than did *Sigmodon hispidus.*

McBee[35] examined G-band karyotypes of *Peromyscus leucopus* from the FS and a control site in order to determine whether differential chromosome banding could increase the resolution obtained in a chromosomal toxicity assay. She identified several classes of chromosomal aberrations from G-bands that were not identified in the previous study of nondifferentially stained chromosomes.[31] Moreover, she found chromosomes 5 and 21 to have expressed significantly higher frequencies of aberrations than other chromosomes. Thus, the use of various chromosome banding procedures[16] holds promise for enhancing the use of chromosome integrity in toxicity assays. It should be noted, however, that differential staining procedures are more time-consuming than standard karyotype analysis.

The observation that cytological studies of natural populations of rodents can be used to demonstrate the mutagenicity of a polluted site[31] has been confirmed by subsequent studies. Thompson[36] also found increased chromosomal aberrations in *Sigmodon hispidus* taken from two hazardous waste sites compared to control site animals. Tice et al.[37] found increased numbers of micronuclei and sister-chromatid exchanges in *Peromyscus leucopus* from a hazardous waste site compared to a control site. Thus, data are now available for these two species from four polluted sites and in each case strong evidence of genetic damage was found. Although relatively few studies have been performed that investigate levels of cytological damage in natural populations,[38] the existing evidence certainly is promising for the future use of these techniques with native rodents as sentinel species.

McBee and Bickham[39] followed the chromosome aberration study of the rodents from the FS with a flow cytometric study of *Peromyscus leucopus* from that site. Once again, evidence was found for genetic damage as a result of exposure to chemical mutagens. In this case, there was an increase in the mean coefficients of variation of DNA content in G_1 cells. Also observed were a few individuals from the FS with apparent aneuploid peaks in their DNA histograms. Therefore, two distinct endpoints (aneuploidy and DNA content variation), both of which can reasonably be concluded to have resulted from mutagen exposure, were shown to have utility for applications in environmental exposure. Aneuploidy, however, has not been an identified result of mutagen exposure in laboratory studies.

Flow cytometric studies were performed on slider turtles exposed to low-level radioactivity at the Savannah River Site (SRS). In the first of these studies,[40] turtles were collected from seepage basins contaminated with radioactive cesium and strontium as well as certain nonradioactive chemical contaminants. Coefficients of variation were significantly increased in blood samples of the turtles taken from the radioactive seepage basins compared to animals taken from nonradioactive control sites. Moreover, a few individuals were identified with apparent aneuploid mosaicism. This study was followed by a flow cytometric analysis of slider turtles inhabiting another radioactive site, pond B, on the SRS. In that study,[41] flow cytometry was performed to better understand the environmental effects of low-level radioactivity. Levels of radionuclide contamination in pond B and total body burdens of radioactive cesium and strontium were considerably lower than for the seepage basins. Also, pond B has no significant levels of nonradioactive chemical contamination. Thus, any genetic effects are likely the result of exposure to radioactivity. As in the study of the seepage basins, significantly increased coefficients of variation were observed in blood cells from turtles from the radioactive pond. In addition, a few animals were observed with apparent aneuploid mosaicism. Thus, both of these endpoints (increased coefficients of variation and aneuploid mosaicism) have been observed repeatedly in environmental studies of animals chronically exposed to mutagens.

Another flow cytometric study of environmental radiation exposure was conducted on previously unexposed ducks placed on pond B. Over the period of 1 year, the internal uptake of radionuclides was monitored along with flow cytometric analysis of blood. Aneuploid mosaicism appeared in some individuals after long exposure time.[42] This study not only confirms and parallels the results of the turtle studies, but also demonstrates the use of blood DNA histograms as a useful nondestructive biomarker.

III. CONCLUSIONS

The procedures described in this chapter have proved to be reliable, accurate, and repeatable both in laboratory and field studies and should be included as routine assays in future biomarker research and in environmental biomonitoring. Moreover, these procedures are readily applicable to the nondestructive sampling and assessment of blood. We have already successfully used flow cytometric assays on blood samples from fish, reptiles, birds, and mammals — thus illustrating the broad applicabilities of these procedures.

ACKNOWLEDGMENTS

I thank Mike Smolen and Rebecca Bickham for help in the preparation of the figure. The Center for Biosystematics and Biodiversity is funded by a grant from the National Science Foundation. This manuscript is contribution no. 12 from the Center for Biosystematics and Biodiversity, Texas A&M University.

REFERENCES

1. Lyne, T. B., Bickham, J. W., Lamb, T., and Gibbons, J. W., The application of bioassays in risk assessment of environmental pollution, *Risk Anal.*, 12, 361, 1992.
2. Kligerman, A. D., Bryant, M. F., Doerr, C. L., Halperin, E. C., Kwanyuen, P., Sontag, M. R., and Erexson, G. L., Interspecies cytogenetic comparisons: studies with X-radiation and bleomycin sulfate, *Environ. Mol. Mutagen.*, 19, 235, 1992.
3. Bickham, J. W., Flow cytometry as a technique to monitor the effects of environmental genotoxins on wildlife populations, in *In Situ Evaluations of Biological Hazards of Environmental Pollutants,* Sandhu, S. S., Lower, W. R., de Serres, F. J., Suk, W. A., and Tice, R. R., Eds. Plenum Press, New York, 1990, p. 97.
4. Depledge, M. A., The rational basis for detection of the early effects of marine pollutants using physiological indicators, *AMBIO,* 18, 301, 1989.
5. McCarthy, J. F. and Shugart, L. R., *Biological Markers of Environmental Contamination,* Lewis Publishers, Boca Raton, FL, 1990.
6. McCarthy, J. F., Halbrook, R. S., and Shugart, L. R., *Conceptual Strategy for Design, Implementation, and Validation of a Biomarker-Based Biomonitoring Capability,* Oak Ridge National Laboratory/TM-11783, Oak Ridge, TN, 1992.
7. Peakall, D., *Animal Biomarkers as Pollution Indicators.* Ecotoxicology Series 1, Depledge, M. H. and Sanders, B., Eds. Chapman & Hall, London, 1992.
8. Shugart, L. R., McCarthy, J. F., and Halbrook, R. S., Biological markers of environmental and ecological contamination: an overview, *Risk Anal.* 12, 353, 1992.

9. Hsu, T. C., *Cytogenetic Assays of Environmental Mutagens,* Allanheld, Osmun Publishers, Totowa, NJ, 1982.

10. Heddle, J. A., A rapid *in vivo* test for chromosome damage, *Mutat. Res.,* 18, 187, 1973.

11. Schmid, W., Chemical mutagen testing on *in vivo* somatic mammalian cells, *Agents Actions,* 3, 77, 1973.

12. Schmid, W., The micronucleus test, *Mutat. Res.* 31, 9, 1975.

13. Leonard, A., Decat, G., and Fabry, L., The lymphocytes of small mammals: a model for research in cytogenetics?, *Mutat. Res.,* 95, 31, 1982.

14. de Boer, P., van Buul, P. P. W., van Beek, R., van der Hoeven, F. A., and Natarajan, A. T., Chromosomal radiosensitivity and karyotype in mice using cultured peripheral blood lymphocytes, and comparison with this system in man, *Mutat. Res.,* 42, 379, 1977.

15. Erexson, G. L. and Kligerman, A. D., A modified mouse peripheral blood lymphocyte culture system for cytogenetic analysis, *Environ. Mutagen.,* 10, 377, 1987.

16. Verma, R. S. and Babu, A., *Human Chromosomes. Manual of Basic Techniques,* Pergamon Press, New York, 1989.

17. Baker, R. J., Haiduk, M. W., Robbins, L. W., Cadena, A., and Koop, B. F., Chromosomal studies of South American bats and their systematic implications, in *Mammalian Biology in South America,* Vol. 6, Mares, M. A. and Genoways, H. H., Eds. Special Publication Series, Pymatuning Laboratory of Ecology, University of Pittsburgh, Pittsburgh, 1982, p. 303.

18. Erexson, G. L., Kligerman, A. D., Bryant, M. R., Sontag, M. R., and Halpern, E. C., Induction of micronuclei by x-radiation in human, mouse, and rat peripheral blood lymphocytes, *Mutat. Res.,* 253, 193, 1991.

19. Erexson, G. L., Kligerman, A. D., Halpern, E. C., Honore, G. M., and Allen, J. W., Micronuclei in binucleated lymphocytes of mice following exposure to gamma radiation, *Environ. Mutagen.,* 13, 128, 1989.

20. Shapiro, H. M., *Practical Flow Cytometry,* 2nd ed., Alan R. Liss, New York, 1988.

21. Bickham, J. W., Sawin, V. L., Burton, D. W., and McBee, K., Flow-cytometric analysis of the effects of triethylenemelamine on somatic and testicular tissues of the rat, *Cytometry,* 13, 368, 1992.

22. Maier, P. and Schawalder, H. P., A two-parameter flow cytometry protocol for the detection and characterization of the clastogenic, cytostatic and cytotoxic activities of chemicals, *Mutat. Res.,* 164, 369, 1986.

23. Otto, F. J. and Oldiges, H., Flow cytogenetic studies in chromosomes and whole cells for the detection of clastogenic effects, *Cytometry,* 1, 13, 1980.

24. Otto, F. J., Oldiges, H., Gohde, W., and Jain, V. K., Flow cytometric measurement of nuclear DNA content variations as a potential *in vitro* mutagenicity test, *Cytometry,* 2, 189, 1981.

25. Bickham, J. W., Sawin, V. L., McBee, K., Smolen, M. J., and Derr, J. N., Further studies of the effects of triethylenemelamine on somatic and testicular tissues of the rat using flow cytometry, *Cytometry,* in press.

26. Evenson, D. P., Baer, R. K., and Jost, L. K., Long-term effects of triethylenemelamine exposure on mouse testis cells and sperm chromatin structure assayed by flow cytometry, *Environ. Mol. Mutagen.* 14, 79, 1989.

27. Lawley, P. D., and Brookes, P., Molecular mechanism of the cytotoxic action of difunctional alkylating agents and resistance to this action, *Nature (London),* 206, 480, 1965.

28. Matter, B. E. and Generoso, W. M., Effects of dose on the induction of dominant-lethal mutations with triethylenemelamine in male mice, *Genetics,* 77, 753, 1974.

29. Rutledge, J. C., Cain, K. T., Cacheiro, N. L. A., Cornett, C. V., Wright, C. G., and Generoso, W. M., A balanced translocation in mice with a neurological defect, *Science,* 231, 395, 1986.

30. Deaven, L. L., Application of flow cytometry to cytogenetic testing of environmental mutagens, in *Cytogenetic Assays of Environmental Mutagens,* Hsu, T. C., Ed., Allanheld, Osmun Publishers, Totowa, NJ, 1982, p. 325.

31. McBee, K., Bickham, J. W., Donnelly, K. C., and Brown, K. W., Chromosomal aberrations in native small mammals *(Peromyscus leucopus* and *Sigmodon hispidus)* at a petrochemical waste disposal site. I. Standard karyology, *Arch. Environ. Contam. Toxicol.,* 16, 681, 1987.

32. Atlas, E. L., Donnelly, K. C., Giam, C. S., and McFarland, A. R., Chemical and biological characterization of emissions from a fire-person training facility, *Am. Ind. Hyg. Assoc. J.,* 46, 532, 1985.

33. Brown, K. W., Final Report to the Firemen's Training School of the Texas Engineering Extension Service. A Plan to Minimize the Volume of Runoff Water which must be Treated and Disposed of and an Assessment of the Feasibility of Land Disposal of the Runoff Water and Sludge, Soil and Crop Sciences Department, Texas A&M University, College Station, TX, 1980.

34. Brown K. W. and Donnelly, K. C., Mutagenic potential of water concentrates from the effluent of a waste oil storage pond, *Bull. Environ. Contam. Toxicol.,* 28, 424, 1982.

35. McBee, K., Chromosomal aberrations in native small mammals *(Peromyscus leucopus)* at a petrochemical waste disposal site. II. Cryptic and inherited aberrations detected by G-band analysis, *Environ. Toxicol. Chem.,* 10, 1321, 1991.

36. Thompson, R. A., Schroder, G. G., and Conner, T. H., Chromosomal aberrations in the cotton rat *(Sigmodon hispidus)* exposed to hazardous waste, *Environ. Mol. Mutagen.,* 11, 359, 1988.

37. Tice, R. R., Ormiston, B. G., Boucher, R., Luke, C. A., and Paquette, D. E., Environmental biomonitoring with feral rodent species, in *Short-Term Bioassays in the Analysis of Complex Environmental Mixtures,* Vol. 5, Sandhu, S. S., Demanine, D. M., Mass, M. J., Moore, M. M., and Mumford, J. L., Eds., Plenum Press, New York, 1987, p. 175.

38. McBee, K. and Bickham, J. W., Mammals as Bioindicators of Environmental Toxicity, in *Current Mammalogy,* Vol. 2, Genoways H. H., Ed., Plenum Press, New York, 1990, p. 37.

39. McBee, K., and Bickham, J. W., Petrochemical related DNA damage in wild rodents detected by flow cytometry, *Bull. Environ. Contam. Toxicol.,* 40, 343, 1988.

40. Bickham, J. W., Hanks, B, G., Smolen, M. J., Lamb, T., and Gibbons, J. W., Flow cytometric analysis of the effects of low level radiation exposure on natural populations of slider turtles *(Pseudemys scripta), Arch. Environ. Contam. Toxicol.,* 17, 837, 1988.

41. Lamb, T., Bickham, J. W., Gibbons, J. W., Smolen, M. J., and McDowell, S., Genetic damage in a population of slider turtles *(Trachemys scripta)* inhabiting a radioactive reservoir, *Arch. Environ. Contam. Toxicol.,* 20, 138, 1991.

42. George, L. S., Dallas, C. E., Brisbin, I. L., and Evans, D. L., Flow cytometric DNA analysis of ducks accumulating ^{137}Cs on a reactor reservoir, *Ecotoxicol. Environ. Saf.,* 21, 337, 1991.

CHAPTER 7

Hemoglobin Adducts*

Lee R. Shugart

TABLE OF CONTENTS

* The submitted manuscript has been authored by a contractor of the U.S. Government under contract No. DE-AC05-84OR21400. Accordingly, the U.S. Government retains a nonexclusive, royalty-free license to publish or reproduce the published form of this contribution, or allow others to do so, for U.S. Government purposes.

I. INTRODUCTION

Since deoxyribonucleic acid (DNA) in animal tissues, including that of humans, is sometimes difficult to obtain, ancillary molecules may serve as a surrogate. Damage to the surrogate molecule (protein adducts, for example) may provide evidence of DNA alterations provided protein damage is proportional to DNA damage in the target tissue/organ. Blood proteins such as hemoglobin and serum albumin have proved useful in this context.[1,2] Proteins adducts are useful for quantifying past exposure because they are stable under biological conditions, easily obtained for biomonitoring, exhibit a relationship to DNA adducts, and provide evidence of exposure over long periods of time.

II. EXPOSURE MONITORING

A. Background

Traditionally, exposure to environmental pollutants has been estimated from direct measurements of pollutant concentration in various environmental media such as air, water, soil, and food. For example, a classical method for assessing occupational exposure to volatile chemicals in industry is to measure the concentration of such chemicals in ambient air. Ambient monitoring of most environmental media has obvious drawbacks because it does not account for: (1) intermittent, continuous, or multiple sources of exposure; (2) routes of exposure (i.e., skin, lung, ingestion, etc.); or, (3) organismal variation. Since only that portion of the total exposure dose that is actually absorbed by the organism is relevant to potential health risk incurred by the organism, some *in vivo* measure of exposure is needed on which to base more realistic estimates of risk.

B. Biological Monitoring

The estimation of the risk resulting from exposure to environmental contaminants requires knowledge of the dose and pharmacokinetics of the chemicals. Biological monitoring represents an effort to estimate the *in vivo* dose of xenobiotics from measurements of biological markers (biomarkers). Biological monitoring entails making measurements of one or more of the following classes of biomarkers: (1) parent compounds or their metabolites in biological media; (2) reaction products with cellular macromolecules such as DNA and protein; and (3) biological endpoints such as altered enzyme levels or clastogenic events.

C. Biomarkers of Exposure

Once a chemical has been absorbed it is distributed via the general circulation to all tissues in the body of an organism. Most xenobiotics have relatively short biological half-lives, and their concentrations in body fluids therefore, change rapidly. Some may be excreted while others may be partitioned into storage tissues.

Measurements of the concentration of exogenous agent absorbed by an organism can serve as biomarkers of exposure. A systemic dose in an individual can be obtained by quantitation of the chemical or its metabolites in body fluids such as blood, urine, feces, sputum, or milk; however, a major drawback to this approach is that the analysis must be performed shortly after exposure, that is, prior to the clearance of the metabolites from the body unless, of course, it is partitioned elsewhere. For example, solvents such as benzene are measurable in blood and expired air, as are polychlorinated biphenyls in adipose tissue and breast milk and heavy metals in hair, teeth, and nails. However, before most environmental chemicals exert a biological effect, they must be metabolized to reactive forms, redistributed to target tissue(s), and react with critical macromolecules.

D. Macromolecular Adducts

Adducts are stable complexes of reactive chemicals that attach themselves through covalent binding to cellular macromolecules. The adverse health effects of most environmental contaminants are the result of covalent binding to physiologically important molecules. Organophosphate poisoning and chemically induced cancer are examples indicative of covalent binding to a protein (cholinesterase) and DNA, respectively. Adducts potentially can represent the most direct and biologically relevant indicator of exposure as well as the risk from that exposure.

It has to be assumed that most DNA adducts are intrinscially damaging events. This assumption might not be entirely correct since adducts may be readily repaired by excision without further consequences. Conversely, adducts on nucleotides in inert regions of DNA may persist, but not produce adverse effects. It should be noted that DNA adducts still provide evidence of specific exposure that has passed all of the toxicokinetic barriers. The detection of DNA adducts in environmental samples provides a means of investigating the qualitative and quantitative relationships between the formation of DNA adducts, subsequent DNA alterations, and resulting lesions in target tissues. Furthermore, characterization of specific DNA adducts may ultimately lead to the identification of a group of genotoxic chemicals of significant environmental concern.

III. HEMOGLOBIN ADDUCTS

A. Xenobiotic Metabolism

The rationale underlying the strategy for determining levels of compounds which become covalently bound to cellular macromolecules as biomarkers of exposure is based on our current understanding of the mode of action of genotoxic compounds.[3,4] The metabolism of chemicals is a relatively complex phenomenon. Although xenobiotic metabolism most often results in detoxification and elimination of the foreign compound, cellular detoxification may not always be this straightforward. Many compounds — most notably chemical carcinogens, organophosphorus insecticides, and certain cytotoxic agents — are "activated" by metabolism resulting in highly reactive electrophilic metabolites. These derivatives can undergo attack by nucleophilic centers in macromolecules such as lipids, proteins, DNA and ribonucleic acid (RNA) which often results in cellular cytotoxicity. Reactive metabolites formed in hepatocytes, for example, may react directly with macromolecules in the liver or may diffuse into the blood that perfuse that organ and bind to serum proteins or hemoglobin.

The binding of a chemical with DNA can cause formation of altered bases that can be repaired, be innocuous, or result in alterations which become fixed and are transmitted onto daughter cells. Current research suggests that the reaction of chemicals with DNA and ensuing changes which can result are causative of cancer.

B. Adduct Formation

Given the fact that genotoxic agents exert their activity through irreversible reactions with nucleophilic atoms, the amount of such reaction products will be proportional not only to the in vivo concentration of the electrophile, but also to the time this concentration is maintained. Therefore, the amount of metabolite bound (*in vivo* dose) provides a reliable dosimetric basis on which to assess the risk connected with exposure to a genotoxic compound.[4,5] However, for DNA adducts to be useful as a biomarker of exposure, it should occur in a tissue/cell that is accessible for analysis, be stable enough to accumulate over time, and be detectable by suitable analytical methods. Highly efficient DNA repair mechanisms may compromise these requirements. Rates of DNA repair can vary between adducts, tissue, and organisms. In addition, cell proliferation (hyperplasia), a common response to cytotoxic insult, may dilute the adduct to a nondetectable level.

Because it meets a number of essential requirements, hemoglobin has been proposed as a surrogate for DNA for estimating the in vivo dose of chemicals subsequent to exposure.[2,6-9] First, it has reactive nucleophilic sites and the reaction products with electrophilic agents are stable. Over 60 compounds to date have been shown to yield covalent reaction products with hemoglobin in animal experiments.[4,10,11] These compounds include representatives of most of the important classes of genotoxic chemicals currently known. No mutagenic or cancer-initiating compound has failed to produce covalent reaction products with hemoglobin. Second, hemoglobin has a well-established life span, is readily available in large quantities in humans and animals, and its concentration is not subject to large variation. Third, modification of hemoglobin has been shown to give an indirect measure of the dose to the DNA in cells which are potential targets for genotoxic agents.[1,7]

For the purpose of estimating exposure to potentially harmful environmental chemicals and the assessment of the effect to ecological systems and human health, biological monitoring of hemoglobin-bound metabolites, therefore, represents a novel approach. Although these adducts have no putative mechanistic role in carcinogenesis, they do relate quantitatively to activation and exposure since they approximate the systemic dose (*in vivo* dose) of a chemical. By way of target-organ dose, they can be a measure of the chemical genotoxic potency as well.

Furthermore, since the adducts which form with hemoglobin are stable and have life spans equal to that of the circulating erthyrocyte, quantitation of these adducts can be used to integrate the dose obtained from chronic lowlevel types of exposure such as occur on release of pollutants around toxic waste sites.

IV. ENVIRONMENTAL MONITORING

A. Background

The detection and measurement of adducts to hemoglobin has been seriously studied in humans exposed to hazardous occupational chemicals for some time, and a considerable amount of scientific literature exists for this subject.[10,11] As early as 1983, Ehrenberg et al.[1] examined the metabolism and subsequent damage to DNA by ethylene oxide using protein adducts as a surrogate for DNA adducts. It has not been until relatively recently that similar studies have been attempted in other organisms,[12] especially in relation to exposure to hazardous chemicals in their environment. This appears to be a very fruitful area for scientific investigation that has been restricted mainly by the availability of facile methods for the detection and measurement of protein adducts.

B. Methods

Methods for the detection of protein adducts vary considerably as to their sensitivity, selectivity, expense, and time consumption. However, as each cubic millimeter of blood contains over 10^6 red blood cells (varies with species), a milliliter of blood will have a considerable amount of hemoglobin from which pure globin can be easily obtained for analysis.

Adducted hemoglobin can be separated by electrophoresis since these molecules have a different charge than normal hemoglobin molecules. This method is useful for general monitoring but suffers from the ability to identify the chemical modification responsible. The usual method for identifying specific protein adducts is gas chromatography (GC) alone or in conjunction with selective ion monitoring mass spectrometry (MS). Two procedures are currently used. In one, the entire protein must first be reduced to its constituent amino acids before it can be analyzed. Modified amino acids are separated usually by high-performance liquid chromatography (HPLC), derivatized by perfluoracylation, and analyzed by GC/MS. In the second method, the tedious workup of total protein hydrolysates is eliminated by performing a modified Edman degradation procedure on the protein. In this procedure pentafluorophenyl isothiocyanate is coupled to the N-terminal amino acid of a protein under basic conditions. Subsequent treatment causes the release of the N-terminal amino acid as the pentafluorophenylthiohydantoin derivative which is suitable for analysis by GC/MS. Both procedures are extremely selective but require special analytical instrumentation; the former is time-consuming and costly.

Some chemicals such as benzo[a]pyrene (BaP) and other polycyclic aromatic hydrocarbons (PAHs) are intrinsically fluorescent. Adducts to proteins formed by these chemicals may be detected by fluorescence spectrophotometry. A method

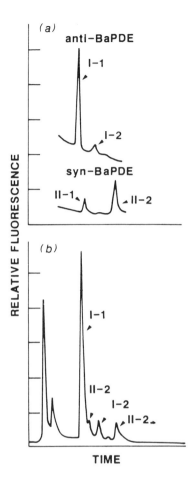

FIGURE 1. HPLC/fluorescence chromatogram of tetrols (a) liberated by hydrolysis of synthetic anti- and syn-B[a]P diolepoxides in water; and (b) liberated by acid hydrolysis hemoglobin isolated from mouse skin exposed to 3 μg benzo[a]pyrene per gram of body weight. Retention times of tetrol I-1, I-2, I-2, and II-1 are 10.5, 12.1, 14.2, and 17.9 min, respectively. (From Shugart, L. R. and Kao, J., *Environ. Health Perspect.,* 62, 223, 1985; Shugart, L. R., Holland, J. M., and Rahn, R., *Carcinogenesis,* 4, 195, 1983. With permission.)

employing this approach has recently been devised.[13,14] Essentially, the technique consists of the acid-induced removal of the diolepoxides of BaP from the hemoglobin (or DNA) as the strongly fluorescent free tetrols which are then separated and quantitated by HPLC/fluorescence analysis. The resulting chromatographic profile (Figure 1) can be used to establish the stereochemical origin of the diolepoxides involved in adduct formation.[14] This method is sensitive and can detect as little as 5 pg of tetrol, but is restricted to the detection of the ubiquitous pollutant, BaP.

C. Studies

As stated above, protein adducts (particularly hemoglobin) as biomarkers to assess environmental contamination have received little attention. The work from

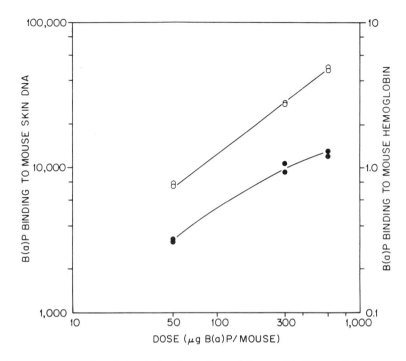

FIGURE 2. Log-log plot relating the log dose of benzo[a]pyrene binding to cellular macromolecules and the log dose of benzo[a]pyrene applied topically to mouse skin. (0): Benzo[a]pyrene binding to DNA expressed as picogram tetrol I-1 per milligram of DNA; (0) Benzo[a]pyrene binding to hemoglobin expressed as picogram tetrol I-1 per milligram hemoglobin. (From Shugart, L. R. and Kao, J., *Environ. Health Perspect.*, 62, 223, 1985; Shugart, L. R., *Toxicology*, 34, 211, 1985. With permission.)

Shugart's laboratory[2,7-9,12-15] will serve to demonstrate the general approach and associated problems.

Laboratory studies by Shugart[7] showed that mild acid hydrolysis of globin preparations from erythrocytes of mice, previously exposed topically to BaP, released tetrols which are detectable by HPLC/fluorescence analysis.[8,13,14] The technique used to isolate the hemoglobin precluded the presence of noncovalently bound BaP or its cellular metabolites. There was a dose-response relationship between the amount of BaP applied to the skin of the mouse and the occurrence of BaP adducts to hemoglobin. The amount of binding to DNA and hemoglobin at various doses of BaP was qualitatively similar (Figure 2). Subsequent investigations[9] showed that BaP administered to the skin of pregnant mice will pass across the placental membrane and bind to the hemoglobin of the progeny.

Field studies[14,16] with terrestrial animals were conducted on a floodplain of a small stream near the U.S. Department of Energy Reservation in Oak Ridge, TN. The floodplain is downstream several kilometers from a point source of industrial pollution, and the stream periodically overflows and deposits sediment. Measured BaP concentrations in the soil are 70 ng/g although 10 times this amount have been reported. Vegetation is abundant and BaP concentrations in plants were less

Table 1. Concentration of Benzo[a]Pyrene Adducts in the Hemoglobin of Animals Trapped on the EFPC Floodplain

Species	Number captured	BaPadducts (ng/g Hb)
White-Footed mouse *(Peromyscus leucopus)*	15	0.00
House mouse *(Mus muscularis)*	1	0.00
Shorttail shrew *(Blarina brevicauda)*	2	0.35
Black rat snake *(Elaphe obsoleta obsolet)*	1	0.00
Snapping turtle *(Chelydra serpentina)*	1	0.16
Norway rat *(Rattus norvegicus)*	1	0.15
Muskrat *(Ondatra zibethica)*	6	0.23

than one tenth of that in the soil. Small animals were trapped at this site to determine the practicality of using hemoglobin as a suitable biomarker of BaP pollution. Data (Table 1) indicated detectable concentrations of BaP adducts to hemoglobin were present in some individuals of species (shorttail shrew and muskrat) which have intimate contact with the sediment or soil. For example, the shorttail shrew burrows into the ground and eats earthworms and insects, while the muskrats feed on invertebrates found in the stream bed and browse on muddy vegetation along the stream bank. BaP adducts were not detectable in herbivores such as the white-footed mouse.

Investigations were extended to aquatic species.[15] Adduct formation in bluegill sunfish exposed to BaP demonstrated that aquatic species are similar to other organisms in that they possess cellular enzymes capable of metabolizing BaP to the ultimate carcinogenic form of the chemical (diolepoxide). A comparison of BaP adduct formation between the mouse and sunfish is given in Table 2. These data indicate that there are essentially no differences between these species as to the levels of BaP adducts found in both DNA and hemoglobin for similiar levels of exposure, suggesting a comparative biochemical mechanism exists in both species for the xenobiotic metabolism of the chemical.

Recently the level of BaP adducts to both hemoglobin and albumin were measured in wild woodchucks *(Moarmota monax)* to assess environmental contamination by PAHs.[17] Animals were collected from a contaminated area in Quebec, Canada and a control area over 100-km distance. Blood was collected and analysis for BaP adducts was by the HPLC/fluorescence method of Shugart.[7] BaP blood protein adducts were significantly higher in animals from the contaminated site, and the authors concluded that these types of measurements were a good biomarker of environmental contamination by PAHs.

V. SUMMARY

Since evidence of DNA adducts in most living organisms is difficult to obtain, DNA adducts in ancillary cells (such as circulating lymphocytes) or adducts on proteins that are easily obtained (such as hemoglobin and serum albumin) may

Table 2. Adduct Formation in the Mouse and Bluegill Sunfish after an Acute Exposure to Benzo[a]Pyrene

Species	Macromolecule	Adduct[a] I-1
Mouse	Skin DNA[b]	588
	RBC Hb[c]	3
Fish	Liver DNA[d]	340
	RBC Hb	2.9

[a] Adduct formation determined 72 hr subsequent to benzo[a]pyrene exposure and expressed as nanograms adduct per gram of macromolecule.
[b] Topical application of benzo[a]pyrene at 3 μm/g body wt.
[c] Topical application of benzo[a]pyrene at 10 μm/g body wt.
[d] Intraperitoneal injection of benzo[a]pyrene at 5 μm/g body wt. Fish were maintained at 13°C.

serve as surrogates. Hemoglobin can serve as a convenient agent for assessing levels of activated chemicals over long periods of time since it has a long half-life, and can be easily purified and analyzed. If protein adduct formation is proportional to DNA adduct formation, then the former can be used to indirectly evaluate the latter. In general, protein adducts are useful for quantifying past exposure because they are: (1) stable under biological conditions; (2) easily obtained for biomonitoring; (3) exhibit a relationship to DNA adducts in the target tissue; and (4) provide evidence of exposure.

ACKNOWLEDGMENT

The Oak Ridge National Laboratory is managed by Martin Marietta Energy Systems Inc. under contract DE-AC05-84OR21400 for the U.S. Department of Energy, Environmental Sciences Division Publication No. 4112.

REFERENCES

1. Ehrenberg, L. E., Moustacchi, S., Osterman-Golkar, O., and Ekman, G., Dosimetry of genotoxic agents and dose/response relationship of their effects, *Mutat. Res.,* 123, 121, 1983.
2. Shugart, L. R. and Kao, J., Examination of adduct formation in vivo in the mouse between benzo[a]pyrene and DNA of skin and hemoglobin of red blood cells, *Environ. Health Perspect.,* 62, 223, 1985.

3. Harvey, R. G., Polycyclic hydrocarbons and cancer, *Am. Sci.,* 70, 386, 1982.

4. Wogan, G. N. and Gorelick, N. J., Chemical and biochemical dosimetry of exposure to genotoxic chemicals, *Environ. Health Perspect.,* 62, 5, 1985.

5. Pereira, M. A. and Chang, L. W., Hemoglobin binding as a dose monitor for chemical carcinogens, in *Banbury Report No. 13,* Bridge, B. A., Butterworth, B. E., and Weinstein, I. B., Eds., Cold Spring Harbor Laboratory, Cold Spring Harbor, NY, 1983, p. 177.

6. Osterman-Golkar, S., Ehrenberg, L., Segerback, D., and Halstrom, I., Evaluation of genetic risks of alkylating agents. II. Hemoglobin as a dose monitor, *Mutat. Res.,* 34, 1, 1976.

7. Shugart, L. R., Quantitating exposure to chemical carcinogens: in vivo alkylation of hemoglobin by benzo[a]pyrene, *Toxicology,* 34, 211, 1985.

8. Shugart, L. R., Quantifying adductive modification of hemoglobin from mice exposed to benzo[a]pyrene, *Anal. Biochem.,* 152, 365, 1985.

9. Shugart, L. R. and Matsunami, R., Adduct formation in hemoglobin of the newborn mouse exposed in utero to benzo[a]pyrene, *Toxicology,* 37, 241, 1985.

10. de Serres, F., Geldhill, B. L., and Sheridan, W., DNA Adducts: Dosimeters to Monitor Human Exposure to Environmental Mutagens and Carcinogens, Vol. 62, Environmental Health Perspective, U.S. Department of Health and Human Services Publication No. 85–218, Research Triangle Park, NC, 1985.

11. Bartsch, H., Hemminki, K., and O'Neill, I. K., *Methods for Detecting DNA Damaging Agents in Humans: Application in Cancer Epidemiology and Prevention,* IARC Scientific Publication No. 89, Lyon, France, 1988.

12. Shugart, L. R., Adams, S. M., Jimenez, B. D., Talmage, S. S., and McCarthy, J. F., Biological markers to study exposure in animals and bioavailability of environmental contaminants, in *ACS Symposium Series 382, Biological Monitoring for Pesticide Exposure: Measurement, Estimation, and Risk Reduction,* Wang, R. G. M., Franklin, C. A., Honeycutt, R. C., and Reinert, J. C., Eds., American Chemical Society, Washington, DC, 1989, p. 86.

13. Rahn, R., Chang, S., Holland, J. M., and Shugart, L. R., A fluorometric-HPLC assay for quantitating the binding of benzo[a]pyrene metabolites to DNA, *Biochem. Biophys. Res. Commun.,* 109, 262, 1982.

14. Shugart, L. R., Holland, J. M., and Rahn, R., Dosimetry of PAH carcinogenesis: covalent binding of benzo[a]pyrene to mouse epidermal DNA, *Carcinogenesis,* 4, 195, 1983.

15. Shugart, L. R., McCarthy, J. F., Jimenez, B. D., and Daniels, J., Analysis of adduct formation in the bluegill sunfish *Lepomis macrochirus* between benzo[a]pyrene and DNA of the liver and hemoglobin of the erythrocyte, *Aquat. Toxicol.,* 9, 319, 1987.

16. Talmage, S. S. and Walton, B. T., *Comparative Evaluation of Several Small Mammal Species as Monitors of Heavy Metals, Radionuclides, and Selected Organic Compounds in the Environment,* Oak Ridge National Laboratory/TM-11605, Oak Ridge, TN, 1990.

17. Blondin, O. and Viau, C., Benzo[a]pyrene protein adducts in wild woodchucks used as biological sentinels of environmental PAH contamination, *Arch. Environ. Contam. Toxicol.,* 3, 310, 1992.

SECTION FIVE

Cellular Biomarkers

CHAPTER 8

An Integrated Approach to Cellular Biomarkers in Fish

Michael N. Moore, Angela Köhler, David M. Lowe, and Michael G. Simpson

TABLE OF CONTENTS

I. INTRODUCTION

The impact of environmental contamination is frequently apparent only after major effects on the populations of marine organisms such as mass mortality and changes in the diversity of ecological communities. If these types of event are to be avoided or ameliorated by effective environmental management, then the level of risk of damage to marine animals or plants needs to be assessed and used to predict the possible consequences before the event occurs. Of the current tools available for this purpose, the biomarker approach appears to offer considerable promise.[1-3]

The underlying basis of all contaminant-induced pathological change is damage to, or perturbation of, living processes at the molecular and subcellular levels of biological organization.[4] Consequently, detection of changes at these most basic levels should in theory provide markers of early onset damage, which if sustained will lead to cell injury and pathology with an eventual deterioration in the health of the population. Our knowledge of these early changes has grown substantially in the past decade, often drawing on the reservoir of understanding of such processes in humans and rodents. The use of these putative biomarkers by marine environmental toxicologists has been relatively limited, and yet the potential power of such markers is probably considerable. Not only do they hold out the prospect of diagnostic predictors of pathological change, but also there is the prospect of biomarkers of exposure/bioavailability for various classes of organic xenobiotics.[5,6] This latter type of biomarker has the prospect of providing rapid and less costly alternatives to routine chemical analytical screening, which would then allow the chemistry to be focused on more specific problems.

Chemical data tell us what is present in the tissues but nothing about whether the xenobiotics or metals are biologically active.[7,8] This latter information can only come from an understanding of the biochemical and cellular processes involved in uptake, detoxication, and excretion (Figure 1). Furthermore, many xenobiotics are rapidly transformed by detoxication/activation processes and will not be detected by conventional analytical chemistry (e.g., polycyclic aromatic hydrocarbons and nitrogen heterocyclics in fish).[6] However, the activated derivatives may be responsible for much of the damage to cellular processes (Figure 1). This argues strongly for a tighter linking of the chemistry with mechanistic biology.

Obviously, it is premature to state that biomarkers are the panacea to detection of problems in the marine environment. However, the biomarker approach could well offer a very powerful set of tools for environmental surveillance with the added potential to provide evidence of exposure to bioavailable contaminants and mechanistic linking of the causative processes. A hypothetical example of this type of functional and hierarchical linking could involve induction of specific gene products such as those for cytochrome P-450 1A1 or multidrug-resistance protein (exposure markers); increases in generation of radicals (the initiators of molecular damage); failure of the antioxidant protection systems to counter radical formation; and damage to the integrity of cellular organelles leading to degeneration of target organs (e.g., liver) resulting in a decline in performance,

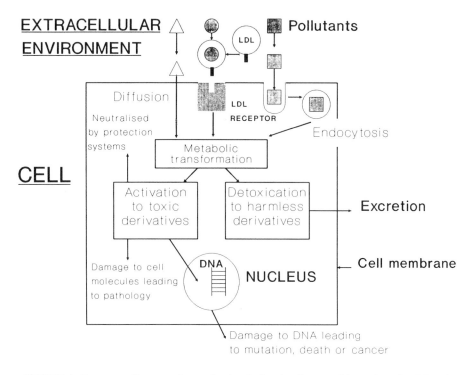

FIGURE 1. Summary diagram of an animal cell showing the possible routes of uptake of pollutant chemicals, their fate within the cell and the possible consequences of their toxic action in processes of cell injury (LDL = low-density lipoprotein).

reproductive impairment, and increased susceptibility to infection and predation (Figure 2).[5,6,8,9]

A number of putative biomarkers are available but generally require further research into their mechanistic basis. However, this should not prevent their use in ecotoxicology as long as they are used circumspectly and the results are not overinterpreted. It must be emphasized that biomarkers will be best utilized as selected batteries of tests rather than individually, where there is a greater likelihood of misinterpretation.[3]

Marine ecotoxicology has been generally slow in adopting available new methods and has relied heavily on LC_{50} type and other lab-based tests. This is perhaps due to the fact that its roots lie in traditional marine biology and ecology, and failure to fully recognize the potential advantages to be gained by interfacing with other sciences such as biochemistry, cell biology, and toxicological pathology.[1] In the instances where such interdisciplinary interactions have occurred, progress has often been rapid both in developing understanding of the processes of toxicity and in the identification of markers of adverse effects on health. An aim of this chapter is to draw the attention of ecotoxicologists to this window of opportunity by outlining some of the possibilities for future exploitation and giving some examples of current prospective biomarkers.

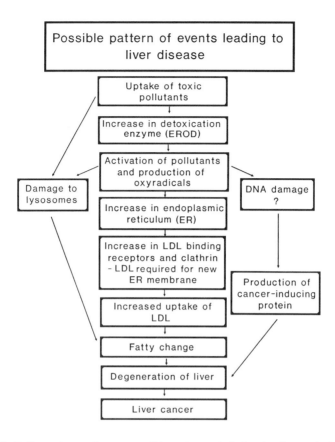

FIGURE 2. Outline scheme of some possible early events in the development of degenerative and neoplastic disease of fish liver (LDL = low-density lipoprotein).

The chapter focuses on environmental impact on marine fish which are an important economic and biological resource and often form part of the human food chain. The main habitats for both juvenile and adult forms of many fish species are estuaries and coastal zones; however, these very areas are likely to be affected by human activities such as urbanization and industrialization. It is conceivable, therefore, that human activities resulting in the contamination of sediments in estuaries and coastal zones will result in the loss of the living resource through chronic toxicity and impairment of reproductive capability or early deaths of both adult and larval fish.[10] Human health may also be at risk through consumption of contaminated fish products.

In recognition of these potential problems, various types of flatfish are increasingly being used as sentinels for the detection of any deterioration in environmental quality. Flatfish live in close association with the sediments, which are often the main repository for toxic chemical pollutants, and hence might be expected to display signs of distress or damage if so exposed. Unfortunately, flatfish are not ideal sentinel animals because they show migratory patterns that make interpretation of observed effects more difficult than in the case of many sedimentary benthic invertebrates, such as bivalve mollusks.[2] Nonetheless, it should still be possible to use rapid biological responses to toxic chemical exposure as markers of biological damage, although longer term pathological changes are obviously more open to interpretation.

The main aim of this chapter is to provide an overview of how a range of molecular cell biological techniques coupled with histopathology has been used in order to obtain a mutitiered picture of molecular and cellular reactions to toxic pollutants in liver cells of the flatfish dab *(Limanda limanda)* from the North Sea.[11,12] The approach is based on the identification of early onset of molecular and cellular changes or biomarkers of the processes of cell injury and attempts to link these to higher levels of biological damage. Liver cells (hepatocytes) are the focus of the study because the liver is an integrator of many functions including detoxication/activation of toxic chemicals, digestion and storage, excretion, and (in females) synthesis of the egg yolk protein. There is also a considerable body of literature on pollutant chemical impact on the cellular pathology of fish liver.[10]

II. NONDESTRUCTIVE BIOMARKERS IN FISH

The main theme of this book is the development of nondestructive biomarkers. In part, this is driven by the necessity not to kill higher mammals and birds during sampling for the assessment of environmental impact. The basis for this approach is founded on both ethical considerations and the frequent need not to overly disturb small and often endangered populations. With fish, in general, these criteria do not apply, although there is obvious value in being able to resample either tagged or experimental fish.

In this context blood samples may be of some limited usefulness; however, they will not provide the integrated picture of cellular pathology that can be obtained from liver samples.[3,10] Liver does appear to be the major target organ for attack by environmental xenobiotics, and any future move to nondestructive sampling will therefore necessitate the successful development of liver biopsy procedures. At present such procedures are unreliable and highly stressful for fish.

III. APPROACH

Molecular cell biology is identifying many possible biomarkers relating to perturbation of cellular processes or defensive reactions involving those systems that protect the cell against hostile environmental agents. These include heat-shock proteins (e.g., chaperonins and metallothioneins); cell surface receptors; cellular oncogenes; neurobiological disturbances; enzymatic machinery for xenobiotic detoxication and activation (biotransformation); and internal membranous components of the cell such as the endoplasmic reticulum, Golgi, transport vesicles and lysosomes, and various building blocks of the cytoskeleton.[13,14]

It is not the intention of this chapter to provide a comprehensive dissertation on all of these components, many of which have been well described in several recent publications.[1] Instead, one particular aspect will be highlighted. This is the pathobiology of cellular membranes, which in fact can embrace many other aspects of pathological and adaptive change.[15]

The cell surface or plasma membrane of liver cells is frequently a major interface with the environment. Pollutants associated with the diet come into contact with the cell surface via the portal system in the case of liver or directly in the case of midut glands. Lipophilic xenobiotics may enter the cell (Figure 1) by diffusion across the plasma membrane or possibly by receptor-mediated endocytosis in association with low-density lipoprotein.[13,16] In addition to being the gateway of entry into the cell, the plasma membrane plays many crucial physiological functions. Not the least of these are endocytosis and exocytosis which regulate the bulk transport of proteins and other macromolecules between the cell and its immediate extracellular environment. Perturbation of these functions by toxic chemicals will have profound effects on the cell and consequently on the organ and animal. Pathological disturbance of endocytic events can involve increases or decreases in specific cell surface receptors (Figure 2), which may result from changes within the cell such as induction of the receptor genes or in the recycling of receptors back to the cell surface via the compartment of uncoupling of receptors and ligands (CURL) component of the endocytic-lysosomal system.[13]

The cell membrane also has many specific proteins associated with it in addition to molecular receptors. These include ion channels, gap junction and desmosomal complexes, adhesion molecules, molecular pumps, and signal transduction systems.[13] As yet there is only limited evidence for pathological alterations in the above systems induced by chemical pollutants; however, there are several consistent patterns emerging. One of these involves disturbance of endocytic processes in fish liver cells (Lowe, personal communication) while the second involves changes in expression of *ras*-protooncogenes, which code for G-type proteins related to signal transduction systems.[17-21]

Reactions of intracellular membranes to toxic pollutants are perhaps better documented than those of the plasma membrane. In particular, the lysosomal system in both mollusks and fish is known to be affected by xenobiotics and metals.[2,22-24] These changes can involve increases in volume; accumulation of lipid and lipofuscin (aging or stress pigment); destabilization of the lysosomal membrane, which impairs function; and inhibition of the lysosomal proton pump (Figure 2).

The endoplasmic reticulum, in addition to being the major site of cellular protein syntesis, is central to many of the enzymic reactions involved in xenobiotic detoxication and activation (Figure 1). These processes involve the cytochromes P-450 and epoxide hydrase, as well as some of the antioxidant protection enzymes.[6,25,26] All of these enzymes are integral components of the endoplasmic reticulum (ER) and some of them can be induced by particular types of xenobiotic. One of the best studied examples is the induction of cytochrome P-450 1A1.[6] In addition to the functional changes and induction of ER-associated proteins, the ER itself can undergo proliferation markedly altering the internal organization of the cell, following exposure to many xenobiotics.[10] Activated derivatives of certain xenobiotics can also be retained within the lipoprotein membranes of the ER where they become involved in self-sustaining futile cycles and give rise to potentially damaging oxyradicals which react with many biological molecules including DNA.[9]

If we view these types of membrane-associated pathological changes holistically, then it becomes possible to see how they can be used as biomarkers not only for indicating exposure to particular types of xenobiotic (i.e., induction of cytochrome P-450 1A1), but also for the possible mechanisms of cell injury (i.e., generation of radicals) and the degree of cell injury (e.g., proliferation of ER, damage to lysosomes, overexpression or activation of cellular protooncogenes; Figure 2).[10,17,18,24]

Histopathology also provides another powerful tool for the assessment of toxiclogical pathology, particularly if coupled with the cellular pathobiological approach described above.[3,10] This coupling permits intracellular changes to be related to a finite number of pathological conditions, which can be quantified if required using the techniques of microstereology and image analysis.[27] Understanding of the early pathological changes involved in the development of contaminant-induced disease will provide the predictive capability that is essential in the biomarker approach. For instance, overexpression of *ras*-oncoprotein may be useful as a potential tumor marker in fish as it already is in mammals, while histologically identifiable foci of cellular alteration are believed to be representative of preneoplastic lesions in livers of fish and indicative of exposure to xenobiotics.[10,18,22,28,29]

IV. APPLICATION OF BIOMARKERS

During March 1990 an international practical workshop was held in Bremerhaven sponsored by the International Council for the Exploration of the Seas (ICES) and the Intergovernmental Oceanographic Commission (IOC). The purpose of this workshop was to test for effects of toxic contaminants on a range of marine animals from the molecular/biochemical level to ecological levels of organization (Figure 3). The work described in this chapter represents a synthesis of five studies carried out as part of the "Molecular and Cellular Pathology" component of this Workshop.[18,23,29-31] For these studies, fish (dab) were sampled along a transect from the chemically contaminated inner German Bight out to the central North Sea.[12]

The basis of all these studies is the use of cellular pathology as a biomarker(s) of exposure to toxic contaminants and of cell injury likely to lead to a deterioration in health of the individual fish. The synthesis is subdivided into three sections, namely, pathological reactions of the endocytic-lysosomal system, pathological reactions associated with the endoplasmic reticulum, and finally histopathology. As indicated in the previous section, the linking theme in all of this work is pathological reactions of membranes as they relate to the processes of cell injury.

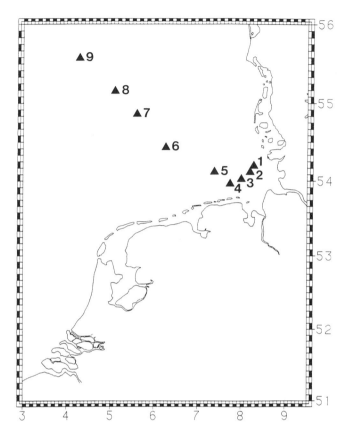

FIGURE 3. Map of the southeastern region of the North Sea showing the locations of the sampling stations for the Bremerhaven Workshop.

A. Pathological Reactions of the Endocytic-Lysosomal System

The endocytic-lysosomal system provides the pathway for bulk transport of macromolecules into the cell. This process involves both nonspecific fluid-phase endocytosis as well as the highly selective receptor-mediated endocytosis for specific macromolecules (e.g., low-density lipoprotein and epidermal growth factor).[13] The lysosomal component of the system is the site of degradative disassembly of the ingested macromolecules by means of a battery (ca. 60) of hydrolytic enzymes, which function optimally in the low pH environment of the lysosomal compartment.[13] This low pH is maintained by means of a membrane-associated ATP-dependant proton pump.[13] The latter system is also believed to be present in the late endosomal compartment; there it presumably facilitates some of the macromolecular sorting and receptor recycling processes, which occur in this prelysosomal component.[13] Another key function of lysosomes is the degradative recycling of intracellular proteins (autophagy), a process that is vital to the normal economy of the cell.[2,13]

Pathological disturbance of the lysosomal system, leading to dysfunction, is a central feature of many diseases.[2] Disturbance of lysosomes also occurs in many

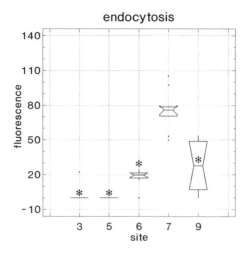

FIGURE 4. Fluorescence intensity (arbitrary units) in isolated liver cells (hepatocytes) for endocytosis of Texas Red labeled albumin. Results are represented by notched-box and whisker plots; significant differences ($p < 0.05$) from the least contaminated station (site 7) are shown by asterisks. Data were assessed using the Mann-Whitney U-test (n = 10 fish per site with 20 measurements per fish). (From Moore, M. N., *Mar Ecol. Prog. Ser.*, 91, 127, 1992. With permission.)

other pathological conditions as a secondary consequence of cell injury.[2] Furthermore, lysosomes are noted for their remarkable ability to sequester and accumulate many toxic metals, as well as many classes of organic xenobiotic.[2]

The lysosomal system in the cells of many aquatic animals is known to be a prime target for toxic chemical pollutants.[2,22-24] There is a growing body of evidence that such lysosomal damage is important in pollutant-induced cell injury and may itself be mediated either by the direct action of some xenobiotics and metals or by the damaging action of reactive radicals produced as a consequence of the biotransformation of xenobiotics and the futile cycling of some of their derivatives.[2,4,9,17,32,33]

A number of tests for damage to the endocytic-lysosomal system were applied to samples taken from the livers of female dab (body length 17–25 cm) caught at each of five sites along a transect running from the plume of the Elbe River to the eastern end of the Dogger Bank in the central North Sea.[23,29,30] These tests included the study of both isolated hepatocytes and tissue sections of liver. Nonspecific endocytosis was tested using the uptake of Texas red-conjugated bovine serum albumin into the endosomal system of the hepatocytes in vitro.[30] Endocytosis was markedly reduced at the most contaminated inshore sites in the southern German Bight in comparison with the offshore "clean" reference site (site 7) as is shown in Figure 4.[30]

In vitro studies on the hepatocytes also included two tests for integrity of lysosomal function.[23] These used the lysosomal retention of the acidotropic probes neutral red and acridine orange, respectively. As with the findings for effects on endocytosis, the results clearly showed a major detrimental impact on the ability of lysosomes to retain the probes in cells taken from the livers of fish caught at

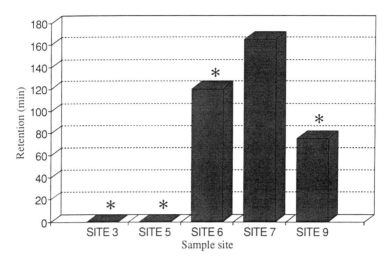

FIGURE 5. Graph showing median values of lysosomal neutral red retention times in hepatocytes from the transect sample sites (* = $p < 0.002$, n = 10, Mann-Whitney U-test, comparing with site 7 which was the least contaminated). (From Lowe, D. M., Moore, M. N., and Evans, B., *Mar. Ecol. Prog. Ser.,* 91, 135, 1992. With permission.)

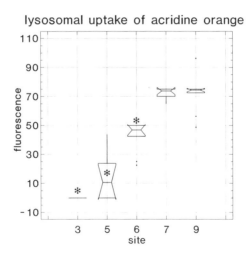

FIGURE 6. Notched-box and whisker plot showing fluorescence of acridine orange uptake in hepatocyte lysosomes from transect sample sites as in Figure 4. (From Lowe, D. M., Moore, M. N., and Evans, B., *Mar. Ecol. Prog. Ser.,* 91, 135, 1992. With permission.)

the inshore contaminated sites (Figures 5 and 6). These results indicate probable damage to the lysosomal proton pump and failure to maintain a sufficiently acid internal environment, as well as possible permeability increases in the lysosomal membrane resulting in leakage of the protonated probe (hydrophilic). The nonprotonated forms of the probes are lipophilic and can permeate biomembranes.[34]

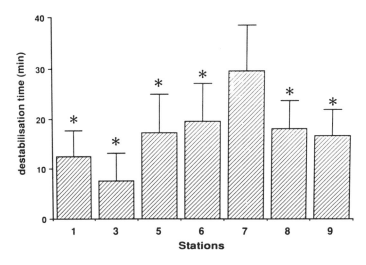

FIGURE 7. Plot of cytochemically determined lysosomal stability in tissue sections of dab liver from the transect sample sites. Values are means and standard and deviation (* = $p < 0.05$, n = 25, Mann-Whitney U-test, comparing with site 7, which was the least contaminated). (From Köhler, A., Deisemann, H., and Lauritzen, B., *Mar. Ecol. Prog. Ser.*, 91, 141, 1992. With permission.)

A cytochemical test for lysosomal membrane stability based on permeability to the histochemical substrate for β-N-acetylhexosaminidase was applied to unfixed cryostat sections of liver tissue.[29,35,36] The findings obtained here fully support the results of the in vitro tests for lysosomal integrity, again showing that lysosomes in cells from fish taken at the inshore contaminated sites were severely affected and their membranes were highly permeable (Figure 7).[29]

Structural changes occurred in the lysosomes of fish from the most contaminated sites.[23,29] These included enlargement and increase in the heterogeneity of lysosomal contents, probably indicative of enhanced autophagy (Figure 8).[29] There was also clear evidence for phospholipidosis; this is an accumulation of membranous phospholipid-derived material believed to be induced by exposure to xenobiotics, particularly those which can give rise to amphiphilic derivatives.[29] Further deleterious changes included the enhanced formation of giant lysosomes or "councilman bodies", which are associated with programmed cell deletion (apoptosis).[29]

The pattern that emerges from the results of the above tests is a clear picture of damage to the endocytic-lysosomal system that is entirely consistent with the known effects of xenobiotic exposure. The results clearly match the gradation of xenobiotic contamination in both sediments and fish livers; and the degree of lysosomal injury described for the inshore sampling sites is likely to lead to irreversible cell damage, liver dysfunction, and severe impairment of the health of the fish.

B. Pathological Reactions Associated with the Endoplasmic Reticulum (ER)

The endoplasmic reticulum (ER) is a multifunctional interconnected system of membranous cisternae found in all eukaryotic cells.[13] Its functions include protein

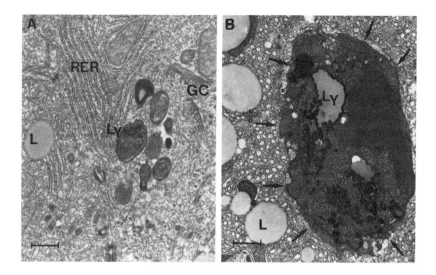

FIGURE 8. (A) Formation of normal sized lysosomes (Ly) near the Golgi complexes (GC) during cell catabolism and macromolecule transport in a dab liver from the offshore reference site. These lysosomes showed "normal" integrity with a membrane stability of 35 min. (B) Dramatically enlarged lysosome ingesting heterogeneous material and cellular components in a liver of dab caught at an inshore site of the Elbe plume. The arrows indicate the boundaries of the lysosome; the membrane was severely injured with a lowered membrane stability of only 3 min. (Bar = 1 μm.) (From Köhler, A., Deisemann, H., and Lauritzen, B., *Mar. Ecol. Prog. Ser.*, 91, 141, 1992. With permission.)

synthesis in the granular or rough endoplasmic reticulum (RER), as well as post-translational modification of proteins including glycosylation of glycoproteins, calcium storage, and detoxication of foreign compounds (xenobiotics) derived either from ingested food stuffs or from exposure to contaminant chemicals.[13] These latter processes are predominantly associated with the smooth endoplasmic reticulum (SER). The endoplasmic reticulum frequently shows marked changes associated with cellular pathology. A characteristic reaction of exposure to many xenobiotics is proliferation (hypertrophy) of the ER, particularly the smooth component.[37]

Pathological changes in the ER have been identified as a potential biomarker of exposure to organic xenobiotics and were tested for in the livers of dab from the North Sea transect.[17,30] Tests included the use of the fluorescent 3,3'-dihexyloxacarbocyanine iodide probe $DiOC_6$ (3), which is relatively specific for ER, in live isolated hepatocytes, and ultrastructural study of liver sections using electron microscopy.[17,30] Complementing these tests, fluorescent probes were used to detect the activity of the biotransformation enzyme 7-ethoxyresorufin-O-deethylase (EROD) and the generation of the oxyradical superoxide O_2^-.[17,30,32] EROD is a detoxication/activation enzyme activity associated with cytochrome P-450 1A1, an ER-associated protein, which is induced by many xenobiotics such as coplanar congeners of PCBs and polycyclic aromatic hydrocarbons (PAHs) like 3-methylcholanthrene.[6]

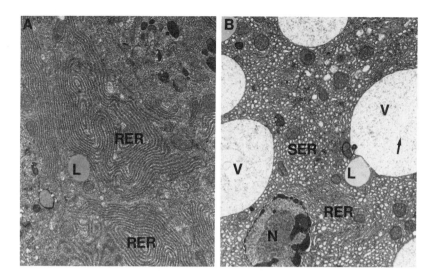

FIGURE 9. (A) Strictly parallel orientation of stacks of granular or rough endoplasmic reticulum (RER) and fields of free ribosomes indicating high protein synthesis in the liver of dab caught at an offshore reference site. (B) Extensive proliferation of smooth endoplasmic reticulum (SER) and loss of rough endoplasmic reticulum (RER) to small bundles in a dab liver caught at the contaminated inshore site 3 close to the mouth of the Elbe River. Note the large vacuoles (V) filled with a fine precipitate (lipoproteins?), fusing with small lipid droplets (L). (Bar = 2 μm.) (From Köhler, A., Deisemann, H., and Lauritzen, B., *Mar. Ecol. Prog. Ser.,* 91, 141, 1992. With permission.)

The results of our study on the livers of dab showed ultrastructural evidence of proliferation of the SER in hepatocytes from fish caught at the inshore contaminated sites (Figure 9).[29] In livers that were severely injured and that contained foci of phenotypically altered cells, there was extensive proliferation of RER in these cells.[29] However, the subject of foci of cellular alteration will be covered in detail in the following section on histopathology.[28] The studies with fluorescent probes on isolated hepatocytes showed a fourfold increase in EROD activity in fish from the contaminated site in the Elbe plume (site 3) as compared with the offshore reference (site 7) (Figure 10).[30] These findings are in good agreement with the results from conventional measurement of EROD activity in fish from the same sites.[38] The probe used to detect ER in the isolated hepatocytes showed a two- to threefold increase in the livers of dab from the inshore sites 3, 5, and 6; and this is in good agreement with the ultrastructural data (Figures 9 and 11).[29,30]

Results obtained from the use of the oxyradical probe dihydrorhodamine 123 indicated an increased generation of superoxide anions at the contaminated inshore sites 3 and 5 (Figure 12).[30] The pattern of oxyradical production showed a strong correlation with both liver PCB content ($r = 0.887$, $p < 0.05$, n = 5) and sediment PAH content ($r = 0.92$, $p < 0.05$, n = 5).[30]

These findings are consistent with the documented effects of many lipophilic xenobiotics on the ER and its associated biotransformation system.[6,37] The proliferation of SER and the increase in EROD activity are indicative of exposure to

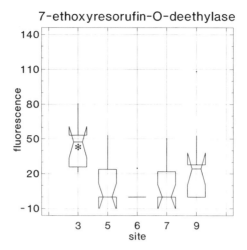

FIGURE 10. Notched-box and whisker plot showing fluorescence of 7-ethoxyresorufin-O-deethylase (EROD) reaction product in isolated hepatocytes from transect sample sites as in Figure 4. (From Moore, M. N., *Mar Ecol. Prog. Ser.,* 91, 127, 1992. With permission.)

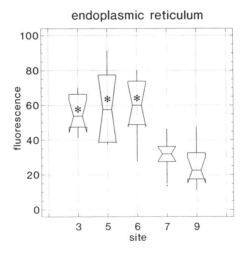

FIGURE 11. Notched-box and whisker plot showing fluorescence of the endoplasmic reticulum probe $DiOC_6$ (3) in isolated hepatocytes from transect sample sites as in Figure 4. (From Moore, M. N., *Mar Ecol. Prog. Ser.,* 91, 127, 1992. With permission.)

xenobiotics, although the possibility of other environmental factors being involved cannot be completely excluded. The increase in oxyradical generation at the most contaminated sites is interpreted as being due to the likely increase in futile cycling of metabolic derivatives of some PAHs in the membranes of the ER, with the consequent generation of oxyradicals.[9]

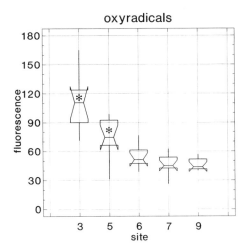

FIGURE 12. Notched-box and whisker plot showing fluorescence of rhodamine 123 produced by the action of superoxide radicals on the oxyradical probe dihydrorhodamine 123 in isolated hepatocytes from transect sample sites as in Figure 4. (From Moore, M. N., *Mar Ecol. Prog. Ser.,* 91, 127, 1992. With permission.)

Therefore, what are the likely consequences for injury and health of the fish? Excessive production of radicals is known to be injurious to cells as they react with lipid biomembranes, proteins, and glycoproteins, as well as with DNA.[9] In this study we did not test for oxidative damage; however, a parallel investigation by Chipman et al.[39] did not reveal any significant increases in oxidative adducts of DNA in livers of fish from the same samples taken at the most contaminated sites. Nonetheless, oxyradicals are known to attack the lysosomal membrane; and in this study there was a strong inverse correlation between lysosomal integrity (measured as retention of acridine orange) and superoxide generation (Spearman rank correlation coefficient = -0.73, $p < 0.00005$, n = 50).[4,30,32] It is therefore not unreasonable to conclude that increased radical production in the liver cells from fish caught at the contaminated sites is contributing to cell injury, and furthermore that this increase in radical generation is a direct consequence of elevated production of metabolic derivatives of the enhanced activity of the biotransformation system in the ER.

C. Histopathological Reactions

Liver cell and tissue structure is an integration of the many biochemical, cellular, and physiological processes occurring within it, as well as any pathological disturbances to these processes. Hence, histopathology provides a potentially powerful tool for the assessment of cell injury by environmental pollutants and in the prediction of higher level consequences of such injury. Extensive histopathological studies have been conducted on the impact of pollutants on fish in North America.[3,10,28,40-42] However, in comparison with these latter, European studies have been rather limited, with investigations of marine fish often tending

Table 1. Summary of Histopathological Findings in the Liver of Non-overtly Diseased Dab (17–27 cm)

Microscopic findings	Site 3	Site 5	Site 6	Site 7	Site 8	Site 9
No. livers examined	20	20	20	19	18	23
No. with:						
Hepatocellular cytoplasmic vacuolation (glycogen and lipid)	17*	10	10	8	10	17
Subcapsular sinusoidal dilatation	1	0	0	1	1	0
Scattered basophilic hepatocytes	3	1	1	1	0	0
Apoptosis	1	0	4	0	2	3
Haemorrhagic foci	1	0	0	0	0	0
Proliferation of melanomacrophage centers	2	2	0	0	1	0
Hepatocyte necrosis	3	1	0	0	0	0
Increased mitotic activity	4	1	0	0	0	0
Infiltration/fibrosis of melanomacrophage centers	0	0	0	0	1	0
Foci of cellular alteration	7*	3	2	0	2	4
Parasitic infection	0	1	1	0	0	5

From Simpson, M.G. and Hutchinson, T., *Mar. Ecol. Prog. Ser.*, 91, 155, 1992.
* Significantly higher than the prevalence at Site 7 by Fisher's Exact Test (*P <0.05)

to follow the protocols recommended by ICES.[24,28,29,31,43-45] Some of these studies have been mainly restricted to gross inspection and have revealed a variety of temporal and spatial trends in patterns of fish disease in the North Sea. This epidemiological type of approach has had a limited degree of detailed histopathologic support.[24,28,29,31,43-45,46] In a recent paper, Moore and Simpson argued strongly for the wider usage of systematic histopathology coupled where possible with other techniques for cellular pathology in order to provide a better mechanistic understanding of the etiology of environmentally induced disease in fish.[3,17,18,23,28-31,46]

A number of changes were seen in the livers of dab from the North Sea transect and are summarized in Table 1.[28] Liver cells from sites 3 and 9, Elbe plume, and Dogger Bank, respectively, had a greater degree of hepatocellular cytoplasmic vacuolation in comparison to other sites (Table 1). Prevalence of this type of vacuolation was significantly greater ($p < 0.05$, Fisher exact test) at site 3 than at the least contaminated site 7. Two types of vacuolation were found: one of these was characteristic of glycogen that had been lost from the sectioned material; the other type was characteristic of lipid vacuolation, and subsequent histochemical tests on cryostat sections confirmed the presence of unsaturated neutral lipid (Oil

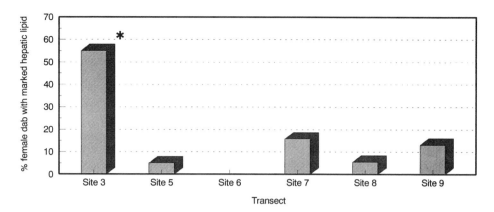

FIGURE 13. Prevalence of marked hepatocellular cytoplasmic lipid vacuolation in non-overtly diseased dab (17–27 cm body length) from the transect sampling sites. Asterisk indicates significant ($p < 0.05$) difference from the prevalence at site 7 (least contaminated) using Fisher's exact test (n = 20). The vacuolation was confirmed as neutral lipid by staining of representative sections with Lillie and Ashburn's Oil Red O. (From Simpson, M. G. and Hutchinson, T., *Mar. Ecol. Prog. Ser.*, 91, 155, 1992. With permission.)

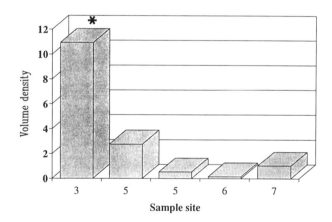

FIGURE 14. Stereological analysis of lipid vacuolation in liver cells of dab from the transect sampling sites. Asterisk indicates significant difference ($p < 0.001$) from the volume density at site 7 (least contaminated) using the Mann-Whitney U-test (n = 10). (From Lowe, D. M., previously unpublished data.)

Red O staining) in these vacuoles. There were significant increases in both the prevalence of lipid vacuolation and lipid content in the livers of dab from the contaminated site 3 compared with the reference site 7 (Figures 13 and 14).[29] Detailed study of this fatty change in methacrylate sections showed that the first signs of pathological processes involved lipid infiltration around blood vessels and lytic cell death (necrosis).[29] These are considered to be indicative of toxic liver injury. This type of change is progressive; and at the next stage of pathogenesis, cell death regularly started around the sinusoids and bile ducts (Figure 15).[29] At

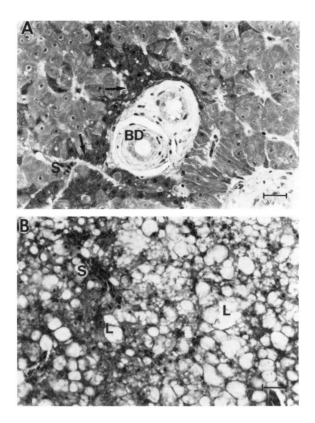

FIGURE 15. Characteristic light microsopic findings of dab liver (*Limanda limanda* L.) caught
at contaminated inshore sites of a North Sea transect. (A) Necrotic cells
(arrows) appear at first around bile ducts (BD) and sinusoids (S) indicating toxic
liver injury. (B) Severe steatosis of dab liver with compression of the cytoplasm
by lipid droplets (L) and complete dissolution of the characteristic parenchymal
structure. (Bar = 20 μm.) (From Köhler, A., Deisemann, H., and Lauritzen, B.,
Mar. Ecol. Prog. Ser., in press. With permission.)

the ultrastructural level, livers affected in this way showed proliferation of SER
and numerous free ribosomes throughout the cytoplasm.[29] The next level of
degenerative change included signs of lysosomal phospholipidosis mentioned
previously, which is considered to be characteristic of a chemically induced
condition.[29] This condition was present in livers from the inshore contaminated site.

Further indication of a degenerative reaction at the contaminated site 3 involved the formation of foci of altered cells.[28,29] These phenotypically distinct
lesions were classified according to the standardized criteria of Harada et al.[48] and
showed the highest prevalence at the contaminated site 3 ($p < 0.05$, n = 20, Fisher
exact test) compared to the reference site (Figure 16).[47] These lesions were
predominantly of the basophilic type and their constituent cells showed extensive
proliferation of granular endoplasmic reticulum (RER) (Figures 16, 17 and 18).[29]
Other changes in these livers included hepatocellular necrosis and prominent

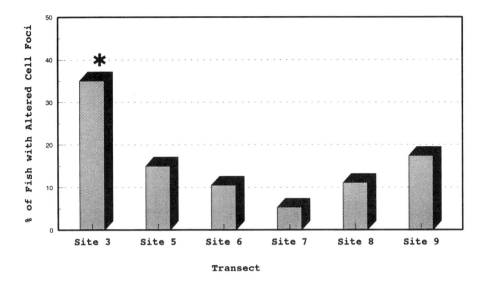

FIGURE 16. Prevalence of altered cell foci in the livers of non-overtly diseased dab (17–27 cm body length) from the transect sampling sites. Asterisk indicates significantly higher value ($p < 0.05$) than the prevalence at site 7 (least contaminated) using Fisher's exact test (n = 19). (From Simpson, M. G. and Hutchinson, T., *Mar. Ecol. Prog. Ser.,* 91, 155, 1992. With permission.)

mitotic activity in hepatocytes (Table 1).[28] Foci of cellular alteration in the livers of vertebrates are generally regarded as putative preneoplastic changes in hepatocytes; these occur spontaneously in laboratory rats, increase in frequency as a function of age, and mostly do not progress to hepatic neoplasia.[47,48] They do increase dramatically in numbers in young rodents exposed chronically to a wide variety of both genotoxic and nongenotoxic carcinogens. Foci of altered cells also occur in both feral fish and laboratory-reared populations.[10] This type of lesion can be induced in fish by exposure to chemical carcinogens and contaminated sediments, and shows considerable similarity to foci induced in rodents following experimental exposure to carcinogenic xenobiotics.[10] The prevalence of foci in dab liver followed the trend of contaminant levels in both livers and sediments.[28,29]

A cautionary note must be introduced at this point in order to emphasize that there is insufficient baseline pathological data on the induction of foci in dab and their relationship, if any, to the development of hepatic neoplasia, to draw conclusions about the significance of foci in liver disease. If dab are to be used effectively as environmental sentinels, then this lack of data needs to be rectified by means of experimental studies in order to establish the pathological significance of foci of altered cells and their consequences for individual health. Such studies are in fact currently in progress by the authors of this chapter. However, it is probably reasonable to conclude that increased prevalence of livers with foci is indicative of exposure to xenobiotics.

In an attempt to get a "handle" on the problem of contaminant-induced preneoplastic change in dab from this study in the North Sea, an immunocytochemical study of *ras*-

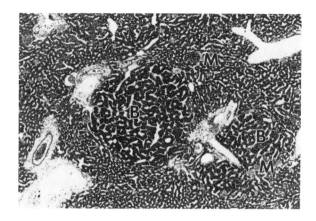

FIGURE 17. Foci of cellular alteration of basophilic type (B) in liver of dab from site 3. Note slight compression of hepatocytes (arrowheads) at the margin of one of the foci. Also, exclusion of melanomacrophage aggregates (M) from the foci. (Hematoxylin-eosin stain; magnification × 40.) (From Simpson, M. G., unpublished data.)

oncoprotein in the liver cells was conducted.[18] A test for *ras*-oncoprotein was developed for use with sections of dab liver, and the results are shown in Figure 19.[18] The role of *ras*-oncoproteins in the earliest stages of carcinogenesis is probably better documented than that for any of the other oncoproteins.[49] Activation or overexpression of *ras*-protooncogenes has been linked with several cases of neoplasia in flatfish, and the results of this study showed a 100% prevalence of the *ras*-oncoprotein in dab caught in the Elbe plume.[18] This finding has since been confirmed in further samples taken at different times of the year (Moore and Köhler, unpublished data).

These histopathologic findings are consistent with the expectations for effects induced by toxic organic xenobiotics. The demonstration of significant degenerative change in many livers from the most contaminated sites would indicate a significant degree of impairment of normal liver function in impacted fish.

V. AN INTEGRATED PERSPECTIVE

The results of this investigation of dab in the North Sea show a considerable degree of internal consistency, as well as clearly following the trend of declining levels of chemical contaminants as the transect moves offshore into the central North Sea.[50] A pattern readily emerges from the data showing lysosomal, cellular, and tissue damage of the type expected due to the toxic action of xenobiotics.[23,28-30] The additional evidence of proliferation of the smooth endoplasmic reticulum in hepatocytes — together with an increase in the activity of the inducible detoxication/activation enzyme EROD — further supports the premise that the fish have been exposed to inducer xenobiotics, which is consistent with the analytical

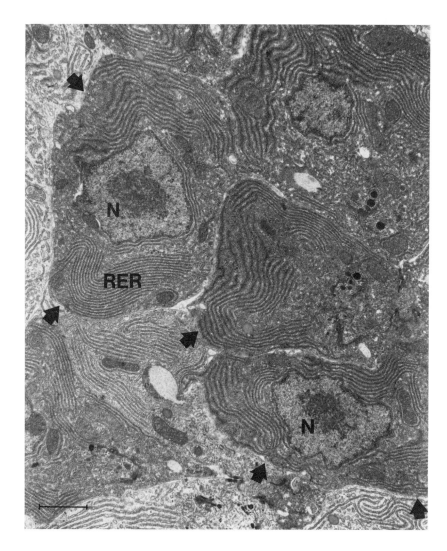

FIGURE 18. Group of phenotypically altered cells with a densely packed rough endoplasmic reticulum (RER) and a notable increase of the nucleus (N)/cytoplasmic ratio. The arrows mark the border of the focus which arises predominantly in areas of necrosis around the blood vessels. (Bar = 2 µm.) (From Köhler, A., Deisemann, H., and Lauritzen, B., *Mar. Ecol. Prog. Ser.*, 91, 141, 1992. With permission.)

chemical data.[29,30,50] Proliferation of SER and induction of EROD also provide a reasonable mechanistic explanation for the enhanced generation of superoxide radicals — through futile cycling — at the most contaminated sites, while radical attack on lysosomes is probably a contributory factor to the clear evidence for lysosomal membrane injury and consequent dysfunction.[9,29,30] The presence of phospholipidosis (accumulation of phospholipid membrane components in lysosomes) in severely damaged livers is also consistent with autophagy of chemically

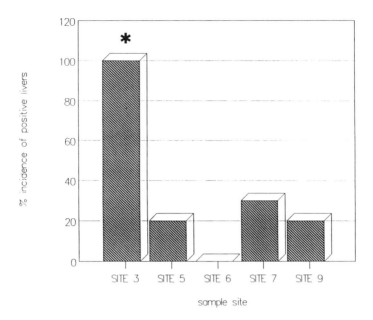

FIGURE 19. Prevalence of livers showing positive immunocytochemical reactions in liver sections with antibody to N-*ras* oncoprotein. The incidence at site 3 is significantly higher than at all other sites, including the least contaminated, site 7 ($* = p < 0.0016$, Fisher exact test, n = 10). (From Moore, M. N. and Evans, B., *Mar. Environ. Res.*, 34, 33, 1992. With permission.)

damaged proliferated endoplasmic reticulum and failure of the functionally impaired lysosomes to degrade the ingested membranes, which results in the characteristic pathological presentation of this storage disease condition.[51]

The presence of foci of altered cells at the most contaminated sites needs to be interpreted carefully. Comparison with the data from other species lends support to the conclusion that the high prevalence of foci is indicative of exposure to xenobiotic pollutants. The relationship of foci to neoplastic change in dab cannot at the present state of our knowledge be determined and requires further experimental investigation. It is, perhaps, worth commenting that neoplastic change in itself is of little significance in ecological terms unless the incidence in the population is extremely high. The value of pursuing the "neoplastic pathway" is that this type of disease is probably indicative of exposure to carcinogenic xenobiotics, although whether this occurs in embryos and larvae associated with xenobiotic-rich surface microlayer or as juveniles and adults in contact with bottom sediment and contaminated prey organisms is still an open question.[52] However, the finding that the prevalences of *ras*-oncoprotein positivity and foci of altered cells in livers of dab are similarly distributed would support the hypothesis that the initial steps in the process are occurring in the adult fish, as the involvement of *ras*-oncoprotein in the process of carcinogenesis is believed to occur at a very early stage.[28,29,49]

This integrated study probably represents the only one to date on such a large scale where a range of early cellular markers of xenobiotic-induced cell injury has been coupled with a detailed histopathologic investigation of the status of the liver in flatfish. The value of the cellular pathological approach is clear from the unequivocal demonstration of liver damage at sites known to be contaminated, and from the fact that the nature of the liver injury is likely to be due to exposure to contaminant xenobiotics. It must be emphasized that early onset cellular markers should be used in an integrated manner and not in isolation; otherwise a distorted interpretation might be derived from the data. If they are used appropriately, then it is possible to rapidly build up an holistic picture of the pattern of liver injury and to identify the probable causative factors, although not as specific single chemical entities. However, there is seldom any call for the latter because most contaminated situations consist of numerous xenobiotics, many of which will exert their toxicities in concert. This argues very strongly for the use of cellular pathology as an integrated multi-tiered biomarker or linked network of biomarkers.

The endpoint in this type of approach is the use of cellular pathology as an integrated predictive biomarker of the health status of individuals in a population.[10,33] It is not for the pathologist to extrapolate to the broader ecological consequences; this lies in the realms of environmental epidemiology and population ecology, and certainly this latter discipline appears to lack the necessary tools for extrapolation at present.[53] We consider that it is sufficient for the reasonable assessment of contaminant impact to demonstrate pathological change of a degree likely either to impair or to be a predictor of impairment of animal function and performance in a significant proportion of a representative sample taken from the population. This in itself is cause for environmental concern without any need to demonstrate ecological effects, which are probably impossible to interpret in terms of causality.[53]

This study has clearly demonstrated that an integrated pathological approach can identify both exposure to xenobiotics and induced cell injury. In order to further strengthen the biomarker approach there is a distinct need to develop cellular markers that will demonstrate whether the degree of pathological change is reversible or irreversible. Only in this way will we derive biomarkers which are truly prognostic as well as being diagnostic.

VI. CONCLUSIONS AND PROSPECTUS

Are we moving towards the era of readily available biomarkers for use in fish? The answer is probably yes; the approach outlined in this chapter and that described elsewhere by Hinton and Lauren indicates that the biomarker concept is both practical and potentially powerful.[10] Diagnosis of human and veterinary disease, together with mammalian toxicology, uses this concept very sucessfully and there is no obvious impediment to its immediate use in fish, so long as it is used circumspectly with full regard to the limitations in the pathological dastabases for many fish species. This latter problem only serves to underline the need for a greater concerted interdisciplinary research effort in identifying biomarkers and verifying the use of the biomarker concept in fish. The biomarker approach has the potential to provide a causative mechanistic basis for both cellular indicators of exposure to pollutants and their damaging effects.[33] As such it provides a bridge between chemical residues in tissues and their consequences, if any, for the individual animal.

So what about developments in the forseeable future? The emergence of effective cellular markers of xenobiotic exposure will reduce the need for comprehensive chemistry, which is costly, time consuming, and frequently limited in scope. This will help in the rapid screening of fish samples for the identification of environmental problems and, hence, permit the chemistry to be more focussed. The use of biomarkers will also facilitate the assessment of whether or not contaminants are biologically available, which cannot be ascertained from chemicals data alone. Finally, it is of paramount importance that experimental studies are closely linked with investigations of field samples in order to establish appropriate criteria for the cellular pathology of fish livers.

ACKNOWLEDGMENTS

Part of this work was funded under contract to the U.K. Department of the Environment as part of its coordinated program of research on the North Sea (PECD Ref. 7/8/137). Results are included from the IOC/ICES Workshop on the Biological Effects of Contaminants in the North Sea, Bremerhaven, Germany, March 1990.

REFERENCES

1. McCarthy, J. F., and Shugart, L. R., *Biomarkers of Environmental Contamination,* McCarthy, J. F. and Shugart, L. R., Lewis Publishers, Boca Raton, FL, 1990, 457 p.
2. Moore, M. N., Lysosomal cytochemistry in marine environmental monitoring, *Histochem. J.,* 22, 187, 1990.
3. Moore, M. N. and Simpson, M. G., Molecular and cellular pathology in environmental impact assessment, *Aquat. Toxicol.,* 22, 313, 1992.
4. Slater, T. F., Biochemical studies on liver injury, in Biochemical Mechanisms of Liver Injury, Slater, T. F., Ed., Academic Press, London, 1979, p. 1.
5. Kurelec, B. and Pivcevic. B., Evidence for a multixenobiotic resistance mechanism in the mussel *Mytilus galloprovincialis. Aquat. Toxicol.,* 19, 291, 1991.
6. Stegeman, J. J. and Lech, J. J., Cytochrome P-450 mono-oxygenase systems in aquatic species: carcinogen metabolism and biomarkers for carcinogen and pollutant exposure, *Environ. Health Perspect.,* 90, 93, 1991.

7. Nott, J. A. and Nicolaidou, A., Transfer of metal detoxification along marine food chains, *J. Mar. Biol. Assoc. U.K.*, 70, 905, 1990.
8. Ugarkovic, D., Kurelec, B., Krca, S., Batel, R., Robitzki, A., Müller, W. E. G., and Schröder, H. C., Alterations in *ras*-gene expression and intracellular distribution of protein kinase C in the sponge *Geodia cydonium* in response to marine pollution, *Mar. Biol.*, 107, 191, 1990.
9. Di Guilio, R. T., Washburn, P. C., Wenning, R. J., Winston, G. W., and Jewell, C. S., Biochemical responses in aquatic animals: a review of determinants of oxidative stress, *Environ. Toxicol. Chem.*, 8, 1103, 1989.
10. Hinton, D. E. and Lauren, D. J., Liver structural alterations accompanying chronic toxicity in fishes: potential biomarkers of exposure, in *Biomarkers of Environmental Contamination*, McCarthy, J. F. and Shugart, L. K., Eds., Lewis Publishers, Boca Raton, FL, 1990, p. 17.
11. De Basio, R., Bright, G. R., Ernst, L. A., Waggoner, A. S., and Taylor, D. L., Five-parameter fluorescent imaging:wound healing of living Swiss 3T3 cells, *J. Cell Biol.*, 105, 1613, 1987.
12. Lohse, J., Distribution of organochlorine pollutants in North Sea sediments, in *Int. Conf. North Sea Pollut. Tech. Strategies Improvement*, IAWPRC, EWPCA, NVA, Amsterdam, 1990, p. 227.
13. Darnell, J., Lodish, H., and Baltimore, D., *Molecular Cell Biology*, Scientific American Books, New York, 1990, 1105 p.
14. Sanders, B., Stress proteins: potential as multitiered biomarkers, in *Biomarkers of Environmental Contamination*, McCarthy, J. F. and Shugart, L. R., Eds., Lewis Publishers, Boca Raton, FL, 1990, p. 165.
15. Trump, B. F. and Arstila, A. V., Cell membranes and disease processes, in *Pathobiology of Cell Membranes*, Vol. 1, Trump, B. F. and Arstila, A. V., Eds., Academic Press, London, 1975, p. 1.
16. Mohammed, A., Eklund, A., Ostlund-Lindqvist, A. M., and Slanina, P., Distribution of toxaphene, DDT and PCB among lipoprotein fractions in rat and human plasma, *Arch. Toxicol.*, 64, 567, 1990.
17. Moore, M. N., Pollutant-induced cell injury in fish liver: use of fluorescent molecular probes in live hepatocytes, *Mar. Environ. Res.*, 34, 25, 1992.
18. Moore, M. N. and Evans, B., Detection of *ras* oncoprotein in liver cells of flatfish (dab) from a contaminated site in the North Sea, *Mar. Environ. Res.*, 34, 33, 1992.
19. McMahon, G., Huber, L. J., Stegeman, J. J., and Wogan, G. N., Identification of a c-Ki-*ras* oncogene in a neoplasm isolated from a Winter Flounder, *Mar. Environ. Res.*, 24, 345, 1988.
20. Pfeffer, S. R., GTP-binding proteins in intracellular transport, *Trends Cell Biol.*, 2, 41, 1992.
21. Wirgin, I., Currie, D., and Garte, S. J., Activation of the K-*ras* oncogene in liver tumours of Hudson River tomcod, *Carcinogenesis*, 10, 2311, 1989.
22. Hinton, D. E., Environmental contamination and cancer in fish, *Mar. Environ. Res.*, 28, 411, 1989.
23. Lowe, D. M., Moore, M. N., and Evans, B., Contaminant impact on interactions of molecular probes with lysosomes in living hepatocytes from Dab *(Limanda limanda)*, *Mar. Ecol. Prog. Ser.*, 91, 135, 1992.

24. Köhler, A., Lysosomal perturbations in fish liver as indicators for toxic effects of environmental pollution, *Comp. Biochem. Physiol.,* 100C, 123, 1991.

25. Jimenez, B. D., Oikari, A., Adams, S. M., Hinton, D. E., and McCarthy, J. F., Hepatic enzymes as biomarkers: interpreting the effects of environmental physiological and toxicological variables, in *Biomarkers of Environmental Contamination,* McCarthy, J. F. and Shugart, L. R., Eds., Lewis Publishers, Boca Raton, FL, 1990, p. 123.

26. Livingstone, D. R., Garcia Martinez, P., Michel, X., Narbonne, J. F., O'Hara, S., Ribera, D., and Winston, G. W., Oxyradical production as a pollution — mediated mechanism of toxicity in the common mussel, *Mytilus edulis* L and other molluscs, *Functional Ecol.,* 4, 415, 1990.

27. Lowe, D. M., Moore, M. N., and Clarke, K. R., Effects of oil on digestive cells in mussels: quantitative alterations in cellular and lysosomal structure, *Aquat. Toxicol.,* 1, 213, 1981.

28. Simpson, M. G., Histopathological studies in the liver of dab *(Limanda limanda)* from a contamination gradient in the North Sea, *Mar. Environ. Res.,* 34, 39, 1992.

29. Köhler, A., Deisemann, H., and Lauritzen, B., Ultrastructural and cytochemical indices of toxic injury in dab liver, *Mar. Ecol. Prog. Ser.,* 91, 141, 1992.

30. Moore, M. N., Molecular cell pathology of pollutant induced liver injury in flatfish: use of fluorescent probes, *Mar. Ecol. Prog. Ser.,* 91, 127, 1992.

31. Simpson, M. G. and Hutchinson, T., Toxicological pathology of the dab *(Limanda limanda)* along pollution gradients in the North Sea, *Mar. Ecol. Prog. Ser.,* 91, 155, 1992.

32. Winston, G. W., Moore, M. N., Straatsburg, I., and Kirchin, M., Decreased stability of digestive gland lysosomes from the common mussel *Mytilus edulis* L. by in vitro generation of oxygen-free radicals, *Arch. Environ. Contam. Toxicol.,* 21, 401, 1991.

33. Mayer, F. L., Versteeg, D. J., McKee, M. J., Folmar, L. C., Graney, R. L., McCume, D. C., and Rattner, B. A., Physiological and nonspecific biomarkers, in *Biomarkers — Biochemical, Physiological and Histological Markers of Anthropogenic Strem,* Huggett, R. J., Kimerle, R. A., Mehrle, P. M., and Bergman, H. L., Eds., Lewis Publishers, Boca Raton, FL, 1992, p. 5.

34. Rashid, F., Horobin, R. W., and Williams, M. A., Predicting the behaviour and selectivity of fluorescent probes for lysosomes and related structures by means of structure-activity models, *Histochem. J.,* 23, 450, 1991.

35. Moore, M. N., Cytochemical demonstration of latency of lysosomal hydrolases in digestive cells in the common mussel, *Mytilus edulis,* and changes induced by thermal stress, *Cell Tiss. Res.,* 175, 279, 1976.

36. Moore, M. N., Cytochemical responses of the lysosomal system and NADPH-ferrihemoprotein reductase in molluscan digestive cells to environmental and experimental exposure to xenobiotics, *Mar. Ecol. Prog. Ser.,* 46, 81, 1988.

37. Stier, A., Membrane fluidity, in *Biochemical Mechanisms of Liver Injury,* Slater, T. F., Ed., Academic Press, London, 1978, p. 219.

38. Renton, K. W. and Addison, R. F., Hepatic microsomal monooxygenase activity and P450IA mRNA in North Sea dab *(Limanda limanda)* from contaminated sites, *Mar. Ecol. Prog. Ser.,* 91, 65, 1992.

39. Chipman, J. K., Marsh, J. W., Livingstone, D. R., and Evans, B., Genetic toxicity in the dab *(Limanda limanda)* from the North Sea, *Mar. Ecol. Prog. Ser.,* 91, 121, 1992.

40. Bodammer, J. E. and Murchelano, R. A., Cytological study of vacuolated cells and other aberrant hepatocytes in winter flounder from Boston Harbor, *Cancer Res.,* 50, 6744, 1990.

41. Black, J. J., Carcinogenicity tests with rainbow trout embryos: a review, *Aquat. Toxicol.*, 11, 129, 1988.
42. Couch, J. A. and Courtney, L. A., Hepatocarcinogenesis in estuarine fish: induced neoplasms and related lesions with comparison to mammalian lesions, *J. Natl. Cancer Inst.*, 79, 297, 1987.
43. Bucke, D., Pathology of fish diseases in the North Sea, in *Environmental Protection of the North Sea*, Newman, P. J. and Agg, A. K., Eds., Heineman Professional Publishing, 1988, p. 286.
44. Bucke, D., Waterman, B., and Feist, S., Histological variations of hepatosplenic organs from the North Sea dab, *Limanda limanda* (L), *J. Fish Dis.*, 7, 255, 1984.
45. ICES, Methodology of fish disease surveys, Dethlefsen, V., Egidias, E., and McVicar, A. H., Eds., Co-operative research report No. 140, ICES, Copenhagen, Denmark, 1986.
46. Köhler, A., Cellular effects of environmental contamination in fish from the River Elbe and the North Sea, *Mar. Environ. Res.*, 28, 417, 1989.
47. Bannasch, P., Enzmann, H., Klimek, F., Weber, E., and Zerban, H., Significance of sequential cellular changes inside and outside foci of altered hepatocytes during hepatocarcinogenesis, *Toxicol. Pathol.*, 4, 617, 1989.
48. Harada, T., Maranpot, R. R., Morris, R. W., Stitzel, K. A., and Boorman, G., Morphological and stereological characterisation of hepatic foci of cellular alteration in control Fischer 344 rats, *Toxicol. Pathol.*, 4, 579, 1989.
49. Rayter, S. I., Iwata, K. K., Michitsch, R. W., Sorvillo, J. M., Vanenzuela, D. M., and Foulkes, J. G., Biochemical functions of oncogenes, in *Onocogenes*, Glover, D. M. and Hames, B. D., Ed., IRL Press, Oxford, 1989, p. 113.
50. Cofino, W., Smeeds, F., de Jong, A. S., Arbenou, A., Boon, J. P., Oostingh, I., Davies, I., Klungosøyr, J., Wilhensen, S., Law, R., Whinnett, J. A., Schmidt, D., and Wilson, S., The workshop chemistry programme, *Mar. Ecol. Prog. Ser.*, 91, 47, 1992.
51. Lüllman-Rauch, R., Drug-induced lysosomal storage disorders, in *Lysosomes in Biology and Pathology*, Vol. 6, Dingle, J. T., Jacques, P. J., and Shaw, I. H., Eds., Elsevier, Amsterdam, 1979, p. 49.
52. Cameron, P. and Berg, J., Morphological and chromosomal aberrations during embryonic development in the dab *Limanda limanda. Mar. Ecol. Prog. Ser.*, 91, 163, 1992.
53. Underwood, A. J. and Peterson, C. H., Towards an ecological framework for investigating pollution, *Mar. Ecol. Prog. Ser.*, 46, 227, 1988.

SECTION SIX

Biomarkers in Eggs

9. Biomarkers in Egg Samples

CHAPTER 9

Biomarkers in Egg Samples

David B. Peakall

TABLE OF CONTENTS

0-87371-648-5/94/$0.00+$.50
© 1994 by Lewis Publishers

I. INTRODUCTION

There are two main reasons for using eggs rather than adult animals for toxicological studies. The first, the rationale for this book, is that while the use of eggs for biomarker studies is not nondestructive as far as the egg itself it concerned, it is far less destructive to the species than using adult organisms. The second reason is that this life stage is often the most sensitive to pollutants and thus needs to be studied in order to assess the impact of a toxicant on the species. Interspecies and interclass comparisons are examined, and the relationship to results found in embryos to those in the adult organisms. Teratogenic studies require the use of embryos and are considered here only from the viewpoint of comparative studies.

Beside the relatively nondestructive nature of using eggs, there are several advantages of using eggs compared to adult organisms. These are:

1. The large clutch size of many species means that a large sample size can be readily obtained. Maintenance has been well worked out for several species, and large numbers can be housed in a small area. Such studies are cost-effective compared to tests on adult organisms.
2. In many species the outer shell is transparent and abnormalities can be observed as they occur in the living egg. In these cases it may be possible for the experiments to be completely nondestructive.
3. The early life stage is frequently the most sensitive stage for effects of chemicals to be seen.
4. For many species collection at the egg stage is easier than the collection of adults.

There are, however, a number of difficulties that have to be overcome before reliable results can be obtained:

1. The embryo needs to be exposed to the chemical; and if artificial means are used, the means of administration should not cause excessive mortality. Exposure via the parent is not usually nondestructive, but can allow the production of a large number of exposed embryos from a single parent.
2. The effect is likely to be dependent on the age of the embryo at the time of exposure. Prepartition exposure is likely to be different to exposure after birth, and the sensitivity of the embryo to toxicants can be expected to change during development.
3. The activity of biomarkers is known to change, often rapidly, during development of the embryo. Thus timing of the measurement can be critical.
4. In field experiments the high natural mortality of eggs can be a serious problem.

Table 1. Clutch-Size of Some Common Test Species

Class	Species	Clutch size	Key references to use as test species
Aves	Japanese quail (*Cortunix cortunix*)	7–12	50
	Mallard (*Anas platyrhynchos*)	10–12	3
	Ring dove (*Streptopelia risoria*)	2	51
Reptilia	Snapping turtle (*Chelydra serpentina*)	15–40	17 52
Amphibia	*Rana pipiens*	10,000	53
	Xenopus laevis	3,000–4,500	31
	Bufo woodhousii	5,000–8,000	26
Pisces	Rainbow trout (*Salmo gairdneri*)	10,000	54
	Bluegill sunfish (*Epomis macrochirus*)	30,000	55

II. CHARACTERISTICS OF EGGS OF DIFFERENT CLASSES

Although there are species in all of the five classes of vertebrates that lay external eggs, in the class Mammalia they are limited to a few species. These species, such as the platypus, have no importance in toxicology. The number of eggs laid by some of the common test species in the remaining four classes is given in Table 1.

The survival of the embryo is, as a first approximation, inversely proportional to the clutch size. However, even in birds with a small clutch size only a modest fraction of eggs result in adults.

III. INTRODUCTION OF CHEMICALS INTO EGGS

Under experimental conditions there are two broad categories of methods to introduce chemicals into eggs. The first is to expose the female and allow transfer to take place before the eggs are laid; the second is to expose the egg to the chemical after laying. The first approach cannot be considered nondestructive to the adult, but especially in the case of fish and amphibians does allow a large number of exposed embryos to be produced from each exposed adult. In field studies the natural exposure route can be used, and this is considered further in Section III.2.

A. Techniques for Avian Eggs

A wide variety of techniques has been used to expose the avian embryo to chemicals. These are listed in Table 2.

Table 2. Techniques of Administration of Chemicals to Avian Eggs

Technique	Comments	Ref.
Total immersion of egg in test solution	Detailed chemical analysis needed to determine embryonic exposure	1 2
Application of chemical to small area of shell	Limited number of chemicals penetrate well; dosage uncertain without analysis	3
Cutting window in shell pipetting onto membranes	Allows accurate placing on membrane; chemicals can be rapidly absorbed	23 5
Injecting into air chamber	Easy technique, chemicals can be rapidly adsorbed from membrane	25
Yolk replacement	Time-consuming; allows chemical to be well mixed	11, 12
Injecting into the yolk	Allows uptake during embryonic development; high mortality at some periods	7 9

1. Immersion of Eggs

Brief (30 sec) immersion of eggs has been used to expose the embryos to a variety of chemicals. David[1] used the technique to study dichlorodiphenyl-trichloroethane (DDT). In this case the DDT was presented as a suspension; uptake was demonstrated by chromatographic analysis. The fact that analysis is needed to quantify the uptake makes it a time-consuming and expensive technique. Hoffman and Albers[2] measured the weight change after immersion and related this to the estimated amount expected after field application by spraying. The technique has the advantage of simplicity and does not have the potential for damaging the embryo, but suffers from uncertainty — unless detailed chemical analyses are made — in the actual uptake of the chemical.

2. Application of the Chemical to the Shell

A series of experiments has been conducted by Albers[3] and Szaro and Albers[4] on the effects of petroleum hydrocarbons on the viability of embryos. They demonstrated that small amounts, in some cases as little as 1 μL, of oil applied to the external shell caused reduced hatching success and that these effects were not due to effects on gaseous exchange. Effects were most marked when the embryos were small; the effects of weathering and dispersants were also studied. The effects caused by the oil clearly demonstrate that it penetrated the shell although detailed chemical analysis or the formation of adducts do not appear to have been undertaken. The technique is easy, but is limited to those compounds that penetrate readily. Detailed studies of uptake would require detailed analysis.

3. Application of Chemicals to Chorioallantoic Membrane

This can be carried out either by cutting a small window or by direct injection in the air chamber. The window technique has been described by Hodach et al.[5]

These workers withdraw 2 mL of albumin with a syringe, which caused the embryo to fall away from the shell membrane. A half inch square piece of shell and membrane were then removed and the chemical placed on the chorioallantoic membrane. The eggs were then sealed with adhesive tape. Injection into the air chamber, located by candling with a high-intensity light source, can be carried out with a syringe without cutting a window. The syringe hole is sealed with wax.[6]

4. Injecting into the Yolk

The technique of injecting chemicals into the yolk sac of chicken *(Gallus gallus)* eggs has been described in detail.[7] A needle was inserted horizontally through the air cell into the yolk. They found that after the elimination of defective eggs (which accounted for 9% of the initial sample) the hatch of control eggs was 95%. They state that most reports refer to injections of chemicals made after the fourth or eighth day of incubation. Because of problems of small size and irregular positioning of the embryo, earlier injection is more difficult. Gilman et al.,[8] in field studies on the herring gull *(Larus argentatus),* found 40% mortality of controls when injections were carried out within the first day of incubation. Highly sophisticated laboratory equipment has been developed to inject small volumes (25–100 µL) into the yolks of chicken eggs.[9,10] This system is dependent on the precise location of the embryo and the use of fine needles and low volumes of materials. Using this technique there was no effect on viability at least down to 24 hr after incubation, although there was some evidence that an effect of corn oil injection was seen on 16-hr embryos.

Yolk replacement has also been used.[11,12] This is a time-consuming technique; only 30 eggs can be treated in a day. The advantage of the technique is that a large volume (up to 5 mL) can be added into the yolk cavity of the egg through an access canal constructed by electrosurgery.[11] The large volume allows a low concentration of the chemical to be used, giving a more uniform exposure of the embryo. Nevertheless, this difficult technique has not been widely used.

B. Techniques for Fish, Amphibians, and Reptiles

The technique for exposing the eggs of fish and amphibians can be very simple. Eggs, within an hour or so of hatching, are transferred to the test vessels. Test water needs to be monitored for pH, temperature, hardness, dissolved oxygen, and toxicant concentration in order to obtain consistent results.[13] Flow-through systems can be used for water soluble compounds.[14] Binder et al.[15] exposed fish eggs to PCBs by incubating them in petri dishes coated with a film of these compounds.

Although Cole and Townsend[16] described the controlled breeding of lizards for experimental studies over a decade ago, comparatively little toxicological work has been conducted on this class of organisms. Some experimental studies have been carried out on alligators *(Alligator mississippiensis);*[17] and field investigations have been carried out on the snapping turtle.[18] No experimental studies appear to have been made on administering chemicals to reptilian eggs.

C. Distribution of Chemicals in Embryos

Swartz[19] examined the effects caused by different solvents in the injection of DDT into the yolk sac. He found that the mortality caused by the solvent alone was low with DMSO and sesame oil, but high with olive oil. The mortality caused by DDT in sesame oil was higher than that in DMSO, although no dose dependence was seen in either case.

The uptake of DDT and PCBs into the embryos of herring gulls following injection into the yolk has been compared to the uptake from natural deposition.[8] These workers found that there was little difference over the entire range of embryonic development. David[1] has studied the uptake of DDT, using the immersion technique, in quail eggs at various stages of development. The uptake was found to be linear, being proportional to the stage of development.

While the technique of injection into the yolk has been widely used, it has been criticized by Walker.[12] Because of problems of distribution of the chemical, he considers that "at best, the injection method appears to a crude tool for determining toxicity." The validity of his criticism does not appear to be independently examined, and the egg injection has remained in widespread use.

D. Field Studies

While the basic techniques of administration of the chemical to the egg are the same for laboratory and field studies, there are some additional problems. The viability of chicken and quail eggs is readily determined by "candling" in the laboratory, but the viability of thick-shelled or heavily pigmented eggs in the field is more difficult. Mineau and Pedrosa[20] devised a portable device that could pick up the embryonic heartbeat. This equipment was demonstrated to have a 95%+ accuracy in determining embryonic viability in the herring gull from day 11 onward.

Another problem of field studies is the natural, and unnatural, predation of eggs. In a large study (916 eggs) involving the injection of organochlorines into herring gull eggs, 179 (19.5%) were lost before hatching.[8] This percentage was lower than that found in the uninjected controls and therefore does not appear to have been due to the experiment. This loss does mean that the sample size has to be sufficiently large to allow for the casualties. There is the further ethical problem of contamination of the environment. One can calculate that the amounts of organochlorines added to the environment in this way are very small; nevertheless, they exist and could certainly pose a problem if large doses or highly toxic materials (e.g., dioxins) were used. A technique to prevent unnatural predation is shown in Figure 1.

Using eggs from areas with different degrees of contamination it is possible to study the effects of these contaminants without administration of additional chemicals. The effects of environmental contamination on the reproduction of the herring gull[21] and the snapping turtle[18] in the Great Lakes have been shown in this way.

FIGURE 1. Unnatural predation.

IV. MORTALITY

There has been a great deal of criticism of the classical LD_{50} test in recent years. A detailed discussion of the value and limitations of this test is outside the scope of this chapter. However, despite all the objections that have been put forward, it is difficult to see how one can handle chemicals without some reasonable knowledge of their toxicity. The idea of using far fewer animals and being prepared to work with values with wider confidence limits is reasonable, particularly in view of wide interspecies variation.

A. Studies With Avian Embryos

Avian embryos have been used extensively for toxicity testing. These studies date back to an extensive series — over 60 — of papers by Férè at the end of the last century. These studies were reviewed by Férè in 1899.[22]

One of the few papers that examines the differences between two methods of delivering the chemicals to the embryo is that of Ridgway and Karnofsky.[23] These workers compared 4 and 8 day of injections of a wide variety of metal salts into the yolk with cutting a window and placing the solution on the chorioallantoic membrane (CAM). They divided the results into four groups:

1. Some elements gave approximately the same LD_{50} irrespective of age of the embryo or the route of administration.
2. Some elements show decreased toxicity at 8 days compared with 4 days, but are toxic by the CAM technique at 8 days. This is interpreted as meaning that the absorption of the salt is more efficient by the CAM route and that as the embryo increases in size more of the toxicant is needed.
3. Some elements are more toxic at 4 days than at 8 days by either route of administration. This suggests that the salt is equally absorbed by both routes and that the embryo becomes less sensitive as it increases in size.
4. Some elements are more toxic by CAM than by the yolk sac route. This was the largest group, accounting for 50% of the salts tested. This difference could be caused by the salt being precipitated or binding to components in the yolk or by the fact that the chemical is more rapidly absorbed from the CAM causing acute effects, whereas absorption from the yolk is slower.

Khera and Lyon[24] examined the toxicity of several organophosphorus insecticides on chicken and duck embryos of various ages by injection into the yolk sac. They found consistent results only at the midpoint of incubation; at earlier times there was a large variation between replications and no clear dose-response relationship. Since these pesticides act by inhibition of esterases, it is likely that these findings are due to changes in the activity of these enzymes in the developing embryo (see Section V.A).

Dunachie and Fletcher[25] examined the toxicity of a range of insecticides by injecting them into chicken eggs in order to assess the likelihood of these materials adversely affecting the hatching of wild bird eggs. These workers found that most compounds did not cause adverse effects at concentrations that could be reasonably expected in the environment. The main exception was deaths of chicks exposed to moderate levels of cyclodienes soon after hatching, if they were also exposed to the stress of decreased food supply.

B. Interclass Comparisons

Since the eggs of fish and amphibians are produced in large numbers, the possibility that embryos from them could be used for the determination of LD_{50} is attractive.

The toxicity of a range of metallic salts to amphibian embryos has been made by Birge et al.[13] and to the avian embryo by Ridgway and Karnofsky.[23] As has already been discussed in Section IV.A. Ridgway and Karnofsky used two different methods of administration. Without making a detailed statistical comparison it is clear that there is little similarity between the ranking of the various metals. A ranking of amphibian LC_{50}s has been made by Harfenist et al.;[26] this shows that few cholinesterase inhibitors are high in the ranking. This is the opposite finding to that of Dunachie and Fletcher.[25]

A much better correlation between fish and avian embryo toxicity was found by van Leeuwen et al.,[27] but the endpoints here were largely teratogenic. These findings are discussed in the next section.

V. TERATOGENIC STUDIES

Teratogenic studies cannot be performed nondestructively. However, there has been a movement away from mammalian and to some extent avian embryos toward the development of screening tests using fish, amphibians, and even invertebrates.

Only in recent years has the use of nonmammalian systems gained favor as screens for teratogens. In 1967 the World Health Organization stated that "the chick embryo contributes greatly to basic embryological knowledge. However, for the screening of drugs for teratogenicity its use is not recommended." More recent work, however, has shown reasonably good correspondence in different test systems. Gebhardt[28] tabulated data on 23 different chemicals and concluded that "substances teratogenic in the chicken usually have the same effect on the rat, mouse or rabbit." Within a decade the chicken had been accepted as a standard system.[29] More recently there has been a further change in approach, and tests based on amphibians are now widely used.

Three nonmammalian systems for the identification of chemical teratogens were put forward by Sabourin et al.[30] This included one embryonic test, that of using the development of *Xenopus* to the swimming tadpole stage. The other assays were head regeneration in flatworms and ability of disassociated cells of *Hydra* to regenerate into intact organisms. The advantage of the *Xenopus* assay is that it allows a variety of endpoints to be assessed. Reasonable agreement was found between the tests for a small range of chemicals.

Recently Dawson and Wilke[31] have put forward an evaluation of the test system called frog embryo teratogenesis assay: *Xenopus* (FETAX) which is based on embryos of the frog, *Xenopus laevis*. These workers tested this system against a number of combinations of chemicals and concluded that it is useful for hazard assessment, especially when malformation is the endpoint considered.

van Leeuwen et al.[27] have compared the 7-day test using zebra fish *(Brachydanio rerio)* with the 30-day rainbow trout test and then compared these tests with the chicken. They found a high degree of correlation ($r = 0.95$) between the two fish tests, and a remarkably good correlation ($r = 0.90$) between the chicken and the two fish tests. The zebra fish spawns repeatedly in the laboratory and produces large numbers of transparent eggs, making early detection of malformation possible. The species lives at a high temperature; and the time from fertilization to yolk absorption is only 11 days, compared to 60 days in cold water trout. These workers consider that the short-term zebra fish test has the same predictive power as the chicken test for mammalian teratogenicity and that both are suitable for screening for teratogens.

Hoffman and Albers[2] examined the embryo toxicity and teratogenicity of a range of insecticides, herbicides, and petroleum derivatives on mallard eggs. Recently, Hoffman[32] reviewed the embryotoxicity and teratogenicity of environmental contaminants to avian eggs. The distinction between embryos and eggs is important here because Hoffman is looking at the effects of contaminants that have been transported across the eggshell after the compound had been applied to the shell, whereas most studies look at the effects of compounds injected into the egg.

VI. ENZYME SYSTEMS

A. Esterases

Esterases are considered in more detail in Chapter 2. Here they are considered only as they apply to the developing embryo. Detailed studies appear to be limited to the avian embryo.

There is a profound difference between the development of cholinesterase (ChE) activity in precocial and altricial birds. In the case of altricial birds the development of activity occurs largely after hatching. Grue et al.[33] examined the changes in nestling starlings and found a threefold increase between day 3 and day 18. No measurements appear to have been made on the ChE activity in the embryos of altricial species, but the activity can be expected to be low.

The development of plasma and brain cholinesterase activity in mallard has been studied by Hoffman and Eastin.[34] They found a fivefold increase in both plasma and brain ChE over the period of day 11 to hatching (day 26). It will be seen that the most rapid increase of plasma ChE was from day 19 to hatching, whereas the increase in brain ChE was more uniform. Measurement of acetylcholinesterase (AChE) activity in whole embryo homogenates of bobwhite quail was made by Meneely and Wyttenbach[35] at day 6, 9, and 12. They found a linear increase over this period.

Brain ChE activity was measured in nestlings of three species of herons.[36] For only one species, the snowy egret *(Egretta thula)*, were measurements made on embryos. In this species measurements made during the last 3 days before hatching showed the linear increase of activity with age extended back to prehatching. In the other two species appreciable activity was found at day 4. Measurable activity can be expected to be found in the late embryos of all of altricial species.

One of the few papers on AChE activity in amphibian embryos is that by de Llamas et al.[37] These workers found that AChE activity started to increase at day 5 of development in the frog *Bufo arenarum* and that this increase was reduced by dieldrin. A dose-dependent inhibition of AChE by malathion was also shown.

In order to measure any differences between normal and exposed embryos it is obviously necessary to match the age of embryos accurately. Since biologically significant AChE inhibition is 50–80%, using mortality as the criterion, it is possible to use embryos for this research. Sublethal effects may be seen at lower doses, but significant behavioral changes are rare below 50% inhibition.[38]

While the direct effects of organophosphates (OPs) and carbamates (CB) on the nervous system are mediated through AChE, the longer term neurological effects are mediated through neurotoxic esterase, also known as neuropathy target esterase (NTE). This enzyme system has been studied extensively by Johnson and co-workers; a recent review of the work of this group has been made.[39] The binding of Ops with NTE leads to organophosphorus compound-induced delayed neuro-toxicity (OPIDN). Although it is assumed that NTE — which is tightly bound to neuronal membranes — has a physiological function, this has not been elucidated. The phylogenetic variations of NTE have been examined by Johnson.[40] In general, the activity is greatest in mammals and birds, then in fish, and finally in amphibians. Many mammals, especially rodents, are resistant to inhibition of NTE. Another important criterion of OPIDN is whether it is caused by a single dose. While the test animal of choice for NTE is the mature chicken, a considerable amount of work has been done on embryos. Some of these studies, particularly those in which both AChE and NTE were studied, are reviewed below.

The relationship between AChE inhibition and skeletal abnormalities for mallard eggs exposed to three Ops was examined by Hoffman and Eastin.[34] All three compounds — malathion, diazinon, and parathion — caused marked inhibition of AChE, but only the highest dose of parathion caused significant increase of skeletal abnormalities. No relationship between the degree of inhibition of AChE and the occurrence of abnormalities was found. The activity of NTE was not examined in this study. In a subsequent paper[41] the effects of the OP EPN (Phenyl phosphonthoic acid-O-ethyl-O-[4-nitrophenyl]ester) were examined on mallard eggs. Although a significant increase in the percentage of abnormal survivors was found at all doses, the occurrence rates were not dose dependent. Significant decrease of both AChE and NTE were also found at all doses. Again clear dose dependence was not established, but NTE was almost completely inhibited at the two higher doses.

In a recent review of the mechanism underlying OPIDN, Carrington[42] concluded that "hundred of compounds have been tested for their ability to inhibit NTE activity both in vivo and in vitro and a very strong correlation between inhibition and the development of OPIDN has been established."

B. Mixed Function Oxidases

Another enzyme system, or rather series of enzyme systems, that have been studied in embryos are the mixed function oxidases (MFOs).

Hamilton et al.[43] measured the aryl hydrocarbon hydroxylase (AHH) activity in the chicken embryo from 3 to 10 days after hatching. For day 3 and 4 measurements were made on the whole embryo, but from day 5 onward (after the appearance of the liver) they were made on this organ. The basal activity in the supernatant was constant throughout the period of measurement and was capable of stimulation after day 6/7 of incubation by tetrachlorobiphenyl (TCB) using a dose of 5×10^6 mol/kg embryo wt. The basal level of the microsomal fraction was also fairly constant, although there was a small peak from day 10 to 14 and a more pronounced peak in the first 8 days after hatching. The microsomal AHH was capable of induction by TCB throughout the period of observation. Typically the activity was increased from 0.3–0.5 to 10–15 nmol/min/mg protein. The chicken embryo shows basal AHH activity comparable to that of the adult at the earliest point at which it was possible to assay activity a full day before the liver began to differentiate. Time response studies conducted on day 17–21 embryos showed that TCB began to induce activity within 3 hr; induction reached maximal values by 6–9 hr and continued for at least 96 hr.

The basal and induced AHH activity in both chicken liver and lung were studied by Hamilton and Bloom[44] over the period day 14 of incubation to day 10 posthatch. The activity pattern in the two organs is quite different. The basal level remains low throughout the liver, whereas the induced level is 15- to 30-fold higher. In the lung the basal activity is high initially and decreases as the embryo grows. In contrast to the liver the induced activity is, apart from a brief diversion at hatching, essentially the same as the basal activity. The authors suggest that these differences between the induced and basal profiles indicate that the changes in basal and induced AHH activity in the two organs are independent of each other. Further, the authors point out that the activities in both tissues are markedly different from those reported in mammalian systems.

A rather different pattern of activity during development was shown by Brunström.[45] He found a substantial increase in activity in the liver between days 5 and 10, after which it remained at a fairly constant level until day 19 and reached a peak 1–5 days posthatching. Treatment of the eggs at day 5 with TCB, injected into the egg, doubled the activity of 7-ethoxycouramin-O-deethylase and increased the activity of AHH 14-fold. Binder and Lech[46] examined the AHH activity in embryos and swim-up fry of lake trout from hatcheries and from contaminated areas of Lake Michigan. Higher activity was found from contaminated lake sites. Additional studies, in which embryos were treated with PCBs, also resulted in an increase of activity. In a subsequent study[15] the induction of AHH activity in the embryos and eleuthero embryos of *Fundulus* by PCBs was investigated. AHH activity in the eleuthero embryos was increased 9-fold within 24 hr of hatching; the dose needed for embryos was considerably higher. In the Japanese medaka *Oryzias latipes* exposed to 2,3,7,8-tetrachlorodibenzo-p-dioxin beginning at the day of fertilization led to induction of benzo(a)pyrene hydroxylase (BaPH) activity on day 5.[47] This coincides with the development of the liver. A number of congeners were examined, and the pattern of induction was similar to that found in mammalian systems.

The induction of the activity of three MFO enzymes and cytochrome P-450 by microliter quantities of Prudhoe Bay crude oil (PBCO) placed on the shells of chicken eggs (day 10–20) has been demonstrated.[48] The enzymes studied were BaPH, naphthalene hydroxylase, and 7-ethoxyresorufin-O-deethylase (EROD). The basal levels of EROD and BaPH were stable over the period of study; some increases were found with cytochrome P-450 and naphthalene hydroxylase. EROD activity was the most responsive, showing an increase of 22–24-fold.

In a subsequent study[49] the effect of the aliphatic, aromatic, and heterocyclic fractions of PBCO on mortality and enzyme induction was examined; it was found that the heterocyclic fraction (compounds of nitrogen, oxygen, and sulfur) although constituting only 7% of the oil, was on an equivalent weight basis as effective as the aromatic fraction in causing induction. The aliphatic compounds were essentially inactive.

Recently studies have been made on the effect of water soluble fractions of crude oil on the induction of cytochrome P-450 IA1 in eggs, larvae, and juveniles of cod.[16] Induction did not occur until several days after hatching, a rapid rise being found 7 to 8 days posthatch; this finding is similar to that for altricial birds.

Although the pattern of change of activity during development needs to be studied carefully in any experiments, the degree of induction of MFOs is sufficiently large to make studies on embryos feasible. Three different techniques have been used on the avian egg — direct placement on the shell, injecting, and cutting windows — to deliver the chemical to the embryo. In fish, exposure by both addition to the incubation media and exposure of the parent have been used. The pattern of induction in the two orders is similar. There do not appear to be studies on the induction of MFOs by chemicals on the embryos of amphibians and reptiles.

VII. SUMMARY AND CONCLUSIONS

Studies of biomarkers in eggs are most advanced in the class Aves, although a number of studies have now been reported on the induction of mixed function oxidase enzymes in fish. It would be valuable to extend these studies on fish to other biomarkers and conduct studies using a range of biomarkers on eggs of amphibians and reptiles. Only with teratogenic studies has the amphibian embryo been seriously investigated. In view of the enormous numbers of eggs produced per female by most species of fish and amphibians, these systems would be valuable for biomarker studies and would allow the transmission of the chemical via the adult to be studied as well direct uptake by the egg. The comparison of these two routes is important, so that the normal route of embryonic exposure that is through the parent can be compared to the experimental procedure of adding chemicals to the incubation media or injecting the chemical directly into the egg.

REFERENCES

1. David, D., Gas chromatographic study of the rate of penetration of DDT into quail eggs at different stages of their development, *Bull. Environ. Contamin. Toxicol.,* 21, 289, 1979.
2. Hoffman, D. J. and Albers, P. H., Evaluation of potential embryotoxicity and teratogenicity of 42 herbicides, insecticides, and petroleum contaminants to mallard eggs. *Arch. Environ. Contam. Toxicol.,* 13, 15, 1984.
3. Albers, P. H., Effects of external applications of fuel oil on hatchability of mallard eggs, in *Fate and Effects of Petroleum Hydrocarbons in Marine Organisms and Ecosystems,* Wolfe, D. A., Ed., Proceedings of a Symposium, Pergamon Press, New York, 1977.
4. Szaro, R. C. and Albers, P. H., Effects of external application of No. 2 fuel oil on common eider eggs, in *Fate and Effects of Petroleum Hydrocarbons in Marine Organisms and Ecosystems,* Wolfe, D. A., Ed., Pergamon Press, New York, 1977, p. 164.

5. Hodach, R. J., Gilbert, E. F., and Fallon, J. F., Aortic arch anomalies associated with the administration of epinephrine in chick embryos, *Teratology*, 9, 203, 1974.

6. Gebhardt, D. O. E. and van Logten, M. J., The chick embryo test as used in the study of the toxicity of certain dithiocarbamates, *Toxic. Appl. Pharmacol.*, 13, 316, 1968.

7. McLaughlin, J., Jr., Marliac, J.-P., Verrett, M. J., Mutchler, M. K., and Fitzhugh, O. G., The injection of chemicals into the yolk sac of fertile eggs prior to incubation as a toxicity test, *Toxicol. Appl. Pharmacol.*, 5, 760, 1963.

8. Gilman, A., Hallett, D., Fox, G. A., Allan, L., Learning, W., and Peakall, D. B., Effects of injected organochlorines on naturally incubated Herring Gull eggs, *J. Wildl. Manage.*, 42, 484, 1978.

9. Kitos, P. A., Wyttenbach, C.R., Olson, K., and Uyeki, E. M., precision delivery of small volumes of liquids to very young avian embryos. II. Description of the injection systems, *Toxicol. Appl. Pharmacol.*, 59, 49, 1981.

10. Wyttenbach, C. R., Thompson, S. C., Garrison, J. C., and Kitos, P. A., Precision delivery of small volumes of liquids to very young avian embryos. 1. Locating and positioning the embryo in ovo, *Toxicol. Appl. Pharmacol.*, 59, 40, 1981.

11. Walker, N. E., Use of yolk-chemical mixtures to replace hen egg yolk in toxicity and teratogenicity studies, *Toxicol. Appl. Pharmacol.*, 12, 94, 1968.

12. Walker, N. E., The effect of malathion and malaoxon on esterases and gross development of the chick embryo, *Toxicol. Appl. Pharmacol.*, 19, 590, 1971.

13. Birge, W. J., Black, J. A., and Westerman, A. G., Evaluation of aquatic pollutants using fish and amphibian eggs as bioassay organisms, in *Animals as Monitors of Environment Pollution*, Neilsen, S. W., Magaki, G., and Scarpelli, D. G., Eds., National Academy of Science, Washington, D.C., 1979, p. 108.

14. Goksoyr, A., Solberg, T. S., and Serigstad, B., Immunochemical detection of cytochrome P450IAI induction in cod larvae and juveniles exposed to a water soluble fraction of North Sea crude oil, *Mar. Pollut. Bull.*, 22, 122, 1991.

15. Binder, R. L., Stegeman, J. J., and Lech, J. J., Induction of cytochrome P-450-dependent monooxygenase systems in embryos and eleutheroembryos of the killfish *Fundulus heteroclitus*, *Chem. Biol. Interact.*, 55, 185, 1985.

16. Cole, C. J. and Townsend, C. R., Parthenogenetic reptiles: new subjects for laboratory research, *Experientia*, 33, 285, 1977.

17. Jewell, C. S. E., Cummings, L. E., Ronnis, M. J. J., and Winston, G. W., The hepatic microsomal mixed-function oxidase (MFO) system of Alligator mississippiensis: induction by 3-methylcholanthrene (MC), *Xenobiotica*, 19, 1181, 1989.

18. Bishop, C. A., Brooks, R. J., Carey, J. H., Ng, P., Norstrom, R. J., and Lean, D. R. S., The case for a cause-effect linkage between environmental contamination and development in eggs of the common snapping turtle *(Chelydra s. serpentina)* from Ontario, Canada, *J. Toxicol. Environ. Health*, 33, 521, 1991.

19. Swartz, W. J., Dissimilarities in the toxic response of early chick embryos to DDT administered in different vesicles, *Bull. Environ. Contam. Toxicol.*, 25, 898, 1980.

20. Mineau, P. and Pedrosa, M., A portable device for non-destructive determination of avian embryonic viability, *J. Field Ornithol.*, 57, 53, 1986.

21. Gilman, A. P., Fox, G. A., Peakall, D. B., Teeple, S. M., Carroll, T. R., and Haymes, G. T., Reproductive parameters and egg contaminant levels of Great Lakes Herring Gulls, *J. Wildl. Manage.*, 41, 450, 1977.

22. Férè, C., Teratogenie experimentale et pathologie generale, *Cinquantenaire de la Societe de Biologie*, Vol. jubilaire, 360, 1899.

23. Ridgway, L. P. and Karnofsky, D. A., The effects of metals on the chick embryo: toxicity and production of abnormalities in development, *Ann. N.Y. Acad. Sci.,* 55, 203, 1952.

24. Khera, K. S. and Lyon, D. A., Chick and duck embryos in the evaluation of pesticide toxicity, *Toxicol. Appl. Pharmacol.,* 13, 1, 1986.

25. Dunachie, J. F. and Fletcher, W. W., An investigation of the toxicity of insecticides to birds' eggs using the egg-injection technique, *Ann. Appl. Biol.,* 64, 409, 1969.

26. Harfenist, A., Power, T., Clark, K. L., and Peakall, D. B., A review and evaluation of the amphibian toxicological literature, CWS Techn. Rep. No. 61, 1989, 222 pp.

27. van Leeuwen, C. J., Grootelaar, E. M. M., and Niebeek, G., Fish embryos as teratogenicity screens. A comparison of embryotoxicity between fish and birds, *Ecotoxicol. Environ. Saf.,* 20, 42, 1990.

28. Gebhardt, D. O. E., The use of the chick embryo in applied teratology, *Adv. Teratol.,* 5, 97, 1972.

29. Collins, F. T. X., Teratological research using in vitro systems. V. Nonmammalian model systems, *Environ. Health Perspect.,* 72, 237, 1987.

30. Sabourin, T. D., Faulk, R. T., and Goss, L. B., The efficacy of three nonmammalian test systems in the identification of chemical teratogens, *J. Appl. Toxicol.,* 5, 227, 1985.

31. Dawson, D. A. and Wilke, T. S., Evaluation of the frog embryo teratogenesis assay: *Xenopus* (FETAX) as a model system for mixture toxicity hazard assessment, *Environ. Toxicol. Chem.,* 10, 941, 1991.

32. Hoffman, D. J., Embryotoxicity and teratogenicity of environmental contaminants to bird eggs, *Rev. Environ. Contamin. Toxicol.,* 115, 39, 1990.

33. Grue, C. E., Powell, G. V. N., and Gladson, N. L., Brain cholinesterase (ChE) activity in nestling starlings: implications for monitoring exposure of nestling songbirds to ChE inhibitors, *Bull. Environ. Contam. Toxicol.,* 26, 544, 1981.

34. Hoffman, D. J. and Eastin, W. C., Jr., Effects of malathion, diazinon, and parathion on mallard embryo development and cholinesterase activity, *Environ. Res.,* 26, 472, 1981.

35. Meneely, G. A. and Wyttenbach, C. R., Effects of the organophosphate insecticides diazinon and parathion on bobwhite quail embryos: skeletal defects and acetylcholinesterase activity, *J. Exp. Zool.,* 252, 60, 1989.

36. Custer, T. W. and Ohlendorf, H. M., Brain cholinesterase activity of nestling Great Egrets, Snowy Egret and Black-crowned Night Herons, *J. Wildl. Dis.,* 25, 359, 1989.

37. de Llamas, M. C., de Castro, A. C., and de D'Angelo, A. M. P., Cholinesterase activities in developing amphibian embryos following exposure to the insecticides dieldrin and malathion, *Arch. Environ. Contamin. Toxicol.,* 14, 161, 1985.

38. Peakall, D. B., Behavioral responses of birds to pesticides and other contaminants, *Res. Rev.,* 96, 45, 1985.

39. Johnson, M. K., Organophosphates and delayed neuropathy — Is NTE alive and well?, *Toxicol. Appl. Pharmacol.,* 102, 385, 1990.

40. Johnson, M. K., The delayed neuropathy caused by some organophosphorus esters: mechanism and challenge, *CRC Crit. Rev. Toxicol.,* 3, 289, 1975.

41. Hoffman, D. J. and Sileo, L., Neurotoxic and teratogenic effects of an organophosphorus insecticide (phenyl phosphonothioic acid-O-ethyl-O-[4-nitrophenyl] ester) on mallard development, *Toxicol. Appl. Pharmacol.,* 73, 284, 1984.

42. Carrington, C. D., Prophylaxis and the mechanism for the initiation of organophosphorous compound-induced delayed neurotoxicity, *Arch. Toxicol.,* 63, 165, 1989.

43. Hamilton, J. W., Denison, D. S., and Bloom, S. E., Development of basal and induced aryl hydrocarbon (benzo[a]pyrene) hydroxylase activity in the chicken embryo in ovo, *Proc. Natl. Acad. Sci.*, 80, 3372, 1983.

44. Hamilton, J. W. and Bloom, S. E., Developmental differences in basal and induced aryl hydrocarbon (benzo[a]pyrene) hydroxylase activity in chick embryo liver and lung in ovo, *Biochem. Pharmacol.*, 32, 2986, 1983.

45. Brunström, B., Activities in chick embryos of 7-ethoxyresorufin O-deethylase and aryl hdyrocarbon (benzo[a]pyrene) and their induction by 3,3',4,4'-tetrachlorobiphenyl in early embryos, *Xenobioticas,* 16, 865, 1986.

46. Binder, R. L. and Lech, J. J., Xenobiotics in gametes of Lake Michigan lake trout *(Salvelinus namaycush)* induce hepatic monooxygenase activity in their offspring, *Fund. Appl. Toxicol.*, 4, 1042, 1984.

47. Wisk, J. D. and Cooper, K. R., Effect of 2,3,7,8-tetrachloro dibenzo-p-dioxin on benzo(a)pyrene hydroxylase activity in embryos of the Japanese medaka *(Oryzias latipes),* Arch. Toxicol., 66, 245, 1992.

48. Lee, Y.-Z., O'Brien, P. J., Payne, J. F., and Rahimtula, A. D., Toxicity of petroleum crude oils and their effects on xenobiotic metabolizing enzymes activities in the chicken embryo in ovo, *Environ. Res.*, 39, 153, 1986.

49. Walters, P., Khan, S., O'Brien, P. J., Payne, J. F., and Rahimtula, A. D., Effectiveness of a Prudhoe Bay crude oil and its aliphatic aromatic and heterocyclic fractions in inducing mortality and aryl hydrocarbon hydroxylase in chick embryo in ovo, *Arch. Toxicol.,* 60, 454, 1987.

50. Hudson, R. H., Tucker, R. K., and Haegele, M. A., Handbook of Toxicity of Pesticides to Wildlife, U.S. Fish & Wildl. Ser. Resource Publ., 153, 1984, p. 90.

51. Peakall, D. B., Lincer, J. L., and Bloom, S. E., Embryonic mortality and chromosomal alterations caused by Aroclor 1254 in ring doves, *Environ. Health Perspect.*, 2, 103, 1972.

52. Hall, R. J., Effects of environmental contaminants on reptiles: a review, *U.S. Fish & Wildl. Serv., Spec. Sci. Rep. Wildl.,* 228, 12, 1980.

53. Cooke, A. S., Tadpoles as indicators of harmful levels of pollution in the field, *Environ. Pollut.,* 25A, 123, 1981.

54. Oikari, A., Holmbom, B., Anas, E., Miilunpalo, M., Kruzynski, G., and Castren, M., Ecotoxicological aspects of pulp and paper mill effulents discharged to an inland water system: distribution in water and toxicant residues and physiological effects in caged fish *(Salmo gairdneri), Aquat. Toxicol.,* 6, 219, 1985.

55. McCarthy, J. F., Jimenez, B. D., Shugart, L. R., and Sloop, F. V., Biological markers in animal sentinels: laboratory studies improve interpretation of field data, in *In situ Evaluation of Biological Hazards of Environmental Pollutants,* Sandhu, S. S., Lower, W. R., de Serres, F. J., Suk, W. A., and Tice, R. R., Eds., Plenum Press, New York, 1990, p. 163.

SECTION SEVEN

Biomarkers in Studies of Endangered Species: Marine Mammals

10. An Improved Whale Biopsy System Designed for Multidisciplinary Research

11. Assessment of Organochlorine Pollutants in Cetaceans By Means of Skin and Hypodermic Biopsies

CHAPTER 10

An Improved Whale Biopsy System Designed for Multidisciplinary Research

R. H. Lambertsen, C. Scott Baker, Mason Weinrich, and William S. Modi

TABLE OF CONTENTS

I. SUMMARY

A new whale biopsy system designed to obtain efficiently a larger quantity of tissue for biological, genetic, toxicological, and pharmacokinetic research was field tested in the southern Gulf of Maine on humpback whales *(Megaptera novaeangliae)*. Powered by pneumatic gun, 21 of 24 (88%) experimental shots hit the study animal. The majority of these experimental shots struck within a small target area at or near the base of the dorsal fin. Out of 20 darts, 19 (95%) fired in the intended design configuration obtained a useful sample. One shot fired experimentally with the biopsy instrument still contained in its sterile wrap did not function. The biopsies obtained were of sufficient size for distribution of 0.5–1.0 cc sample portions to 6–8 different research groups studying population structure, sex distributions, inbreeding coefficients, aging methods, histopathology, genotoxicity, and ecotoxicology. Seven out of nine (78%) samples plated in vitro were viable and showed active cell proliferation 8 to 20 days after culture initiation. The behavioral response to the sampling procedure was examined in detail in 10 animals. A statistical comparison was made of respiratory and behavioral variables recorded for 30 min before and 30 min immediately after the biopsy. At the instant of the biopsy strike 40% of the whales flicked their tails once in what appeared to be a reflexive action. Respiratory variables and estimated swimming speeds did not change significantly. The number of animals which displayed tail slashes and trumpet blows increased, as did the overall frequency of these two displays, but the changes were not statistically significant. When the postbiopsy observation time was broken into two 15-min periods, a majority of all numerical changes in behavior was found to be limited to the first 15 min after the biopsy. The typical behavioral response to the new biopsy system did not appear to be more than that caused by previously approved equipment. The improved trajectory, range, control, and power of the new unit minimized the number of failed biopsy shots decreased disturbance of study animals caused by close vessel approach and repeat biopsy attempts, and increased the range of sea conditions in which sampling effectively could be pursued. Repeat sampling of individual whales for different investigative purposes also was reduced because archival quality samples suitable for multiple research uses were obtained. On these bases it is concluded that a more humane and cost-effective whale biopsy system has been field-proven by a team of experienced personnel which included a skilled marksman. The combined advantages likely will enable nondestructive tissue sampling of all large Cetacea, including fleet species such as the fin whale, *Balaenoptera physalus,* and tough-skinned animals such as the sperm whale, *Physeter catodon.*

II. INTRODUCTION

The detection of high levels of organochlorine pesticides and polychlorinated biphenyls (PCBs) in the blubber and internal organs of seals and certain odontocete species,[1-13] sometimes in association with high rates of premature births,[4,7] reproductive failure,[10] and uterine pathology,[8] has increased concern in recent years about adverse effects of pollution on coastal populations of marine mammals. Concurrently it has been recognized that toxicological studies of marine mammals are important from a human ecological perspective. As living integrators of environmental factors, including pollutants, these species can serve as early warning systems which provide valuable information on the fate and effects of anthropogenic compounds in higher mammals.[14,15] Others have argued persuasively that the measurement of organochlorine metabolites in the blubber of these species holds promise for discriminating population structure and for evaluating age, sex, and species-specific trophic divergences.[16,17]

Compared with seals and odontocetes, the burdens of pollutants carried by mysticetes (baleen whales) are less well known. Pelagic mysticetes taken in past whaling operations, such as fin whales *(Balaenoptera physalus)* and sei whales *(B. borealis),* show detectable concentrations of organochlorine pesticides in their blubber and liver.[16,18] Similarly, in their pioneering studies Taruski et al.[13] found significant concentrations of organochlorines, including PCBs, in blubber biopsies taken from living humpback whales with a harpoon gun two decades ago. In a more recent attempt to assess toxicant levels in a coastal dwelling mysticete, Woodley et al.[19] used the approximately 1 g of blubber obtained by the current standard whale biopsy dart to make measurements of organochlorine contamination of northern right whales *(Eubalaena glacialis).* They found, however, that such a small sample did not provide enough tissue to perform these measurements on an individual basis. Multiple samples had to be pooled to obtain quantifiable results.

Anticipating this problem, we undertook to develop a new whale biopsy system that is specially designed for multidisciplinary studies which include toxicology. The biopsy punch developed obtains a sample of skin and blubber weighing 5–6 g. Surfaces which come in contact with the sample are fabricated of titanium or Teflon™* materials which meet the protocol of the U.S. National Bureau of Standards for archival quality toxicological sampling. To increase both the efficiency and the efficacy of sampling, the 150-lb. draw crossbow previously

---

* Registered trademark of E.I. du Pont de Nemours and Company, Inc., Wilmington, DE.

recommended as a projection unit[20,21] was replaced with a pneumatic delivery mechanism. This weapon was further modified to enable a wide range of control over projection power. A lightweight line with an elastic leader was used as the tether for dart retrieval.

Here we describe this new whale biopsy system and present the results of tests conducted on humpback whales to assess the sampling technology developed. On the basis of our results, it is anticipated that in the hands of appropriately trained personnel this new technology will be especially useful for nondestructive research on large whales which includes high sensitivity measurements of organohalogens in cutaneous adipose tissue. In this context we discuss the eclectic value of the humpback whale as a model species for large-scale ecotoxicological studies in ocean systems. Parallel use of sample portions for cell culture, molecular, genetic, histopathologic, biological, and veterinary medical research also is possible. By these means answers to many persistent questions about contaminant loading in free-ranging whales are put within the reach of modern science.

III. METHODS

A. Whale Biopsy System

Figure 1 gives a mechanical drawing of the new biopsy punch developed for archival quality toxicological sampling of the blubber, including the epidermis, of large whales. According to this design all surfaces of this instrument which come in contact with the sample are constructed of titanium or Teflon. For sampling in which the analysis of heavy metals is not intended, stainless steel can be substituted for certain titanium components (e.g., the internal hook) to facilitate production. The instrument functions by taking a core of blubber to a depth limited by a large diameter flange that serves as a penetration stop. A Teflon half-ball valve, held in place at the back of the tube of the punch with a spring, serves to release back pressure generated during penetration. A hook is pressure-seated into this valve and extends forward into the lumen of the punch to secure the sample. On retrieval of the biopsy, the same hook and valve mechanism serves as a handle to facilitate removal of the biopsy from the instrument and to aid in manipulation of the sample in processing.

Prior to going to sea the individual components of the punch assembly are cleaned in separate lots. Final cleaning and assembly of the instrument is performed in a laminar flow hood by personnel gowned, gloved, and masked in clean room apparel. Each assembled biopsy punch unit is then placed in a 2 mil thick, 4×6 inch, contaminant-free Teflon bag (Clean Room Products, Inc., Ronkonkoma, NY) which is closed around on the valve casing with autoclave tape. The bagged units are then wrapped in lots in clean surgical towels and sterilized using dry heat. After sterilization each instrument in its Teflon bag is ready for mounting on the shaft of the pneumatic gun. The Teflon bag normally is removed by the marksman and his assistant shortly before firing and is replaced once the biopsy sample has

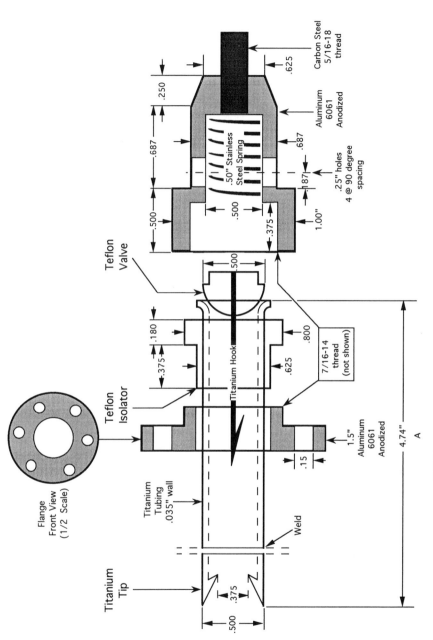

FIGURE 1. Mechanical drawing of whale biopsy instrument.

FIGURE 2. Pneumatic gun used as delivery mechanism.

been retrieved. The biopsy instrument with the sample is transferred to the assistant for subsequent processing using both clean and sterile techniques.

Figure 2 gives an illustration of the pneumatic gun used to project the whale biopsy instrument. The unit — manufactured by Swivel Machine Works, Inc., Milford, CT under the product name of AIRROW — is conveniently operated with compressed air supplied by way of a 10–20 ft long, 5000 psig rated, 1/2-inch diameter hose from a standard steel or aluminum Scuba tank. Inserted in the supply line at the tank end is a 5000-psig rated pressure regulator (Circle Seals Corporation, Anaheim, CA) which permits manual control of projection pressures through normal working range of 1000–3000 psig of the delivery mechanism. This regulator can be worked by the marksman or his assistant to enable biopsy sampling at very close range (10–20 yards) under reduced projection pressures. In our experience, sampling at a range of 25–35 yards can easily be accomplished under a normal working pressure supplied by steel Scuba tanks in the range of 1200–2250 psig. Sampling at maximum practical ranges of 50–60 yards may require projection pressures of 2250–3000 psig and hence aluminum tanks rated for higher pressures.

Retrieval of the dart is by means of a tether tied through one of six small holes drilled primarily for aerodynamic reasons in the flange of the biopsy instrument. In this study we used prestretched, 200-lb test braided polyethylene cord (Spectra 1000), supplied by the manufacturer of the gun. In one case, whale no. 43, this tether broke. To avoid repetition of this problem we recommend a 400-lb test. A 40 yard long leader made from this top quality cord is tied to a 6 yard long, 3/16-inch diameter elastic line wrapped with nylon braid ("bungy cord"). The free end of the elastic cord is secured by a shackle to the main staysail halyard of the research vessel and hoisted part way up the mast. This configuration of the rig enables the retrieval line to be pulled up and away from the marksman by the halyard if the dart remains lodged in a sounding whale. The elastic tether applies a gradually increasing tension to retract the sampling instrument.

Sample processing is accomplished on a clean Teflon anvil using clean sterile knives appropriate to the toxicological research intended. According to standard toxicological protocols for archival quality sampling,[22] the individual responsible for subdividing the samples is masked, gloved, and gowned in clean room apparel. In our study sterile technique also was used. Sample portions meant for toxicological research are placed in contaminant-free Teflon vials (Savillex Corporation, Minnetonka, MN) and then frozen in liquid nitrogen or on dry ice for preservation and transport to the appropriate land-based analytical facility. Samples for toxicological and genetic research in this study were cryopreserved in liquid nitrogen kept during our cruise in a 50 L metal dewar (MG Industries, Valley Forge, PA).

B. Vessels and Field Study

Biopsy sampling was conducted over a 10-day period from a 50 ft long motor ketch equipped with a 95-hp six-cylinder diesel engine in the region of 42°00′00″ to 42°30′00″ N by 70°00′00″ to 70°50′00″ W (southern Gulf of Maine). In most instances the final approach to a study animal was made under sail. However, auxillary power often was used when at a >150-yard from the animal to accelerate maneuvers when the whale was submerged. Sampling was done entirely under power on two calm days. Visual observation and a hand-held video camera were used to document the site where an animal was struck by the biopsy dart. A 20 ft long runabout equipped with a 150-hp outboard engine was used in parallel with the sampling vessel on 7 study days to obtain behavioral data on the whales sampled. Foul weather precluded operation of the second smaller vessel on 3 of the 10 days of the study.

C. Division and Disposition of Biopsies

Each biopsy was divided into several portions within 1 hr of its acquisition. These were placed in appropriate containers, transport media, and/or preservatives for distribution to as many as eight collaborating research laboratories. The aim was to make possible a large number of specialized analytical studies on each whale sampled.

The basic plan for sample division is indicated in Figure 3. Past experience had shown that sample portions of less than 0.5 g and consisting solely of epidermis are sufficent for both mitochondrial DNA analysis[23] and identification of gender using sex-specific molecular probes.[24] Similarly, the tissue which spans the dermoepidermal junction has been found most useful for cell culture, chromosome isolation, and cytogenetics.[21,25] Sample portions from this region therefore were collected in tissue culture media. Because the deeper layers of the dermis and hypodermis consist predominantly of lipid and hence are tissues of choice for analysis of (lipid soluble) organochlorines, these portions were collected for toxicological analyses. All subsequent toxicological methods and results are to be described elsewhere.[26]

D. In vitro Evaluation of Sample Viability and Establishment of Cell Lines

Skin biopsies were processed from nine individual whales for this aspect of the study. The primary explant technique was used in cell line initiation. The procedure employed followed that of Carrel[27] and is summarized more recently by Freshney[28] and Modi et al.[29] Aboard the sampling vessel, sample portions approximately 1 g in size were first washed in 10 mL of 70% ethanol and then in an equal volume

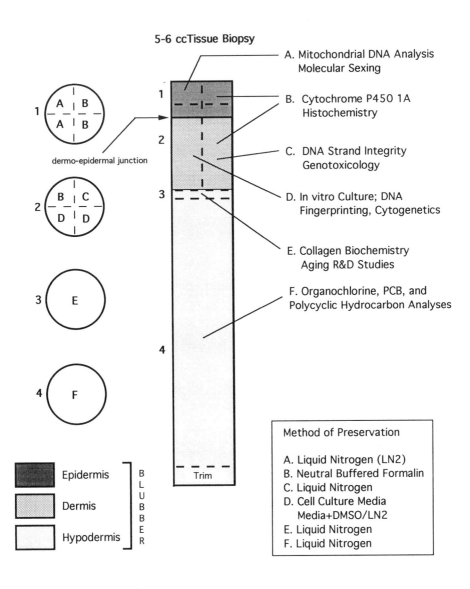

FIGURE 3. Division of whale blubber biopsy.

of Hank's balanced salt solution containing penicillin (100 units/mL), streptomycin (10 µg/mL), and amphotericin (2.5 µg/mL). The washed sample portions were then placed in 100 × 20 mm sterile petri dishes along with 3 mL of complete tissue culture medium (Leibovitz's L15 medium containing 10% fetal calf serum with antibiotics and fungicide as described above). They were then minced into smaller sized pieces using a pair of sterile curved scissors and pipetted into a sterile tube containing 10 mL complete medium, in which they were stored at room temperature until returning to the laboratory 4 to 7 days later. For each individual the minced tissue pieces were then plated into a single 25-cm² tissue culture flask along with 1.5 mL of complete medium and the flasks incubated at 37°C in a 5% CO_2 atmosphere.

E. Behavioral Observation

In light of the well-developed sensory systems of mammals, we hypothesized that biopsy sampling of free-ranging whales — necessarily conducted without the aid of anesthesia — will cause a change in the behavior of the subject animal. In order to allow comparisons with our earlier ethological studies on the responses of whales to biopsy procedures, the protocol of Weinrich et al.[30] was followed. The paired-sample "focal animal" method of Altmann[31] was used. Team members in a 20-ft runabout initially would find a humpback whale and identify it as the focal animal for subsequent study. Whales were identified by natural marks on the dorsal fin and on the ventral surface of the flukes.

For exactly 30 min, data were collected on the behavior of the focal whale and its respiration. During this period the engines of the observation vessel were shut down and neither research vessel approached the animal closer than 500 yards. Three observers were used. One individual observed the focal animal through 10 × 50 power binoculars and called out respirations and behaviors. Two others recorded these observations in code on data sheets. After data had been collected for the 30 min (prebiopsy control period), the sampling vessel was notified by radio that it could approach the animal for a biopsy. On firing of the biopsy dart the postbiopsy observation period was initiated. Behavioral data were then collected for an additional 30 min.

Definitions of the behavior types which we recorded were based on a preexisting ethogram used by the Cetacean Research Unit (CRU) in long-term studies of whale behavior.[30,32] From a possible total of 64 behavior types in the ethogram, 32 were observed during this study. The probability that the number of animals displaying a particular behavior changed significantly from the pre- to the postbiopsy period was tested by means of a binomial distribution, in which the probability of that behavioral display was assumed to be equal in the two samples. The change in the frequency with which a particular behavior was displayed between periods was compared using a Wilcoxon signed rank test.[33] Because the definitions of behaviors in the CRU ethogram are very specific and hence might mask broader trends, three summary variables also were used in the analysis: (1) all aerial behaviors (the sum of lob-tails, belly-up lob-tails, flipper slaps, belly-up flipper slaps, tail breaches, and full-body breaches); (2) all investigative, nonessential behaviors (the sum of belly-up rolls, low spy-hops, high spy-hops, single bubbles, and passes under a vessel); and (3) all forceful behaviors (the sum of tail slashes, trumpet blows, lob-tails, belly-up lob-tails, and tail breaches).

Many behaviors which showed no significant difference between the pre- and postbiopsy periods are not considered in detail in our results. These included back arches, belly-up flipper slaps, belly-up lob-tails, belly-up rolls, bubble cloud behaviors, flipper breaches, flipper in air, flukes, head stands, high spy-hops, lob-tails, low spy-hops, low flukes, passes under a vessel, rumble blows, single bubbles, sounding dive, spinning head breaches, spins, tail breaches, underwater blows, 1/2 flukes, 1/2 rolls, and 1/4 rolls. Those behaviors which showed notable or significant variation were:

- *Back rise* — animal's dorsal surface breaks surface of water while swimming with no accompanying exhalation.
- *Flipper slap* — animal lifts one pectoral flipper into the air, and forcefully hits the ventral surface of the flipper against the surface of the water.
- *Flipper flick* — animal raises the tip of one flipper above the water surface briefly.
- *Hard tail flick* — animal rapidly and forcefully flexes its tail up and down one time during otherwise normal swimming behavior; much spray can be thrown; flukes clear surface of water.
- *Tail flick* — animal flicks the posterior margin of the flukes just above surface of water; flukes come downward in light, splashing motion.
- *Tail rise* — animal slowly straightens its caudal peduncle at the surface during normal upstroke of the flukes.
- *Tail slash* — animal moves tail forcefully from side to side with flukes at or just below the surface.
- *Trumpet blow* — animal makes a loud, broad-band, wheeze-like sound during exhalation at the surface.

F. Estimation of Swimming Speed

Net rate of movement of each of the 10 whales studied ethologically was crudely estimated by timed LORAN-C positions taken by observers aboard the runabout. LORAN-C positions were corrected by a visually estimated distance, and a compass bearing was taken on the whale. Bearings were taken with the boat compass to the nearest 5°. Correction invariably was required during the prebiopsy observation period due to the 500-yard limitation on close approach for this phase of the study. LORAN-C positions thereafter could be obtained without correction at the beginning of the postbiopsy period in association with the close approach to take the biopsy. At the conclusion of each postbiopsy sample, the 20-ft runabout approached within 50 yards of the whale to get a final LORAN-C position.

G. Behavioral Samples

Pre-biopsy focal samples were obtained for 24 whales; however, restrictions such as approachability of the whale for the biopsy strike or decreasing visibility due to darkness limited the total number of 30 min long postbiopsy samples to 10 individuals. Animals which were engaged in prolonged dives at the time of the initial approach of the sampling vessel often were left for other individuals without attempting a biopsy. In addition, five whales sampled were not included in our behavioral analysis because these were sampled on days when the smaller observation vessel could not operate due to foul weather. All animals for which behavior was analyzed were successfully sampled.

In two cases collection of behavioral data was discontinued during the postbiopsy period due to extenuating circumstances. Into the postbiopsy period by 12 min, whale no. 38 — swimming slowy — snagged the drop line of a nearby cod fisherman. The animal reacted to this stimulus by swimming off at high speed (14–16 knots). It was not followed. In a similar case (whale no. 43), an accurately aimed but prematurely fired biopsy dart entered near the dorsal fin of the whale from a direction slightly behind the animal just as its back was maximally arched for a sounding dive. Because this dart penetrated the blubber at a very acute angle, rather than at the desired perpendicular angle to the skin, the flange at its tip jammed. The dart apparently became even more tightly jammed as the whale straightened out its back and tail during its dive.

As whale no. 43 dove the retrieval line stretched and then broke in the middle of its leader. In response the whale bolted and surfaced approximately 500 yards off 221 sec later moving fast (8–10 knots). The observation team followed as best it could in a wind freshening to 20+ knots but the collection of data while attempting to catch up in choppy seas simply was not possible. It was, however, visually confirmed that the dart had dropped off and within 15 min that the animal had ceased its forceful activity.

This atypical case emphasized the need to follow closely the previously recommended procedure for approaching a whale for biopsy sampling.[20] With respect to methods, it is extremely important not to fire before correctly positioning the research vessel *alongside* the animal. Figure 4 illustrates one maneuver for approaching a traveling whale that we have used extensively with considerable success under reefed sails in 20–30 knot winds.

H. Analysis of Behavioral Response

Using the behavioral data obtained in this study we analyzed four respiratory variables for differences in the means between the pre- and postbiopsy periods: (1) the number of respirations ("blows") per surface interval; (2) the length of time between each respiration ("blow interval"); (3) the length of time the animal spent at the surface ("surface interval"); and (4) the time spent below the surface during each dive ("dive time"). Dive time was defined as the period of submergence following typical sounding behavior (a prominent, high arching of the back and tail, often followed by tail flukes being brought well above the water surface). Only full dives or surfacings were included in our analysis of surface time, dive time, and number of respirations per surfacing. That is, if a focal sample was initiated after a surface interval had begun or if the focal sample ended during either a dive or a surfacing, data for the dive or surface interval were discounted. To examine the respiratory pattern that the whale was using before and after the biopsy procedure, a surface-to-dive-time ratio was calculated by dividing total surface time by total dive time for each focal sample. The Wilcoxon sign rank test[33] was used for comparisons between mean values of each respiratory variable recorded in pre-and postbiopsy focal samples.

In order to examine the duration of behavioral reactions, postbiopsy focal samples were further broken into two 15-min intervals: one that lasted from the moment of the biopsy strike until 16 min later, and one from 16 to 30 min after the strike. Respiratory and behavioral variables were tested for variation between control and response periods as described above.

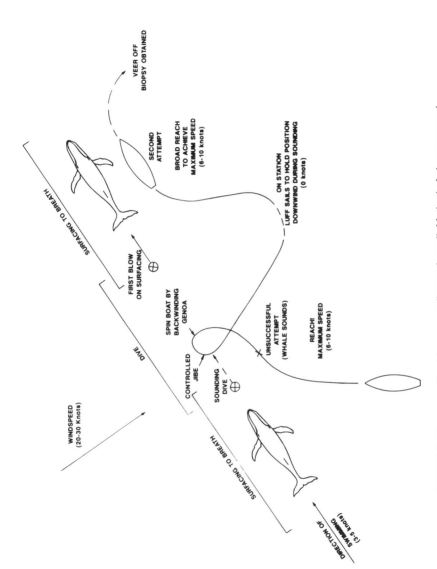

FIGURE 4. Heavy weather biopsy sampling under sail. Method of close approach.

Nonparametric tests were used for all statistical comparisons because of the small sample size. Furthermore, to increase sensitivity we considered two distinct levels of p values: (1) standard statistical significance, defined a priori at $p \leq 0.05$; and (2) a nonstandard and arbitrary measure, which we call notability, where $0.05 < p < 0.17$. Tests results which showed a "notable but not statistically significant change at the 95% confidence level nonetheless are suggestive of a trend in behavioral or respiratory variation between comparative periods. These identify specific changes which are more likely to approach statistical significance with a larger sample size.

IV. RESULTS

A. Range and Accuracy

Figure 5 depicts the precise distribution of the strikes recorded. Of 24 experimental shots fired, a total of 21 (88%) struck within a restricted target area just below the dorsal fin. The maximum range at which a sample was successfully taken, which also was the maximum range of any experimental shot made in this study, was estimated at 35–36 yards. The trajectory of this and all other experimental shots was essentially flat. Because of the design of the pneumatic mechanism, the biopsy dart did not travel in a parabolic arc as do ballistic projectiles — such as the bolt from a crossbow — but instead like a rocket. On the basis of these results and earlier land-based practice tests which found flat trajectories to the maximum range tested of 40 yards,[34] it was the opinion of our marksman that effective sampling range could be extended to 40 yards. The time required for careful aiming in relation to the time that a whale spends at the surface was anticipated to be the practical limiting factor on accuracy. Unlike ballistic delivery, such as those employed by Arnason,[36] Brown,[34] Lambertsen,[20] Palsboll et al.,[37] Taruski et al.,[13] and Whitehead et al.,[39] there was no major imprecision inherent to the performance of the AIRROW projection unit.

B. Efficiency and Efficacy

In the first of our experimental shots (whale no. 1) the biopsy instrument was intentionally fired while still contained in the 2 mil thick Teflon bag used as a sterile clean wrap. Although this was not the intended design configuration for firing, we wanted to determine whether the wrapped dart would function by piercing through the bag on impact. When this dart was retrieved, the bag had a very small hole in it. The biopsy instrument was still in the bag and a sample had not been obtained.

Thereafter all biopsy darts were fired unwrapped in the intended design configuration. In the resultant 20 experimental strikes, 19 obtained samples of sufficient volume for distribution to six to eight research groups. Thus a 95% success rate was achieved with respect to number of samples obtained vs number of strikes made (Table 1). The samples obtained consisted of cylindrical cores of integumentary tissue ≥5–6 cc in volume (Figures 1 and 3).

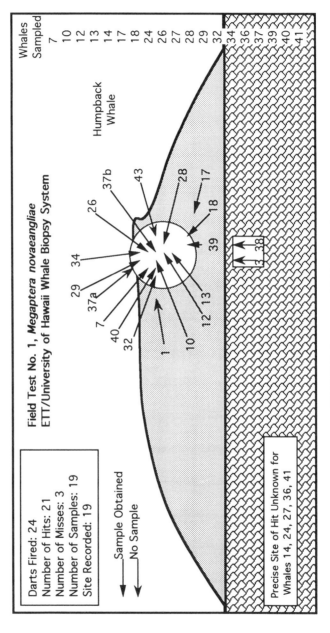

FIGURE 5. Summary of field test no. 1.

Table 1. Efficiency and Efficacy of Newly Developed Whale Biopsy System in Comparison with Standard[a] Instrument

Instrument	No. hits/ No. firings	No. samples/ No. hits	No. misses/ No. firings	No. samples/ No. firings[b]	No. firings/ (5 cc tissue)
Modified[c] standard instrument	0.42	0.63	0.58	0.26	>38.5[d]
Standard[a,e] instrument[29]	0.81	0.87	0.19	0.71	7.0
New system[f] (this study)	0.88	0.95	0.12	0.83	<1.2

[a] Standard instrument takes a biopsy sample of approximately 1 cc vol using a tethered dart fired from a 150-lb draw crossbow.[20]

[b] "Firings" refer to those made in intended design configuration.

[c] Based on data from Whitehead et al.[39] Study reports skin sampling or attempted sampling of sperm whales using standard biopsy dart fired with 50-lb draw crossbow; 5 small samples (<4 mm²) were obtained in 19 firings, 8 of which struck the animal.

[d] Calculation assumes that a sample (<4 mm²) gives a maximal sample volume of <0.5 cc.

[e] Based on data from Weinrich,[38] who describe skin sampling or attempted sampling of 146 humpback whales using standard biopsy instrument.

[f] New system takes biopsy of 5–6 cc vol using a tethered biopsy dart fired by a pneumatic delivery mechanism.

In terms of numbers of shots fired per unit volume of sample obtained, these results represent at least a 5.8-fold improvement over the standard unit used in our hands on humpback whales[38] and at least a 32.1-fold improvement over the standard unit, as modified and used by Whitehead et al.[39] in sperm whales. We have not calculated other significant managerial indicators of improved efficiency and efficacy, such as the savings in fuel and personnel costs that could be achieved through the use of the new instrument. The figures in Table 1, however, are instructive in providing an initial basis to make such calculations.

C. Operational Capability

In an effort to extend previous capabilities for effective work at sea, attempts were made to continue biopsy sampling under weather conditions that would have precluded the use the standard, crossbow-powered system. Experience has shown that that system does not function well in winds greater than 12–15 knots.

We found that at apparent wind velocities of less than 25 knots measured by a masthead anometer, all or most aspects of the technology worked very well. However, with the equipment configuration initially used, sampling became impossible once winds reached greater strength. The problem encountered at 25–27 knots was that the retrieval line was blown out of its storage cannister, an aluminum cylinder (5-inch depth, 18-inch diameter), secured with U-bolts to the life rail of the boat. Subsequently this problem was solved by using a deep-bottomed cannister for line stowage. This maintained the retrieval line in a completely workable condition under wind conditions of 25–30 knots. In the present study 30 knot winds were the maximum encountered in daylight hours.

These findings indicated that biopsy sampling with the system developed does not need to be limited by moderately strong (15–30 knot) winds which would have precluded use of the previously described unit. This is a significant result because the number of effective working days that can be expected per sampling cruise has been increased.

D. Sample Processing

All biopsy samples obtained were divided and processed within 1 hr of the sample being taken. These were appropriately preserved for subsequent distribution to several different research groups, as indicated in Figure 3.

E. Sample Viability and Establishment of Cell Lines

Sample portions plated in vitro were examined daily and after 8 to 20 days were observed to be proliferating around the margins of adherent tissue pieces in seven of the nine flasks (78% viability). Three to four days after first appearing, the cells in two flasks died. Approximately 2 weeks after the first viable cells were observed, the proliferating cells in the five remaining flasks became confluent. At this time these cultures were harvested and transferred into 75-cm^2 flasks. Cell expansion was continued, and six vials each containing about 10^6 cells subsequently were cryogenically frozen in complete medium containing 10% dimethylsulfoxide for each of the five specimens. These vials are now stored in liquid nitrogen at the U.S. National Institute of Health for use in future studies which require living cetacean cell lines.

F. Pre- and Postbiopsy Behavioral Comparisons

1. Respiratory Variables

None of the five respiratory variables we examined showed significant differences between the pre- and postbiopsy observation periods (Table 2). Wilcoxon sign rank tests gave the following results: (1) mean number of respirations per surfacing: $z = -1.2$, $p = 0.21$; (2) mean dive time: $z = -1.36$, $p = 0.17$; (3) mean surfacing time: $z = -0.41$, $p = 0.67$; (4) mean interbreath interval: $z = -1.24$, $p = 0.21$; and (5) mean surface-to-dive-time ratio: $z = -0.05$, $p = 0.95$.

2. Behavioral Variables

None of the 32 behaviors noted in this study showed a statistically significant difference in the number of animals displaying the behavior between the prebiopsy control and postbiopsy observation periods. Thus, when averaged over one-half hour, the data collected failed to confirm the hypothesis that experimental darting will induce a change in the behavior of subject animals.   However, several behaviors showed notable variation (see Section III for explanation): (1) the number of whales displaying back rises changed from 7 to 3 ($p = 0.16$); (2) flipper

Table 2. Mean Values of Respiratory Variables in Humpback Whales in Prebiopsy → Postbiopsy Periods

Whale	No. of respirations per surfacing	Dive time (sec)	Surfacing time (sec)	Interbreath interval (sec)	Surface-to-dive-time ratio
7	3.6 → 4.6	186 → 232	163 → 118	50 → 30	0.83 → 0.51
14	18.0 → 2.5	249 → 176	776 → 293	43 → 0	6.23 → 1.75
17	0.0 → 3.3	184 → 177	0 → 213	132 → 92	2.25 → 0.89
18	3.3 → 7.7	334 → 221	44 → 451	29 → 47	0.23 → 2.83
25	5.5 → 5.3	331 → 159	358 → 135	74 → 31	0.85 → 0.81
28	3.8 → 3.3	360 → 146	81 → 105	25 → 40	0.22 → 0.76
32	3.4 → 5.3	271 → 253	53 → 95	18 → 26	0.27 → 0.28
36	4.3 → 7.0	310 → 322	46 → 181	33 → 39	0.41 → 0.53
37	3.5 → 5.7	256 → 276	106 → 228	44 → 33	0.41 → 0.48

Note: See Section III.F for explanation of calculations.

flicks, from 3 animals to 0 ($p = 0.12$); (3) hard tail flicks, from 0 animals to 4 ($p = 0.06$); (4) tail flicks, from 2 animals to 5 ($p = 0.22$); and (5) tail slashes, from 2 animals to 5 ($p = 0.16$). While the numbers of animals displaying inquisitive and forceful behaviors were almost identical between pre- and postbiopsy periods (3 and 3, 7 and 9, respectively), the number of animals which displayed aerial behavior increased from 3 to 6 ($p = 0.25$).

The frequency with which any single behavior was recorded showed no significant difference between the pre- and postbiopsy periods. However, a number of behaviors showed notable differences: (1) back rises decreased in frequency ($z = -1.52$, $p = 0.12$); (2) flipper flicks decreased in frequency ($z = -1.82$, $p = 0.07$); (3) flipper slaps decreased in frequency ($z = -1.60$, $p = 0.11$); (4) hard tail flicks increased in frequency ($z = -1.82$, $p = 0.07$); and (5) trumpet blows increased in frequency ($z = -1.60$, $p = 0.11$). No notable or statistically significant difference in frequency was found in the three summary variables: (1) inquisitive behaviors: $z = -0.73$, $p = 0.46$; (2) aerial behaviors: $z = -0.94$, $p = 0.34$; and (3) forceful behaviors: $z = -0.29$, $p = 0.76$.

3. Estimated Swimming Speed

No statistically significant difference between pre- and postbiopsy periods was found in estimated net rate of movement ($z = -0.05$, $p = 0.95$).

G. Comparisons Within the Postbiopsy Period

1. Respiratory Variables

No significant difference or otherwise notable difference was found in any of the respiratory variables when comparing data recorded 0–16 and 16–30 min after the biopsy. Wilcoxon sign rank tests gave the following values: (1) mean number of respirations per surfacing: $z = -0.63$, $p = 0.52$; (2) mean dive time: $z = -0.41$, $p = 0.67$; (3) mean surfacing time: $z = -0.29$, $p = 0.76$; (4) mean interbreath interval: $z = -0.17$, $p = 0.85$; and (5) mean surface-to-dive-time ratio: $z = -1.07$, $p = 0.28$.

2. Behavioral Variables

No behavior showed a significant difference in the number of animals displaying that behavior when comparing the periods 0–16 and 16–30 min after the biopsy. Four behaviors, however, showed a notable variation: (1) the number of animals displaying hard tail flicks changed from 4 to 0 ($p = 0.06$); (2) trumpet blows, from 5 animals to 2 ($p = 0.16$); (3) tail flicks, from 3 animals to 0 ($p = 0.12$); and (4) tail slashes, from 4 animals to 1 ($p = 0.15$). Of the three summary variables, none showed significant variation: (1) inquisitive behaviors: 2 animals in the first period, 2 in the second ($p = 0.68$); (2) aerial behaviors: 4 animals in the first period, 1 in the second ($p = 0.18$); forceful behaviors: 8 animals in the first period, 5 in the second ($p = 0.29$).

The trumpet blow was the only behavior that showed a significant difference in frequency of occurrence between the 0–16 and 16–30 min postbiopsy period (z

$= -1.96$, $p = 0.05$). In the context of this change, notable differences in frequency between these two periods were found in three behaviors: (1) hard tail flicks decreased ($z = -1.82$, $p = 0.07$); (2) tail flicks decreased ($z = -1.60$, $p = 0.11$); and (3) tail slashes decreased ($z = -1.61$, $p = 0.10$). Among summary variables, the frequency of inquisitive behaviors was not found to change ($z = -0.44$, $p = 0.65$), but the frequency of both forceful behaviors and aerial behaviors decreased notably in the 16–30 min postbiopsy period (forceful behaviors: $z = -1.77$, $p = 0.07$; aerial behaviors: $z = -1.82$, $p = 0.06$).

H. Unusual Cases

Worthy of separate mention are four cases that were not included in the above analysis because of incomplete behavioral data or because a biopsy sample was not obtained. In whale no. 1 (see above) the dart was intentionally fired while still contained in its sterile wrap. This whale reacted with an immediate tail flick and began to trumpet blow at a greater frequency than in the prebiopsy period. In whale no. 38 the shot was considered a miss because it fell approximately 0.5 yard short and did not obtain a sample (see Figure 5), although the dart may have touched the whale after traveling through water. This whale reacted with a hard tail flick, trumpet blew 6 times, and tail slashed three times. Both cases seem significant because the disturbance reactions observed were merely to close approach of the research vessel and/or to the touch or nearby splash of the biopsy dart. This indicates that efforts to minimize the number of attempts or firings made per sample obtained are of considerable importance.

The third unusal case involved whale no. 41, which was successfully sampled. This animal showed little overt reaction outside of a tail flick at the instant the dart struck. Thereafter it continued the slow (1–2 knot) surface swimming that it had been engaged in immediately prior to the biopsy strike. However, 12 min later it snagged the drop line of a cod fisherman in the area and evidently became alarmed. It immediately increased its swimming speed and began a series of short dives, each punctuated by a loud trumpet blow. As it disappeared into the distance, we estimated it to be swimming 14–16 knots. This reaction to the fishing line was greater than any we observed in this study to biopsy sampling.

The fourth unusual case, whale no. 43, was the only strong reaction to biopsy darting seen in this study. Although instructive, this reaction was not considered representative of the normal biopsy procedure because the dart had been fired prematurely from a position behind, rather than alongside, the whale (see Section III).

V. DISCUSSION

In this study a previously approved sampling instrument was redesigned to enable supravital acquisition of archival quality samples of whale blubber in sufficient volume for multidisciplinary studies which examine many chemical analytes and biological variables. Our technical aim was to create a cost-effective means to allow baseline toxicological measures for statistically significant numbers of whales representing normal, free-living populations. Such data are essential to scientific efforts to better understand human impacts on marine ecosystems and to better interpret toxicological data obtained from marine mammals found moribund or dead of unknown causes. It is clear that existing data on contaminant burdens in dead beached animals would hold greater meaning if compared with equivalent data from animals currently alive.

The larger intention was to address in a substantive way internationally recognized needs for multidisciplinary studies required to better understanding the changing environmental system of the earth.[40]

Evidence that baleen whale populations once exerted a *dominant* control on the dynamics of marine ecosystems, prior to their exploitation by man, underscores the essential nature of research on their populations.[41] Baseline data obtained at the relatively low population densities which characterize most species of baleen whales today will be needed if we are to grant future generations the ability to measure, and understand, critical growth processes in marine ecosystems.[18,42] As these powerful species recover, their role as controlling agents in ocean biosystems naturally will return toward a state of real ecological dominance.

It is clear that with its natural preference for coastal habitats the humpback whale could be disproportionately impacted by anthropogenic compounds introduced to ocean systems by rivers; coastal runoff; wind; ocean dumping; and various industrial activities including oil, gas, and mineral exploitation. With respect to broader issues in ecotoxicology it is also notable that studies of contaminant burdens in the humpback whale promise unusual perspective. Given the ability of researchers to photographically identify individuals of this species;[24,43-50] to estimate the age of known individuals on the basis of long-term sighting histories;[51-53] and to sex individuals using cytogenetic, molecular, and morphological techniques,[23,24,54] process-oriented oceanographic research can be pursued in this species over time and space by means of the extraordinarily powerful method of cohort analysis. Biological and pathological variables can be measured repeatedly in known individuals over the course of seasons or years and across oceanic areas spanning thousands of kilometers.

The rapidly improving status of our knowledge of the zoogeography of this species likewise promises important ecotological resolution on hemispheric, regional, and finer scales. In major ocean basins, discontinuous patterns of distribution indicate that humpback whales aggregate into regional subpopulations.[55,56] Presently some 13 management stocks of this species, each thought to use a distinct breeding ground, are recognized.[57] Although a detailed review of humpback whale zoogeography is beyond the scope of this chapter, the fidelity of known individuals to particular feeding regions, with low rates of interchange between regions,[47,57,58] promises a reasonable foundation upon which to build ecotoxicological perspective. For example, long-term field observation now suggests sorting of the southeast Alaska feeding subpopulation into a series of small geographic neighborhoods heavily utilized by particular individuals. Indeed, certain humpback whales are known to have returned to the specific locale of Glacier Bay and the adjacent waters of Icy Strait to feed over as many as 16 consecutive summers.[60] These findings indicate that ecotoxicological sampling strategies focused on this species could be designed on rather fine geographic scales to more fully evaluate the ecological fate of hazardous effluents arising from specific, isolated sources. One related essential point is that the design and interpretation of such ecotoxicological studies will be greatly facilitated by the commensurate ability to identify known reference populations and even "reference individuals" utilizing virtually pristine feeding areas.

The ability of the biopsy system developed to take a core of blubber approximately 10 cm deep also increases the range of biological variables that can be evaluated in the laboratory. In their studies of lipid stratification in blubber from fin whales taken by commercial whalemen, Aguilar and Borrell[17] found that the percent lipid of the outer blubber layer is essentially constant. However, the percent lipid content in the inner blubber layer varied substantially, both on a seasonal basis and with reproductive status. They found that the slope of the gradient in lipid content across the inner and outer blubber layers reflects the state of fattening of the individual and hence can be used as a sensitive indicator of nutritive condition. The biopsy instrument described here — which unlike previously described systems does sample the inner blubber layer — enables the same type of information to be obtained from living whales. One therefore can foresee the possibility of using the phenomenal integating power of these species to monitor efficiently important aspects of the dynamics and productivity of large marine ecosystems. Moreover, the new possibility of measuring contaminant burdens and body condition simultaneously suggests an approach to the challenging problem of detecting adverse systemic effects of anthropogenic toxicants on the health of these gigantic species.

Against this background, the results of our present study may prove beneficial to international efforts to establish comprehensive baseline information on at least one species of baleen whale. There is essentially no doubt that the new whale biopsy system developed and tested is more effective than any previously described. Certainly the range of sea conditions in which biopsy sampling can be conducted has been increased. Of equal or greater importance, the level of disturbance induced by normal use of the new system evidently is no greater than that caused by the less effective equipment currently available. Like the previously described instrument,[20,30,38] those reactions detected typically were of low or moderate intensity and tend to be short lived. In contrast, the improved trajectory, range, control, and power of the new system all served to minimize the number of failed biopsy shots. This decreases disturbance to study animals that is caused by close vessel approach and repeat biopsy attempts. This improvement is particularly significant because repeated biopsy attempts in sequence have the potential to induce very marked increases in the force and duration of behavioral reactions, possibly causing adverse systemic effects mediated through an acute stress response.[29] Improvement is also gained because any future need for repeat sampling of individual whales to supply different investigations can be reduced because archival quality samples of sufficient volume for many research uses are obtained.

On these several bases we conclude that a more humane and cost-effective whale biopsy system has been field-proven and can now be recommended for wider use by appropriately trained personnel. We strongly advise special training to ensure familiarity and competence with the pneumatic delivery mechanism, which has unusual trigger performance,[61] and to guarantee instrument sterility. Special training also is essential to ensure that both the marksman and the helmsman understand the absolute requirement both for proper positioning of the research vessel and for proper timing of the firing of the dart in order to achieve the intended functioning of the system. Finally, to acquire high quality, uncontaminated samples for archival purposes, special training, and considerable practice with both clean and sterile techniques are needed for proper handling of samples. When used by a team of individuals having all these proficiencies and skills, the improved biopsy system we describe can reasonably be expected to enable high-quality sampling of the integument of all large Cetacea, including fleet species such as the fin whale, *Balaenoptera physalus,* and tough-skinned animals like the sperm whale, *Physeter catodon.* In any application to environmental monitoring, the effective working length of the biopsy punch (length "A," 4.74 in., see Figure 1) should be anatomically scaled to be less than or equal to the expected thickness of the blubber.

ACKNOWLEDGMENTS

Field work was conducted in compliance with Marine Mammal Research Permit No. 675 issued to C.S. Baker by the National Marine Fisheries Service, National Oceanic and Atmospheric Administration, U.S. Department of Commerce. The authors express their gratitude to Mr. Stephen Bryant, who served as our marksman; Dr. Robert Schaper, who assisted with the sampling; Ms. Gentry Holloway, Ms. Courtney-Paige Morrison, Ms. Anne Newcombe, and Mr. Mark Schilling, who assisted in the collection of behavioral data; and Mr. Scott Coe and Ms. Elaine Sacco, who assisted with vessel handling. Components of this study were conducted under contract with the Pacific Biomedical Research Center, University of Hawaii with financial support from the U.S. National Science Foundation (Project BSR 90006); and under Contract 14–35–0001–30553 between Ecosystems Technology Transfer, Inc. and the Minerals Management Service, U.S. Department of Interior. Primary aspects of this study were a result of responsibilities assumed by R.H. Lambertsen under a fellowship with the Committee on the Challenges of Modern Society, Scientific Affairs Division, North Atlantic Treaty Organization, Brussels, Belgium. Views expressed are those of the authors and do not necessarily reflect those of the Committee or NATO member countries.

REFERENCES

1. Aguilar, A., Organochlorine pollution in sperm whales, *Physeter macrocephalus*, from the temperate waters of the eastern North Atlantic, *Mar. Pollut. Bull.*, 14, 349, 1983.
2. Aguilar, A., Relationship of DDE/ΣDDT in marine mammals to the chronology of DDT input into the ecosystem, *Can. J. Fish Aquat. Sci.*, 41, 840, 1984.
3. Cockroft, V. G., DeKock, A. C., Lord, D. A., and Ross, G. J. B., Organochlorines in bottlenose dolphins, *Tursiops truncatus*, from the east coast of South Africa, *S. Afr. J. Mar. Sci.*, 8, 207, 1989.
4. Delong, R. L., Gilmartin, W. G., and Simpson, J. G., Premature births in Californian sea lions: associated with high organo-chlorine pollutant levels, *Science*, 181, 1168, 1973.
5. Gaskin, D. E., Holdrinet, M., and Frank, R., Organochlorine pesticide residues in harbour porpoises from the Bay of Fundy region, *Nature (London)*, 233, 499, 1971.
6. Gaskin, D. E., *The Ecology of Whales and Dolphins*, Heinemann Educational Books, Plaistow, NH, 1982, 459 pp.
7. Gilmartin, W. G., DeLong, R. L., Smith, A. W., Sweeney, J. C., DeLappe, B. W., Risebrough, R. W., Griner, L. A., Dailey, M. D., and Peakall, D. B., Premature parturition in the California sea lion, *J. Wildl. Dis.*, 12, 104, 1976.
8. Helle, E., Olsson, M., and Jensen, S., PCB levels correlated with pathological change in seal uteri, *Ambio*, 5, 261, 1976.
9. Muir, D. C. G., Ford, C. A., Stewart, R. E. A., Smith, T. G., Addison, R. F., Zinck, M. E., and Beland, P., Organochlorine contaminants in belugas, *Delphinapterus leucas*, from Canadian waters, *Can. Bull. Fish. Aquat. Sci.*, 224, 165, 1990.
10. Reijnders, P. J. H., Reproductive failure in harbour seals feeding on fish from polluted coastal waters, *Nature (London)*, 324, 456, 1986.
11. Reijnders, P. J. H., Pespectives for studies of pollutin in cetaceans, *Mar. Pollut. Bull.*, 17, 31, 1986.
12. Tanabe, S., Watanabe, S., Kan, H., and Tatsukawa, R., Capacity and mode of PCB metabolism in small cetaceans, *Mar. Mamm. Sci.*, 4, 103, 1988.
13. Taruski, A. G., Olney, C. E., and Winn, H. E., Chlorinated hydrocarbons in cetaceans, *J. Fish. Res. Board Can.*, 32, 2205, 1975.
14. Martineau, D., Lagace, A., Masse, R., Morin, M., and Beland, P., Transitional cell carcinoma of the urinary bladder in a beluga whale *(Delphinapterus leucas)*, *Can. Vet. J.*, 25, 297, 1985.
15. Reijnders, P. J. H., Ecotoxicological perspectives in marine mammalogy: research principles and goals for a conservation policy, *Mar. Mamm. Sci.*, 4, 91, 1988.
16. Aguilar, A., Using organochlorine pollutants to discriminate marine mammal populations: a review and critique of the methods, *Mar. Mamm. Sci.*, 3, 242, 1987.
17. Frank, R., Ronald, K., and Braun, H. E., Organochlorine residues in harp seals *(Pagophilus groenlandicus)* caught in eastern Canadian waters, *J. Fish. Res. Bd. Canada*, 30, 1053, 1973.
18. Lambertsen, R. H., Disease biomarkers in large whale populations of the North Atlantic and other oceans, *Biomarkers of Environmental Contamination*, McCarthy, J. E. and Shugart, L., Eds., Lewis Publishers, Boca Raton, FL, 1990, p. 395.
19. Woodley, T. H., Brown, M. W., Kraus, S. D., and Gaskin, D. E., Organochlorine levels in North Atlantic right whale *(Eubalaena glacialis)* blubber, *Arch. Environ. Contam. Toxicol.*, 21, 141, 1991.

20. Lambertsen, R. H., A biopsy system for large whales and its use for cytogenetics, *J. Mamm.,* 68, 443, 1987.

21. Lambertsen, R. H. and Duffield, D. A., Biopsy Studies of the Humpback Whale, *Megaptera novaeangliae.* Final Contract Report from the University of Florida to the Northeast Fisheries Center, National Marine Fisheries Service, National Oceanic and Atmospheric Administration, U.S. Department of Commerce, Woods Hole, MA, 1987, p. 1.

22. Becker, P. R., Wise, S. A., Koster, B., and Zeisler, R., Alaskan Marine Mammal Tissue Archival Project: A Project Description Including Collection Protocols. National Bureau of Standards, Center of Analytical Chemistry, U.S. Department of Commerce, NBSIR 88–3750, 1988, 46 pp.

23. Baker, C. S., Lambertsen, R. H., and Palumbi, S. R., Extraction and identification of mitochondrial (mt)DNA from the epidermal tissue of individually identified humpback whales, 7th Biennial Conf. Biol. Mar. Mammals, Miami, FL, Abstr., 1987.

24. Baker, C. S., Lambertsen, R. H., Weinrich, M. T., Calambokidis, J., Early, G., and O'Brien, S. J., Molecular genetic identification of the sex of humpback whales *(Megaptera novaeangliae), Rep. Int. Whaling Comm. (Special Issue),* 13, 105, 1991.

25. Lambertsen, R. H., Baker, C. S., Duffield, D. A., and Chamberlin-Lea, J., Cytogenetic determination of sex among individually identified humpback whales, *Can. J. Zool.,* 66, 1243, 1988.

26. Lambertsen, R. H., Baker, C. S., Weinrich, M. T., Modi, W. S., A baseline method to evaluate chemico-biological interactions involving cetacea, with description of an improved whale biopsy system designed for multidisciplinary research. Final report to the Minerals Management Service, U.S. Dept. of the Interior, Anchorage, AK from Ecosystems Technology Transfer, Inc., Philadelphia, PA, 1993, p. 1.

27. Carrel, A., On the permanent life of tissues outside of the organism, *J. Exp. Med.,* 15, 516, 1912.

28. Freshney, R. I., *Culture of Animal Cells: A Manual of Basic Techniques,* Alan R. Liss, New York, 1983, 295 pp.

29. Modi, W. S., Nash, W. G., Ferrari, A. C., and O'Brien, S. J., Cytogentic methodologies for gene mapping and comparative analyses in mammalian cell culture systems, *Gene Anal. Tech.,* 4, 75, 1987.

30. Weinrich, M. T., Lambertsen, R. H., Belt, C. R., Schilling, M. R., and Iken, H. J., Behavioral reactions of humpback whales to biopsy procedures, *Fish. Bull.,* 90, 90, 588, 1992.

31. Altmann, J., Observational study of behavior: sampling methods, *Behaviour,* 49, 227, 1974.

32. Weinrich, M. T., Schilling, M. R., and Belt, C. R., Evidence for acquisition of a novel feeding behavior: lobtail feeding in humpback whales, *Megaptera novaeangliae, Anim. Behav.,* 44, 1059, 1992.

33. Zar, J. H., *Biostatistical Analysis,* Prentice-Hall, Englewood Cliffs, NJ, 1984.

34. Lambertsen, R. H., Documentation of Practice Session to Evaluate Safety and Efficacy of a Modified Whale Biopsy System to acquire Blubber Samples for Multidisciplinary Studies, unpublished report from Ecosystems Technology Transfer, Inc. to Office of Protected Resources, National Marine Fisheries Service, National Oceanic and Atmospheric Administration, U.S. Department of Commerce, Silver Springs, MD, 1991, p. 1.

35. Brown, M. W., Kraus, S. D., and Gaskin, D. E., Reaction of North Atlantic right whales *(Eubalaena glacialis)* to skin biopsy sampling for genetic and pollutant analysis, *Rep. Int. Whaling Comm. (Special Issue),* 13, 81, 1991.

36. Arnason, U., Bellamy, H., Eythorsson, T., Sigurjonsson, J., and Widegren, B., Conventionally stained and C-banded karyotypes of a female blue whale, *Heriditas,* 102, 251, 1985.

37. Palsboll, P. J., Larsen, F., and Hansen, E. S., Sampling of skin biopsies from free-ranging large cetacean at West Greenland: development of new biopsy tips and bolt designs, *Rep. Int. Whaling Comm. (Special Issue),* 13, 71, 1991.

38. Weinrich, M. T., Lambertsen, R. H., Baker, C. S., Schilling, M. R., and Belt, C. R., Behavioural reactions of humpback whales *(Megaptera novaeangliae)* in the southern Gulf of Maine to biopsy sampling, *Rep. Int. Whaling Comm. (Special Issue),* 13, 91, 1991.

39. Whitehead, H., Gordon, J., Mathews, E. A., and Richard, K. R., Obtaining skin samples from living sperm whales, *Mar. Mamm. Sci.,* 6, 316, 1990.

40. Anon., NATO and global environmental change, *Q. Bull. NATO Sci. Comm. Comm. Challenges Mod. Soc.,* 30, 2, 1990.

41. Valiela, I., *Marine Ecological Processes,* Springer-Verlag, New York, 1984, p. 1.

42. Grubb, P. J. and May, R. M., Comments on the sustainable biosphere initiative, *Conserv. Biol.,* 5, 548, 1991.

43. Baker, C. S., The Population Structure and Social Organization of Humpback Whales *(Megaptera novaeangliae)* in the Central and Eastern North Pacific, Ph.D. dissertation, University Microfilms International, Ann Arbor, MI, 1985.

44. Hammond, P. S., Mizroch, S. A., and Donovan, G. P., Individual recognition of cetaceans: use of photo-identification and other techniques to estimate population parameters, *Rep. Int. Whaling Comm. (Special Issue),* 12, 1990.

45. Katona, S., Baxter, B., Brazier, O., Kraus, S., Perkins, J., and Whitehead, W., Identification of humpback whales by fluke photographs, *Behavior of Marine Animals — Current Perspectives in Research, Vol. 3, Cetaceans,* Winn, H. E. and Olla, B. L., Eds., Plenum Press, New York, 1979, p. 33.

46. Mattila, D. K., Clapham, P. J., Katona, S. K., and Stone, G. S., Humpback whales on Silver Bank, 1984; population composition and habitat use, *Can. J. Zool.,* 67, 281, 1989.

47. Mattila, D. K. and Clapham, P. J., Humpback whales *(Megaptera novaeangliae)* and other cetaceans on Virgin Bank and in the northern Leeward Islands, 1985 and 1986, *Can. J. Zool.,* 67, 2210, 1989.

48. Perry, A., Mobley, J. R., Baker, C. S., and Herman, L. M., Humpback Whales of the Central and Eastern North Pacific: A Catalog of Individual Identification Photographs. Sea Grant Miscellaneous Report, UNIHI-SEAGRANT-MR-88–02, 1988.

49. Stone, G. S., Florez-Gonzalez, L., and Katona, S., Whale migration record, *Nature (London),* 346, 705, 1990.

50. Straley, J. M., Fall and winter occurrence of humpback whales *(Megaptera novaeangliae)* whales in southeastern Alaska, *Rep. Int. Whaling Comm. (Special Issue),* 12, 319, 1990.

51. Clapham, P. J. and Mayo, C. A., The attainment of sexual maturity in two female humpback whales, *Mar. Mamm. Sci.,* 3, 279, 1987.

52. Clapham, P. J. and Mayo, C. A., Reproduction and recruitment in individually identified humpback whales *(Megaptera novaeangliae)* observed in Massachusetts Bay: 1979–1985, *Can. J. Zool.,* 65, 2853, 1987.

53. Jurasz, C. M. and Palmer, V. P., Censusing and Establishing Age Composition of Humpback Whales, *Megaptera novaeangliae,* Employing Photodocumentation in Glacier Bay National Monument, Alaska, report to the National Park Service, Anchorage, AK, 1981, p. 1.

54. Glockner-Ferrari, D. A. and Venus, S. C., Identification, growth rate and behavior of humpback whales *(Megaptera novaeangliae)* cows and calves in the waters of Maui, Hawaii, *Communication and Behavior of Whales,* Payne, R., Ed., Westview Press, Boulder, CO, 1983, p. 1.

55. Baker, C. S., Herman, L. M., Perry, A., Lawton, W. S., Straley, J. M., Wolman, A. A., Kaufman, G. D., Winn, H. E., Hall, J. D., Reinke, J. M., and Ostman, J., Migratory movement and population structure of humpback whales *(Megaptera novaeangliae)* in the central and eastern North Pacific, *Mar. Ecol. — Prog. Ser.,* 31, 105, 1986.

56. Baker, C. S., Palumbi, S. R., Lambertsen, R. H., Weinrich, M. T., Calambokidis, J., and O'Brien, S. J., The influence of seasonal migration on the distribution of mitochondrial DNA haplotypes in humpback whales, *Nature (London),* 344, 238, 1990.

57. Anon., Humpback Whale Draft National Recovery Plan, Office of Protected Species, National Marine Fisheries Service, National Oceanic and Atmospheric Administration, U.S. Department of Commerce, Washington, DC, 1990.

58. Katona, S. K. and Beard, J. A., Population size, migration and substock structure of the humpback whale *(Megaptera novaeangliae)* in the western North Atlantic ocean, *Rep. Int. Whaling Comm. (Special Issue),* 12, 295, 1990.

59. Perry, A., Baker, C. S., and Herman, L. M., Population characteristics of individually identified humpback whales in the central and eastern North Pacific; a summary and critique, *Rep. Int. Whaling Comm. (Special Issue),* 12, 307, 1990.

60. Baker, C. S., Perry, A., and Vequist, G., Humpback whales of Glacier Bay, *Whalewatcher,* Fall Issue, 13, 1988.

61. Bryant, S., personal communication, 1991.

CHAPTER 11

Assessment of Organochlorine Pollutants in Cetaceans by Means of Skin and Hypodermic Biopsies

Alex Aguilar and Asunción Borrell

TABLE OF CONTENTS

I. INTRODUCTION

Interest in the extent and effects of pollutants in cetaceans has been growing in the last few years. According to the Science Citation Index database (Institute for Scientific Information, Philadelphia, PA), the number of indexed publications addressing this subject has multiplied by a factor of five from the early '70s to the present, a much higher growth rate than that observed for publications on other comparable taxonomic groups. The reason for this increased interest stems from the fact that cetaceans share some particularities which make them especially useful for the understanding of certain biological and ecological processes to which the dynamics of organochlorines are bound, but also that their unique position in the marine ecosystem has put them among the most vulnerable targets of the effects of such pollutants.

On the one hand, cetaceans are long living, are mobile within rather vast home ranges which in many cases are predictable, and are very often situated at the top of the long and complex marine food webs, usually relying on a wide range of prey organisms. These characteristics combine to make cetaceans good integrators of mid- or long-term changes in marine pollution loads in wide bodies of water. Because bioaccumulative pollutants reach them after having been dispersed or diluted in the water and having been incorporated into trophically basal organisms and transferred through successive food web steps, they summarize and average local variations or point sources of pollution to which the typical sessile, low trophic organisms are subjected. However, cetaceans are not useful for discerning short-term changes in marine environment quality or monitoring pollutants which are rapidly degraded by marine biota or nonliving processes.

On the other hand, top predator cetacean species inhabiting polluted regions accumulate extremely high pollution loads, which increase progressively through their life span in males, and pass into the fetal tissues and milk in reproductive females, contaminating their offspring. Moreover, the capacity of at least some cetaceans to metabolize certain PCB forms has been found to be much lower than that of birds and other mammals.[1,2] Therefore, it is not surprising that the highest organochlorine loads ever recorded in a living organism were found on small cetaceans living in heavily polluted bodies of water, such as belugas from the St. Lawrence estuary,[3] striped dolphins from the Mediterranean Sea,[4] bottlenose dolphins from the eastern U.S.,[5] and common and bottlenose dolphins from the Californian coasts.[6]

Because of the obvious difficulties involved in keeping in captivity a number of cetaceans large enough to produce statistically reliable results and also because of ethical and legal constraints on conducting invasive toxicological experiments in them, evaluation of the effects of high concentrations of organochlorines in the tissues of these animals relies mostly on indirect evidence and has been the subject of much controversy. Organochlorine residue levels lower than those typically supported by some cetacean species have been associated with reproductive failure,[7-11] development of uterine tumors,[12,13] and alterations in skeletal development[14,15] in several species of pinnipeds — a group that is taxonomically distant from cetaceans but which nevertheless shares food resources, habitat degradation, and many life history traits. Moreover, because of their capacity to depress the immune system of mammals,[16-19] organochlorine pollutants (especially PCBs) have also been implicated in triggering virulent epizootics that have recently devastated some seal and dolphin populations inhabiting polluted waters.[4,20]

Finally, organochlorines can be used to gain understanding of certain biological traits of cetaceans, such as home ranges and population segregation[21,22] or main reproductive strategies.[23]

However, monitoring of organochlorine pollutant loads in cetaceans is difficult, and most studies in the past have relied on specimens found stranded on the shore. Most species are highly inaccessible; when dead, they sink or reach the shore in a decomposed state; and when these difficulties are overcome, representative sampling of the animal's body load of organochlorines is plagued with difficulties because of their large size. In recent years, sampling of free-ranging, supposedly healthy cetaceans by means of biopsy darts at a distance has become increasingly used. This technique solves or circumvents some of the difficulties posed by the sampling of beach-stranded specimens, but creates others.

The present chapter reviews the potentials and difficulties involved in monitoring organochlorine pollutant loads in cetaceans through biopsy techniques, and presents some new data concerning the questions raised in the last few years as to the reliability and practical usefulness of this technique. The unpublished results on tissue organochlorine residue levels used here were obtained following standard analytical techniques as detailed by Aguilar and Borrell.[24] PCB congener composition in the samples was determined by means of individual congener standards and of a precalibrated Aroclor 1260 standard.

II. WHY USE BIOPSIES?

Until the mid-1980s, most studies of organochlorine pollutants in cetaceans had been performed on specimens caught by commercial whaling operations or found stranded on beaches. The latter group was composed of individuals that either had died from natural causes or had been incidentally caught by fisheries directed to other marine organisms. The reliability of these specimens to monitor incidence of organochlorine pollution in the overall population has been repeatedly questioned for at least three main reasons.

First, these samples may not be representative of the overall population. Thus, neither incidental nor direct catches of cetaceans provide unbiased samples in terms of sex and age composition, two variables which strongly affect the tissue organochlorine concentration of cetaceans. Depending on the type of operation, especially if commercial, certain components of the population are not accessible to sampling. For example, regulations for exploiting large whales forbid the taking of juvenile and lactating individuals and, for sperm whales, restrict or forbid the killing of females.[25] The composition of samples coming from incidental catches may also contain substantial bias,[26-28] although in this case the nature of this bias is more variable and usually unpredictable. Such an effect also hinders collections of samples from stranded cetaceans because natural mortality is not homogeneous for all ages or reproductive status, but appears to be higher for neonates, calves, senile individuals, and near-term females.[25,29,30] Geographic segregation of certain population components, known to occur in many species,[22,27,31,33] may even accentuate the inaccessibility of certain individuals to shore-based sampling. These biases in sample composition could be taken into account if the age and reproductive state of the individuals sampled are determined, but consideration of the specimen's biological data is often lacking in pollutant studies.

Second, the reason for the decease of a stranded cetacean is often difficult to establish, but may be significant for the levels of pollutants carried by a given specimen. In animals that have died from natural causes, the occurrence of disease is to be expected and may involve impairment of function of important organs for the excretion or metabolization of pollutants. This may lead to an artifactual alteration of organochlorine loads, which have often been recognized to hinder interpretation of organochlorine concentrations in naturally stranded specimens.[3,34] Although no comparative study of organochlorines between stranded and live-captured cetaceans has been reported, Bergman et al.[35] found no correlation between age and residue levels in a sample of seals from the Baltic, although a correlation was found for a sample of apparently healthy, slaughtered individuals from the same population.

Moreover, altered physiological or nutritional conditions may produce substantial variation in tissue organochlorine concentrations. For example, lipid mobilization from the blubber, the hypodermic fat cover of cetaceans, leads to an increase in the organochlorine concentration in this tissue.[3,21,36] However, measurement of nutritional condition is difficult for many species and is quite often not taken into account when interpreting organochlorine pollutant levels in stranded cetaceans.

Third, specimens in a poor state of conservation or decomposed, as many — if not most — stranded cetaceans are, may show considerably altered residues of organochlorines. Borrell and Aguilar[37] studied variation of dichlorodiphenyl-trichloroethane (DDT) and polychlorobiphenyl (PCB) concentrations in the blubber and muscle of a stranded striped dolphin that had been left outdoors for 55 days; and found considerable changes caused by evaporation of water, volatilization of both lipids and organochlorines, and degradation by microorganisms. Thus, they found that the water content of the tissues fluctuated widely throughout the study according to variations in the weather and the degree of humidity of the atmosphere, for which reason concentrations of organochlorines expressed on a fresh weight basis (FWB) were considerably affected. On a lipid basis (LB), both DDT and PCB concentrations in blubber and muscle decreased significantly during the period of study, reaching values which were much lower than those originally present in the tissues. However, the pattern of change in concentrations was not identical for all the compounds, and their relative concentrations thus varied throughout the experiment. Especially, the concentrations of p,p'-dichlorodiphenyldichloroethylene (DDE) and p,p'-dichloropdiphenyldichloroethane

(TDE) — by-products of microorganism degradation of p,p′-DDT — significantly increased in relation to the total DDT concentrations both in blubber and muscle over the period of time studied. These results were consistent overall with similar studies on other biological materials and preservation conditions.[38]

These three main flaws — unrepresentative composition of the sample, uncontrolled effects of disease on individual pollutant levels, and alteration of concentrations caused by decomposition — are extremely difficult or impossible to solve. These flaws indicate that stranded cetaceans and, to a lesser extent, commercially caught cetaceans are of questionable reliability for monitoring wild populations.

In contrast, biopsies do not suffer from such drawbacks and are potentially a much more reliable means of measuring the organochlorine levels of wild populations. Thus, biopsy sampling can be scattered over the whole range of distribution of a given population or species, and the possibility of selecting target individuals ensures full and balanced representation of different size classes and sexes. Biopsies are taken from free-ranging, apparently healthy individuals; and incidence of disease or altered physiological or nutritional states should not be expected to be higher than the normal rates for the population. Finally, the samples obtained with a biopsy dart can be adequately preserved and stored immediately after collection, so that their state of conservation is optimal.

III. TECHNIQUES FOR BIOPSY SAMPLING OF CETACEANS

Cetacean skin is hairless, devoid of glands and accessory structures, and extremely smooth in comparison to that of terrestrial mammals. Below the skin, fat layer or blubber forms a cover of variable thickness (1–40 cm depending on the species) around the body. Blubber is important for thermoregulation, but in most species it is also the main site for fat storage. Because organochlorines are highly lipophilic and concentrate in fatty tissues, the bulk of the organochlorine load of cetaceans is therefore located in the blubber compartment. For example, in striped dolphins, organochlorines contained in the blubber have been estimated to represent between 90 and 98% of the total organochlorine body load.[39]

The ease with which the skin of cetaceans can be penetrated with a cutting device and the location of the blubber layers containing organochlorines near the surface greatly facilitate the collection of biopsies from free-ranging individuals at a distance and, in this way, the relatively noninvasive monitoring of their organochlorine residue levels.

The first biopsy dart tested on cetaceans was a modification of one originally developed for obtaining a flesh sample of the Loch Ness monster. This dart, essentially consisting of a punch equipped with internal barbs to ensure retrieval of the sample, was successfully used by Winn et al.[40] in the West Indies, where they collected a limited number of skin and blubber samples from humpback whales, sperm whales, and spotted dolphins. Some of these samples were later used to determine organochlorine pollutant levels.[41] However, the passing of the

U.S. Marine Mammal Protection Act, which placed restrictions on the use of potentially harmful techniques on cetaceans, set back the development of biopsy devices. In 1982, a biopsy dart equipped with a butterfly valve retrieval system was introduced to monitor organochlorine pollutant loads in striped, common and bottlenose dolphins off the coast of Spain,[42] but it was not until the second half of the '80s that biopsy darts were fully developed.

The reason for this late evolvement was the need for the organizations managing cetacean populations to apply newly developed genetic techniques to ascertain the structure of populations and stocks of large whales. The tissues of these whales, however, had become less accessible because of the abrupt reduction in catches caused by the moratorium on whaling initiated by the International Whaling Commission in 1985. A variety of propulsive devices and dart tips, most of them designed to retrieve skin, have been tested since 1988.[43] Although these were aimed at genetic studies on large whales, the same techniques could be deployed equally on medium-sized or small cetaceans and used for pollutant analyses. Table 1 details the surveys conducted to date on cetaceans with the specific objective of studying pollutants, as well as the characteristics of the devices and techniques used.

A biopsy dart is typically composed of a hollow tip 2–10 mm diameter, mounted on a shaft and secured to the boat by a tether or equipped with a float to facilitate recovery once it falls off the whale after impact. A hole drilled in the base of the tip prevents pressure buildup inside the biopsy chamber as it penetrates the dermal tissues. Penetration is limited by flanges or a collar located at the base of the tip. The tissue sample is usually extracted from the back of the dart. The most important variations between biopsy devices relate to the type of system used to ensure retrieval of the sample inside the hollow tip. Some rely on internal barbs or prongs,[40,44-46] butterfly valves,[42] constrictions on the inner sides of the hollow tip,[47,48] or internal hooks.[49,50] In some cases, there is also a piece of stainless steel wire spiraled up the outside of the tip to impart a twist to the dart as it enters the whale, thus improving penetration.[45,51]

Typically, biopsy tips collect samples weighing 0.3–1 g, containing both skin and blubber. This sample is enough in most cases for correct organochlorine analyses using standard techniques. No detailed comparative study of the efficiency of these tips is available, but most authors claim a success in sample retrieval of about 40–70% of correct hits. However, considerable variation should be expected in this regard taking into consideration the size of the biopsy, firing device, and species subject to study. The skin of sperm whales, for example, is tougher than that of baleen whales; and that of the latter, more than that of dolphins. Alternative biopsy darts composed of a protruding tip made of abrasive materials to collect only skin have also been developed,[52,53] but have not yet demonstrated their utility in the field.

In early studies, darts were fired by conventional harpoon guns,[40] but these were too powerful and noisy; and in more recent work they were substituted by crossbows, compound bows, spear guns, or (for collecting biopsies from bow-riding dolphins) long poles with the biopsy tip attached to their end.[43]

Table 1. Summary of Surveys Monitoring Organochlorine Levels in Cetaceans Using Hypodermic Biopsy Techniques

Species	Area	Delivery system	Type of biopsy tip	Ref.
Humpback and sperm whales	West Indies	Harpoon gun	Barbs	40, 41
Common and striped dolphins	N.E. Atlantic	Air gun	Butterfly valve	75
Fin whales	Mediterranean	Crossbow	Barbs	90, 91
Striped dolphins	Mediterranean	Pole	Barbs	91
Right whales	N.W. Atlantic	Crossbow	Internal hook	49, 60
Striped dolphins	Mediterranean	Spear gun	Butterfly	76

The effects of biopsy sampling on cetaceans are generally considered to be negligible. If properly done, collection of a small hypodermal biopsy is probably painless, and reaction of the animal to darting is minimal and essentially short term. Indeed, many field-workers have observed to their surprise that cetaceans react more strongly to the noisy splash of a missing dart breaking the water surface near them, than to a clean hit in their own body. However, repeated attempts to obtain a biopsy from a given animal may elicit avoidance or even aggressive behavior, a possibility which should be taken into account from both a procedural and safety perspective.[43] Despite its unusual thickness (1–4 mm), cetacean skin has a large germinative layer which permits high epidermal cell proliferation rates and healing is considered to be as fast as in man or other mammals.[54,55] Thus, the small size of the wound produced with standard darts is unlikely to produce significant physical trauma to the animal sampled.

IV. BLUBBER AS TARGET TISSUE FOR BIOPSIES

Blubber appears to be the most convenient tissue for assessing organochlorine pollution in cetaceans. Not only is it the body compartment that contains by far the largest quantity of organochlorines, but also it is easily accessible in live animals; and because of its high lipid content, a small sample is usually enough for a precise determination of these pollutants using standard techniques, even in species or populations with a low exposure to organochlorines.

However, blubber is a rather heterogeneous tissue, and neither lipids nor organochlorines are distributed uniformly within the blubber compartment. This poses some difficulties when using biopsy techniques for monitoring organochlorines in cetaceans, because the levels detected may not be a true reflection of organochlorine loads. Methodological research in this regard has shown that organochlorine surveys relying on biopsies may be distorted by two types of heterogeneities: that among blubber strata and that among body locations.

Blubber, which in large whales can be up to 40 cm thick, is a clearly layered tissue. In mysticetes and the sperm whales, in which the subject has been studied in great detail, the outer stratum appears to be essentially devoted to thermoregulation and its lipid content is fairly constant throughout the year, hardly reflecting changes in body condition. The inner stratum, in contrast, has a significant role in the dynamics of fat storage in whales and its lipid richness fluctuates according to the individual's energy demands. The midstratum is a transition between the external and internal layers.[56-59] This stratification applies not only to the lipid content of the blubber layers, but also to the composition of these lipids in each layer. Thus, triglycerides from the outer strata contain more fatty acids of longer chain length than those from the inner ones.[57]

Obviously, these variations have consequences for the quantity and quality of organochlorines that are deposited in the various blubber strata. Expressing concentrations on an extractable lipid basis rather than on a fresh weight basis reduces the differences between layers caused by stratification of the lipids, but does not eliminate them. In a survey conducted on fin whales, residue levels of all organochlorines calculated on a lipid basis and the tDDT/PCB ratio were significantly higher in the outer layer. The reasons for these differences were not ascertained, although the differential role of the various strata in the fattening cycle, the heterogeneous lipid composition of blubber throughout its depth, and the likely difference in turnover rates of pollutants in the various layers were considered to be probable causes.[23]

Such heterogeneities in organochlorine composition between blubber strata have been recognized as a drawback for monitoring pollutants by means of biopsies in large cetaceans.[23,36,60] The only solution appears to be the development of biopsy tips capable of penetrating the whole blubber thickness, a technique that may ensure an equal representation of all the blubber layers in the analytical determination of organochlorines.[61] In small cetaceans the blubber cover is much thinner (usually between 0.5 and 3–4 cm); and a chunk containing all layers is therefore easy to collect and, indeed, collected by most biopsy tips currently in use.

The second difficulty, the heterogeneity in organochlorine concentrations among blubber from different body locations, has also attracted a number of studies in stranded or commercially caught small cetaceans.[39,62-64] The results from these studies indicate that differences between body sites are small and probably reflect analytical variation rather than actual differences among tissue locations. The only regions in which blubber showed significant deviations from the rest of the body were the head (including the melon) and the dorsal peduncle, locations in which lipid content and composition is known to differ markedly from the rest of the body.[36] However, the head region is never targeted when collecting biopsies in order to preserve the integrity of the animal, and the peduncle is also usually avoided because its high content in fibrous tissue impairs or prevents penetration of the biopsy tip. Therefore, body location does not appear to be a major concern in biopsy collection from small cetaceans, provided that the dart is aimed at the trunk region of the animal.

Unfortunately, however, the information available in this respect for large whales is extremely limited. In contrast to small-sized cetaceans, the lipid content of the blubber of mysticetes and the sperm whale increases markedly from head to tail and from the ventral to the dorsal regions;[59,65] and the quality of lipids constituting the blubber in these different body regions also appears to vary.[56,57]

It is also unfortunate that no studies on variation of organochlorine concentration among body locations are available for large whales; thus the importance of sampling-site consistency in the determination of organochlorine levels in these animals is unclear. In contrast to the small dolphins and porpoises, the studies on variation in lipid content and composition suggest that this may be a major source of variation in pollutant concentrations and is a potential source of bias or heterogeneity. Therefore, when sampling large cetaceans, care should be taken to collect biopsies from the same body site and to record precisely when this does not happen in order to take this into account when interpreting analytical outputs.

V. HOW REPRESENTATIVE OF THE MAIN BODY TISSUES IS BLUBBER?

As seen above, because organochlorines are highly lipophilic, the bulk of the organochlorine load of cetaceans is located in the blubber, the main site for fat storage in these animals. Therefore, a biopsy of blubber will be a fairly good indicator of the total amount of organochlorines carried by a given specimen. However, the extent to which blubber is truly representative of the other body tissues or organs remains undetermined.

Many surveys of organochlorines in cetaceans include the analytical screening of various tissues in order to determine the variation patterns between them and the blubber. The conclusions drawn from these studies are, however, quite contradictory, mostly because of the lack of an appropriate sample size to examine these differences statistically.

Furthermore, comparison between tissues depends on how organochlorine concentrations are calculated. Figures 1 and 2 compare concentrations of tDDT and PCBs in 12 different tissues from a fin whale caught by a whaling operation off northwestern Spain in 1985, calculated both relative to the fresh weight of the sample (FWB) and relative to the amount of lipids extracted (LB). On a FWB, concentrations differ greatly from one tissue to another, largely because variable lipid content of a tissue greatly influences its capacity to retain organochlorines. When concentrations in the same tissues are calculated on a lipid basis, differences are greatly reduced. This partly explains why analyses using concentrations expressed on a FWB[6,66] usually failed to find correlations between tissue residue levels and concluded that blubber was poorly representative of the other organs.

Calculated on a lipid basis, however, blubber concentrations much resemble those from other tissues. Table 2 details the levels of significance of correlations between concentrations in blubber expressed in this way and those in other tissues in three cetacean species.[67] As can be seen, correlations are highly significant in most cases, indicating that blubber concentrations do maintain a correspondence with those in other body tissues. Indeed, this was to be expected if we consider

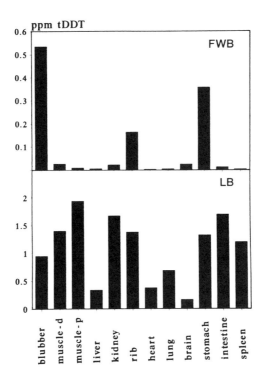

FIGURE 1. Concentrations of tDDT in different tissues from a fin whale calculated both relative to the fresh weight of the sample (fresh weight basis [FWB]) and relative to the amount of lipids extracted (lipid basis [LB]).

that organochlorines follow a compartmental model of the mammillary type, in which blood acts as a central compartment freely distributing organochlorines among the other body compartments. An application of this model to the distribution of organochlorines in cetaceans led to the conclusion that in most cases the movement of these pollutants within the body is restricted only by their association with lipids, the quality of which — on the other hand — determines their relative abundance in a given tissue or organ.[36] This explains why, contrary to studies based on concentrations calculated on a FWB, those using lipid-based calculations usually conclude that blubber is a good indicator of organochlorine residue levels in other tissues.[2,36,39,63]

However, proportionality is not synonymous with equality. Thus, even calculating concentrations on a lipid basis, residue levels in blubber are not identical to those in other tissues. Table 3 details the *"distribution factors,"* calculated as defined by Massé et al.[68] as the ratio of the organochlorine concentration in the blubber to that measured in other tissues, for the same sample collection used for Table 2. The closer the distribution factor is to unity, the more similar the organochlorine concentration of a given tissue is to that of blubber. Most tissues present distribution factors which do not deviate significantly from unity overall, although muscle and bone appear to present higher residue levels than blubber;

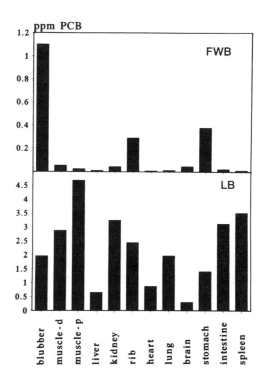

FIGURE 2. Concentrations of PCB in different tissues from a fin whale calculated both relative to the fresh weight of the sample (FWB) and relative to the amount of lipids extracted (LB).

while liver, kidney, skin and milk tend to present slightly lower ones. These results are consistent with a similar study conducted by Duinker et al.[2] on fewer body tissues.

Such differences in distribution factors may relate to the capacity of a given tissue to retain organochlorines which, in turn, strongly depends on the composition of the lipids constituting it and, especially, on the polarity affinities between lipids and pollutants. Thus, Aguilar,[36] Massé et al.,[68] and Kawai et al.[63] have shown that tissue levels of highly chlorinated PCBs, DDTs, and certain Hexachlorocydohexane (HCH) isomers are related to the richness of the tissue in triglycerides and nonesterified fatty acids (both of neutral character). Conversely, lowly chlorinated PCB congeners and HCH isomers are favorably retained by the highly polar phospholipids. Such affinities may also explain the low concentrations of organochlorines and the proportionally high quantity of lower chlorinated biphenyls usually found in the brain and nervous system — tissues which are extremely rich in phospholipids — as compared to those of blubber, milk, or other tissues rich in neutral lipids.[2,36,39,63]

The only exception to this general rule appears to be the liver. Although the lipid composition of the liver is not very different from that of muscle or lung,[63] the distribution factors typically found in this organ are lower than those calculated for

Table 2. Level of significance of correlations between concentrations (expressed on a LB) of tDDT and PCBs in different tissues from fin whales caught in the Atlantic waters off N.W. Spain in 1983, pilot whales from the Faroe Island in 1987–1988, and striped dolphins from the western Mediterranean in 1990–1991

Species	N	Muscle/ Blubber	Liver/ Blubber	Kidney/ Blubber	Bone/ Blubber	Skin/ Blubber	Milk/ Blubber
DDTs							
Fin whale	26	N.S.	**	**	**	—	**
Pilot whale	19	**	**	*	—	—	**
Striped dolphin	25	**	**	**	—	**	**
PCBs							
Fin whale	26	N.S.	*	**	**	—	**
Pilot whale	19	**	**	**	—	—	**
Striped dolphin	25	**	**	**	—	**	**

Source: Borrell, unpublished data.[67]
Note: N.S.: nonsignificant correlation; *: correlation significant at $p < 0.05$; **: correlation significant at $p < 0.01$.

Table 3. Distribution Factors of Concentrations (Expressed on an LB) of tDDT and PCBs in Different Tissues from the Cetacean Sample of Table 2

Species	N	Muscle	Liver	Kidney	Bone	Skin	Milk
DDTs							
Fin whale	26	1.1	0.5*	0.9	1.2	—	0.5
Pilot whale	19	0.7	0.5*	0.6	—	—	0.9
Striped dolphin	25	0.9	0.7	0.5*	—	0.8	0.5*
PCBs							
Fin whale	26	1.4	0.7*	1.0	1.3	—	0.6
Pilot whale	19	1.5*	0.7*	0.9	—	—	1
Striped dolphin	25	1.2	0.9	0.9	—	0.9	0.8

Note: An asterisk indicates that pollutant concentration in a given tissue was statistically different, according to a *t*-test, to that determined in blubber.

the rest of tissues. Thus, values depicted in Table 3 ranged 0.5–0.9 and the *t*-test performed between tDDT and PCB concentrations in liver and those in blubber yielded significant differences in four out of the six instances tested. These differences do not appear to be caused by dissimilarities in lipid content or composition, or by any protective barrier limiting access of pollutants to the hepatic compartment; instead they appear to be caused by the role of this organ in the detoxification process, which apparently produces a localized abatement of organochlorine concentrations. Thus, liver is the main center for degradation of xenobiotics, and the highest concentration of the enzymes involved in the metabolism of organochlorines is found within the hepatic tissue.[69,70]

This active role in detoxification also explains why the abundance of metabolized forms of organochlorines is higher in the liver than in the rest of body tissues or organs. For example, screening of cetacean tissues for organochlorines almost invariably shows higher percentages of DDE in the liver than in the blubber, muscle, kidneys, or blood.[2,36,39,66]

In conclusion, blubber biopsies can be used to infer organochlorine residue levels in other tissues, provided that the lipid content, lipid composition, and distribution factors of the latter in relation to blubber have been previously determined. However, for tissues with different lipid composition or metabolic function, such as brain or liver, this inference can be seriously biased, even when this information is available.

VI. BLUBBER AS INDICATOR OF ORGANOCHLORINE BODY LOADS

The total amount of organochlorine compounds carried by an organism, or "body load," is a potentially useful parameter to measure the extent of reproductive transfer in females; to study whole body dynamics of accumulation of pollutants in individuals subjected to changes in weight either by fat mobilization or deposition, or by growth; and to establish patterns of bioaccumulation between trophic levels, among others.

However, measuring body loads in cetaceans is cumbersome. Cetaceans are massive animals and pollutants are inhomogeneously distributed among the various body compartments, each of considerable volume. For this reason, measurement of body loads requires the determination of pollutant concentrations in the volumetrically greater tissues and subsequent calculations of the absolute amounts through the weight of these tissues.

This explains why these calculations so far have been properly performed only for a few species: striped dolphins,[39] Stejneger's beaked whales,[71] fin whales,[67] and pilot whales.[67] These studies have indicated that blubber loads represent between 75 and 98% of total body loads, depending on the species and compound, and have justified that blubber loads have been used almost indistinctly to body loads.[72,73]

Therefore, knowledge of the concentration of pollutants in a biopsy of blubber permits indirect calculation of total body loads, provided that previous information on organochlorine distribution among tissues and weight of these tissues is available for the species.

VII. THE SKIN ALTERNATIVE

As seen above, blubber is the most convenient tissue for assessing organochlorine pollution in cetaceans. However, some biopsy darts are especially designed to collect only skin; and even when using those prepared to penetrate the hypodermis, sidelong hits often yield dermal tissues only. Moreover, in small tropical species with a thin blubber cover, blubber biopsies may be too invasive. In these situations, skin may be a good alternative to blubber as an indicator tissue.

Cetacean skin measures 2–4 mm in most species and is, therefore, 10–20 times thicker than that of terrestrial mammals. Moreover, its germinal epithelium contains prominent lipid droplets[74] which are potentially capable of retaining, like any other body lipid compartment, measurable quantities of lipophilic xenobiotics such as organochlorines. As we have seen above, accepting that organochlorines distribute in the body compartments of mammals following simple laws of dilution in neutral lipids according to a mammillary model,[36] skin — like any other tissue with no intrinsic metabolic activity — may be used as proper indicator of organochlorine body loads. However, some imbalance may be expected if the overall polarity of lipids found in skin differs from that of triglycerides, the main constituents of blubber.

To examine this possibility, we studied the relationship between organochlorine concentration in the blubber and in the skin of 20 striped dolphins sampled in the western Mediterranean Sea. This population is heavily polluted by organochlorines, and the usual levels are in the range of 100–300 ppm for tDDT and 200–400 ppm for PCBs.[75] For both tissues concentrations were calculated on a lipid basis, that is, in relation to the total amount of the lipids extracted. Expressed in this way, the pollutant concentration in any given tissue is not subject to variation in chemical composition between tissues.

Figure 3 shows the results of this comparison for PCBs and total DDT (tDDT = p,p'-DDE + p,p'-DDT + o,p'-DDT + p,p'-TDE). In both cases the correlation between the levels of skin and blubber is very high, and the small variance observed around the regression lines may be explained by analytical or sampling variability.

However, for any given concentration in the blubber, the corresponding values for the skin were about 20% lower, both for tDDT and PCBs. This suggests that skin has a slightly lower capacity than blubber for retaining organochlorines. The reason for this difference is unknown, although a difference in lipid qualitative composition between the two tissues may account for it.

Figure 4 shows the relative contribution (expressed as a percentage of the total) of the different PCB congeners and the various DDT forms in the blubber and skin of the same dolphins. The variation in relative contribution of each compound of the two tissues is insignificant and apparently does not follow any definite trend that might be associated with a variation in polarity of the tissue lipid fraction. Moreover, a *t*-test did not reveal significant differences between the tDDT/PCB ratio of either tissue.

This indicates that skin is as good as blubber for monitoring organochlorine compounds in cetaceans, although it may yield slightly lower residue levels than those present in blubber. Apparently, the relative proportion of the different compounds is not very different in the two tissues. However, because of the low lipid content of the skin (mean:0.116; SD:±0.05 in our sample) and consequently

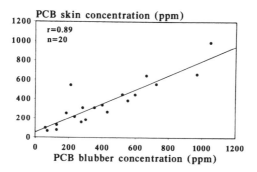

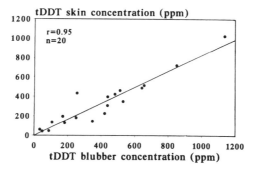

FIGURE 3. Relationship between tDDT and PCB levels in skin and those in blubber of striped dolphins (concentrations expressed on a lipid basis).

the lower absolute quantities of pollutants per unit of fresh weight, organochlorines may be undetectable in skin of animals exposed to low levels of pollution. Thus, assessment of organochlorines in skin samples from North Atlantic fin whales, a population carrying tDDT and PCB blubber levels usually in the range 0.5–1 ppm lipid basis, was impracticable because the amount of organochlorines extracted from a skin biopsy sample of usual size (0.05–0.2 g) was below analytical detection limits.

VIII. MONITORING SIGNIFICANT BIOLOGICAL VARIABLES

The biological characteristics of the individual sampled are necessary to correctly interpret organochlorine residue levels. Within a given population composed of individuals in principle similarly exposed to pollutants, the range of variation in concentrations is usually between twofold and fourfold the mean for the population. Although many factors may account for at least part of this variation, four variables have been identified as significant: induction of enzymatic activity, age, sex, and nutritional state.

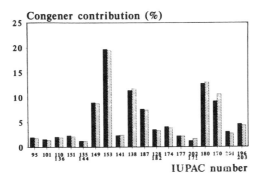

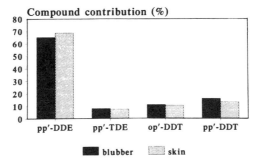

FIGURE 4. Relative contribution (expressed as a percentage of the total) of the different PCB congeners and the various DDT forms in the blubber and skin of striped dolphins.

Organochlorines and other xenobiotic compounds have the capacity of inducing the activity of certain enzymes. Measuring the levels of that induction is interesting for at least two reasons. First, a rise in enzyme activity is likely to increase the decomposition rate of some degradable pollutants and therefore affect their relative abundance in relation to the total pollutant load. Apparently because of this effect, the concentration of p,p'-DDE relative to that of p,p'-DDT was found to correlate positively to total DDT concentrations in fin and sei whales.[76] Second, variation in the activity of enzymes is a more reliable indicator of the effect of pollutants on the organism than their mere concentration levels in the tissues.[77] Enzyme levels are usually measured in organs with a high metabolic activity such as the liver or brain, or in the blood, where they are found at relatively high concentrations. Enzymes also occur in tissues with a lower metabolic activity, such as blubber or skin, but their lower concentrations in these tissues make the analytical determination more complex. While the activity of enzymes in the liver of cetaceans is well documented,[78-80] the only study on this regard conducted so far on cetacean skin is that of Fossi et al.,[81] which presents levels of mixed function oxidase activity in skin biopsies of striped dolphins and fin whales from the Mediterranean. They found that the levels in striped dolphins heavily polluted by organochlorines were substantially higher than those in fin whales, which carry only moderate levels of these pollutants. However, data from further species are still needed to validate the use of this technique in cetaceans.

The influence of age and sex on organochlorine body loads is indeed interconnected by the effect that the reproductive events have on the dynamics of pollutants in individuals. In all cetacean species studied so far, the intake of organochlorines in juvenile females and in males of any age exceeds excretion and degradation and, consequently, concentrations build up progressively with age. In adult females, in contrast, gestation and lactation permits a discharge of a substantial part of the organochlorines contained in their body; as a consequence, concentrations of these pollutants in reproductively active females stabilize or even decrease with age.[21] This pattern of accumulation of organochlorines explains the differences between sexes and, to a lesser extent, between individuals of the same gender but of different age. For this reason, proper monitoring of pollutants in a population needs to take age and sex into account.

In a skin biopsy, gender can be determined by karyotyping of fibroblasts cultured in vitro from dermal tissues and subsequent determination of presence or absence of the male-specific Y chromosome. This technique has been successfully applied to sexing skin biopsies of several cetacean species by Arnason et al.,[82] Mathews,[83] and Lambertsen et al.[84] An alternative method consists of revealing highly sex-specific restriction-fragment patterns in a human Y chromosome clone by hybridization to DNA from a skin biopsy of the cetacean under study. The clone hybridized contains a sequence coding an important protein for certain sex-specific functions in males. This technique — successfully applied by Brown et al.,[85] Baker et al.[86] and Andersen and Grandl[87] to seven cetacean species — has the advantage of not requiring time-consuming in vitro culturing and can be used with poorly preserved tissues. However, its validity in every cetacean species still needs to be proved. Other researchers have employed analysis of sex chromatin bodies,[40] but Mathews[83] found inconsistencies in the counts of bodies suggesting that the technique is not accurate.

Determination of age, on the contrary, is not possible from a biopsy of skin or blubber, and requires the examination of inaccessible tissues such as teeth, ear plugs, bones, or eye lenses. However, body length can be used as a rough indicator of age in young animals, especially in species in which growth is protracted, such as the sperm whale and some medium-sized whales. Whitehead and Gordon[88] reviewed techniques for measuring whales at sea, and Gordon[89] recently developed a simple photographic technique for the unbiased estimation of body length in live cetaceans.

Nutritional condition is the third variable substantially affecting organochlorine tissue concentrations. Cetaceans, and very especially those which engage in long migrations and are seasonal feeders, rely heavily on their fat stores. Because organochlorines are highly lipophilic and dissolve in lipids, changes in the volume of fat present in the body (i.e., of solvent for the organochlorines) affect pollutant tissue concentrations. Increase in organochlorines following fat mobilization and dilution of their concentrations after expansion of the body fat reserves are events that have been commonly observed in humans and other mammals,[69] and may occur in cetaceans.[21] Therefore, the nutritive condition of the specimens sampled needs to be known with some accuracy in order to interpret the pollutant concentrations observed correctly.

Fortunately, blubber is also the main body site for energy storage in cetaceans, and its lipid richness reflects the condition of the individual. However, deposition of lipids in the blubber of large whales is stratified,[56] and the outer layers — the most accessible to biopsy darts — are essentially thermoregulatory, react only mildly to variations in nutritive condition, and are therefore only slightly indicative.

Because lipids from the inner strata are more readily mobilized or deposited, the pattern of variation of lipid content between the outer and the inner strata can be used as a reliable indicator of the nutritive condition of the whale under study.[57,58] However, the determination of this pattern requires a blubber biopsy containing all the blubber strata which, as mentioned before, can only be taken using darts capable of penetrating the whole blubber thickness. The lipid content of the inner blubber layers alone can also be used as a condition index;[58] but, again, a deep biopsy is needed.

Small cetaceans also rely on the lipids accrued in the blubber for energy storage. Although some stratification also appears to exist in their blubber, the small thickness of the hypodermis makes the collection of a whole thickness biopsy considerably easier. Nutritional state can therefore be more readily ascertained in them than in their larger size congeners.

ACKNOWLEDGMENTS

This work was made possible thanks to partial funding granted by the Instituto para la Conservación de la Naturaleza (ICONA), the Comisión Interministerial de Ciencia y Tecnología (project NAT91-1128-CO4-02), and the Dirección General de Política Ambiental (MOPT) of Spain.

REFERENCES

1. Tanabe, S., Watanabe, S., Kan, H., and Tatsukawa, R., Capacity and mode of PCB metabolism in small cetaceans, *Mar. Mamm. Sci.*, 4(2), 103, 1988.
2. Duinker, J. C., Hillebrand, M. T. J., Zeinstra, T., and Boon, J. P., Individual chlorinated biphenyls and pesticides in tissues of some cetacean species from the North Sea and the Atlantic Ocean; tissue distribution and biotransformation, *Aquat. Mammal.*, 15(3), 95, 1989.
3. Martineau, D., Béland, P., Desjardins, C., and Lagacé, A., Levels of organochlorine chemicals in tissues of beluga whales *(Delphinapterus leucas)* from the St. Lawrence Estuary, Québec, Canada, *Arch. Environ. Contam. Toxicol.*, 16, 137, 1987.

4. Borrell, A. and Aguilar, A., Pollution by PCBs in striped dolphins affected by the western Mediterranean epizootic, in *Proc. Mediterranean Striped Dolphin Mortality Int. Workshop*, Pastor, X. and Simmonds, M., Eds., Greenpeace Mediterranean Sea Project, Palma de Mallorca, Spain, 1991, p. 121.

5. Geraci, J. R., Clinical Investigation of the 1987–88 Mass Mortality of Bottlenose Dolphins Along the U.S. Central and South Atlantic Coasts, Final report to (U.S.) N.M.F.S., U.S. Navy and the Marine Mammal Commission, 1989, 63 pp, (unpublished).

6. O'Shea, T. J., Brownell, R. L., Clark, D. R., Walker, W. A., Gay, M. L. and Lamont, T. G., Organochlorine pollutants in small cetaceans from the Pacific and South Atlantic Oceans, November 1968-June 1976, *Pest. Mon. J.*, 14(2), 35, 1980.

7. Le Boeuf, B. J. and Bonnell, M. K., DDT in California sea lions, *Nature (London)*, 234, 108, 1971.

8. Helle, E., Olsson, M. and Jensen, S., DDT and PCB levels and reproduction in ringed seal from the Bothnian Bay, *Ambio*, 5, 188, 1976.

9. Olsson, M., PCB and reproduction among Baltic seals, *Finn. Game Res.*, 37, 40, 1978.

10. Reijnders, P. J. H., Organochlorine and heavy metal residues in harbour seals from the Wadden Sea and their possible effects on reproduction, *Neth. J. Sea Res.*, 14, 30, 1980.

11. Addison, R. F., Organochlorines and marine mammal reproduction, *Can. J. Fish. Aquat. Sci.*, 46, 360, 1989.

12. Helle, E., Olsson, M. and Jensen, S., PCB levels correlated with pathological changes in seal uteri, *Ambio*, 5, 261, 1976.

13. Baker, J. R., Pollution-associated uterine lesions in grey seals from the Liverpool Bay area and the Irish Sea, *Vet. Rec.*, 125, 303, 1989.

14. Bergman, A., Olsson, M., and Reiland, S., High frequency of skeletal deformities in skulls of the Baltic grey seal, *Int. Counc. Explor. Sea*, doc CM/1986/N:15, 7 pp, 1986.

15. Zakharov, V. M. and Yablokov, A. V., Skull asymmetry in the Baltic grey seal: effects of environmental pollution, *Ambio*, 19(5), 266, 1990.

16. Loose, L. D., Pitman, K. A., Bentitz, K. F., and Silkworth, J. B., Polychlorinated biphenyl and hexachlorobenzene induced humoral immunosuppression, *J. Reticuloendo. Soc.*, 22, 253, 1977.

17. Thomas, P. T. and Hinsdill, R. D., Effect of polychlorinated biphenyls on the immune responses of Rhesus monkeys and mice, *Toxicol. Appl. Pharmacol.*, 44, 41, 1978.

18. Brouwer, A., Reijnders, P. J. H., and Koeman, J. H., Polychlorinated biphenyl (PCB)-contaminated fish induces vitamin A and thyroid hormone deficiency in the common seal *(Phoca vitulina)*, *Aquat. Toxicol.*, 15, 99, 1989.

19. Vos, J. G. and Luster, M. I., Immune alterations, in *Halogenated Biphenyls, Terphenyls, Naphthalenes, Dibenzodioxins and Related Products*, Kimbrough, R. D. and Jensen, A. A., Eds., Elsevier, Amsterdam, 1989, chap. 10, p. 295.

20. Simmonds, M. and Johnston, P. A., Seals, sense and science, *Mar. Pollut. Bull.*, 20(11), 580, 1989.

21. Aguilar, A., Using organochlorine pollutants to discriminate marine mammal populations: a review and critique of the methods, *Mar. Mamm. Sci.*, 3(3), 242, 1987.

22. Aguilar, A., Use of trace compounds to discriminate home ranges in cetaceans, *Rep. Int. Whaling Comm.*, 38, 125, 1988.

23. Subramanian, A. N., Tanabe, S., and Tatsukawa, R., Use of organochlorines as chemical tracers in determining some reproductive parameters in Dalli-type Dall's porpoise *Phocoenoides dalli, Mar. Environ. Res.*, 25(3), 161, 1988.

24. Aguilar, A. and Borrell, A., Heterogeneous distribution of organochlorine contaminants in the blubber of baleen whales: implications for sampling procedures, *Mar. Environ. Res.*, 31, 275, 1991.

25. Anon., International Convention for the Regulation of Whaling, 1946. Schedule, International Whaling Commission, Histon, Cambridge, U.K., 1991, p. 27.

26. Kasuya, T., Reconsideration of life history parameters of the spotted and striped dolphins based on cemental layers, *Sci. Rep. Whales Res. Inst. Tokyo*, 28, 73, 1976.

27. Kasuya, T., The life history of Dall's porpoise with special reference to the stock off the Pacific coast of Japan, *Sci. Rep. Whales Res. Inst. Tokyo*, 30, 1, 1978.

28. Lockyer, C. H., Goodall, R. N. P., and Galeazzi, A. R., Age and body length characteristics of *Cephalorhynchus commersonii* from incidentally-caught specimens off Tierra del Fuego, *Rep. Int. Whaling Comm., (Special Issue)*, 9, 103, 1988.

29. Kraus, S. D., Rates and potential causes of mortality in North Atlantic right whales *(Eubalaena glacialis), Mar. Mamm. Sci.*, 6(4), 278, 1990.

30. Aguilar, A., Calving and early mortality in the western Mediterranean striped dolphin, *Stenella coeruleoalba, Can. J. Zool.*, 69, 1408, 1991.

31. Ohsumi, S., Sexual segregation of the sperm whale in the North Pacific, *Sci. Rep. Whales Res. Inst. Tokyo*, 20, 1, 1966.

32. Swartz, S. L., Gray whale migratory, social and breeding behavior, *Rep. Int. Whaling Comm., (Special Issue)*, 8, 207, 1986.

33. Cubbage, J. C. and Calambokidis, J., Size-class segregation of bowhead whales discerned through aerial stereophotogrammetry, *Mar. Mamm. Sci.*, 3(2), 179, 1987.

34. Gaskin, D. E., Frank, R., and Holdrinet, M., Polychlorinated biphenyls in harbor porpoises *Phocoena phocoena* (L.) from the Bay of Fundy, Canada and adjacent waters, with some information on chlordane and hexachlorobenzene levels, *Arch. Environ. Contam. Toxicol.*, 12, 211, 1983.

35. Bergman, A., Olsson, M., and Reutergardh, L., Lowered reproduction rate in seal populations and PCBs. A discussion of comparability of results and a presentation of some data from research on the Baltic seals, *Int. Counc. Explor. Sea*, doc CM/1981/ N:10, 18 pp, 1981.

36. Aguilar, A., Compartmentation and reliability of sampling procedures in organochlorine pollution surveys of cetaceans, *Res. Rev.*, 95, 91, 1985.

37. Borrell, A. and Aguilar, A., Loss of organochlorine compounds in the tissues of a decomposing stranded dolphin, *Bull. Environ. Contam. Toxicol.*, 45, 46, 1990.

38. Stickel, W. H., Stickel, L. F., Dyrland, R. A., and Hughes, D. L., Comparison of methods of preserving tissues for pesticide analysis, *Environ. Monit. Assess.*, 4, 113, 1984.

39. Tanabe, S., Tatsukawa, R., Tanaka, H., Maruyama, K., Miyazaki, N., and Fujiyama, T., Distribution and total burdens of chlorinated hydrocarbons in bodies of striped dolphins *(Stenella coeruleoalba), Agric. Biol. Chem.*, 45(11), 2569, 1981.

40. Winn, H. E., Bischoff, W. L., and Taruski, A. G., Cytological sexing of cetacea, *Mar. Biol.*, 23, 343, 1973.

41. Taruski, A. G., Olney, C. E., and Winn, H. E., Chlorinated hydrocarbons in cetaceans, *J. Fish. Res. Board Can.*, 32(11), 2205, 1975.

42. Aguilar, A. and Nadal, J., Obtención de biopsias hipodérmicas de cetáceos en libertad, *Invest. Pesq.*, 48(1), 23, 1984.

43. Anon., Report of the *Ad-hoc* Working Group on the Effect of Biopsy Sampling on Individual Cetaceans, *Rep. Int. Whaling Comm.*, 41, 201, 1991.

44. Lambertsen, R. H., A biopsy system for large whales and its use for cytogenetics, *J. Mammal.*, 68(2), 443, 1987.

45. Mathews, E. A., Keller, S., and Weiner, D. B., A method to collect and process skin biopsies for cell culture from free-ranging gray whales *(Eschrichtius robustus), Mar. Mamm. Sci.*, 4(1), 1, 1988.

46. Palsbøll, P. J., Larsen, F., and Hansen, E. S., Sampling of skin biopsies from free-ranging large cetaceans in West Greenland: development of new biopsy tips and bolt designs, *Rep. Int. Whaling Comm. (Special Issue)*, 13, 71, 1991.

47. Kato, H., Hiroyama, H., Fujise, Y., and Ono, K., Preliminary report of the 1987/1988 Japanese feasibility study of the special permit proposal for Southern Hemisphere minke whales, *Rep. Int. Whaling Comm.*, 39, 235, 1989.

48. Kato, H., Fujise, Y., Yoshida, H., Nakagawa, S., Ishida, M., and Tanifuji, S., Cruise report and preliminary analysis of the 1988/89 Japanese feasibility study of the special permit proposal for Southern Hemisphere minke whales, *Rep. Int. Whaling Comm.*, 40, 289, 1990.

49. Brown, M. W., Kraus, S. D., and Gaskin, D. E., Reaction of North Atlantic right whales *(Eubalaena glacialis)* to skin biopsy sampling for genetic and pollutant analysis, *Rep. Int. Whaling Comm. (Special Issue)*, 13, 81, 1991.

50. Weinrich, M. T., Lambertsen, R. H., Baker, C. S., Schilling, M. R., and Belt, C. R., Behavioural responses of humpback whales *(Megaptera novaeangliae)* in the southern Gulf of Maine to biopsy sampling, *Rep. Int. Whaling Comm. (Special Issue)*, 13, 91, 1991.

51. Medway, W., Evaluation of the Safety and Usefulness of Techniques and Equipment Used to Obtain Biopsies from Free-Swimming Cetaceans, Report MMC-82/01 to the U. S. Marine Mammal Commission, 1983, p. 14.

52. Aguilar, A., unpublished data.

53. Whitehead, H., Gordon, J., Mathews, E., and Richard, K. R., Obtaining skin samples from living sperm whales, *Mar. Mamm. Sci.*, 6(4), 316, 1990.

54. Brown, W. R., Geraci, J. R., Hicks, B. D., St. Aubin, D. J., and Schroeder, J. P., Epidermal cell proliferation in the bottlenose dolphin *(Tursiops truncatus), Can. J. Zool.*, 61, 1587, 1983.

55. Bruce-Allen, L. J. and Geraci, J. R., Wound healing in the bottlenose dolphin *(Tursiops truncatus), Can. J. Fish. Aquat. Sci.*, 42, 216, 1985.

56. Ackman, R. G., Hingley, J. H., Eaton, C. A., Sipos, J. C., and Mitchell, E. D., Blubber fat deposition in mysticeti whales, *Can. J. Zool.*, 53, 1332, 1975.

57. Lockyer, C. H., McConnell, L. C., and Waters, T. D., The biochemical composition of fin whale blubber, *Can. J. Zool.*, 62, 2553, 1984.

58. Aguilar, A. and Borrell, A., Patterns of lipid content and stratification in the blubber of fin whales *(Balaenoptera physalus), J. Mammal.*, 71(4), 544, 1990.

59. Lockyer, C. H., Body composition of the sperm whale, *Physeter catodon*, with special reference to the possible functions of fat depots, *Rit. Fiskideildar*, 12(2), 1, 1991.

60. Woodley, T. H., Brown, M. W., Kraus, S. D., and Gaskin, D. E., Organochlorine levels in North Atlantic right whale *(Eubalaena glacialis)* blubber, *Arch. Environ. Contam. Toxicol.*, 21(1), 141, 1991.

61. Lambertsen, R. H., this volume.

62. Calambokidis, J., Chlorinated hydrocarbons in harbor porpoise from Washington, Oregon, and California: regional differences in pollutant ratios, *NMFS Southwest Fish. Cent. Admin. Rep.*, LJ-86–35C, 1986, p. 29.

63. Kawai, S., Fukushima, M., Miyazaki, N., and Tatsukawa, R., Relationship between lipid composition and organochlorine levels in the tissues of striped dolphin, *Mar. Poll. Bull.,* 19(3), 129, 1988.

64. Borrell, A., Aguilar, A., Corcuera, J., and Monzón, F., Distribution of organochlo-rines in tissues and organs of the franciscana, *Pontoporia blainvillei,* in *European Research on Cetaceans,* Evans, P. G. H., Aguilar, A., and Smeenk, C., Eds., 4, 111, 1990.

65. Lockyer, C. H., McConnell, L. C., and Waters, T. D., Body condition in terms of anatomical and biochemical assessment of body fat in North Atlantic fin and sei whales, *Can. J. Zool.,* 63, 2328, 1985.

66. Alzieu, C. and Duguy, R., Teneurs en composés organochlorés chez les Cétacés et Pinnipèdes frequentant les côtes françaises, *Oceanologica Acta,* 2(1), 107, 1979.

67. Borrell, A. unpublished data.

68. Massé, R., Martineau, D., Tremblay, L., and Béland, P., Concentrations and chro-matographic profile of DDT metabolites and polychlorobiphenyl (PCB) residues in stranded beluga whales *(Delphinapterus leucas)* from the St. Lawrence Estuary, Canada, *Arch. Environ. Contam. Toxicol.,* 15, 567, 1986.

69. Hayes, W. J., *Toxicology of Pesticides,* Williams & Wilkins, Baltimore, 1975, p. 580.

70. Matthews, H. B. and Kato, S., The metabolism and disposition of halogenated aromatics, *Ann. N.Y. Acad. Sci.,* 320, 131, 1979.

71. Miyazaki, N., Nakamura, I., Tanabe, S., and Tatsukawa, R., A stranding of *Mesoplodon stejnegeri* in the Maizuru Bay, Sea of Japan, *Sci. Rep. Whales Res. Inst. Tokyo,* 38, 91, 1987.

72. Abarnou, A., Robineau, D., and Michel, P., Contamination par les organochlorés des dauphins de Commerson des îles Kerguelen, *Oceanol. Acta,* 9(1), 19, 1986.

73. Cockcroft, V. G., De Kock, A. C., Lord, D. A., and Ross, G. J. B., Organochlorines in bottlenose dolphins *Tursiops truncatus* from the east coast of South Africa, *S. Afr. J. Mar. Sci.,* 8, 207, 1989.

74. Geraci, J. R., St Aubin, D. J., and Hicks, B. D., The epidermis of odontocetes: a view from within, in *Research on Dolphins,* Bryden, M. M. and Harrison, R., Eds., Clarendon Press, Oxford, 1986, 1.

75. Aguilar, A., Borrell, A., and Perrin, W. F., Differentiation of Population Units of Striped and Common Dolphins in the Eastern North Atlantic Ocean and Western Mediterranean Sea, Report to the U.S.-Spain Joint Committee for Scientific and Technological Cooperation, 1987, p. 34.

76. Borrell, A. and Aguilar, A., Variations in DDE percentage correlated with total DDT burden in the blubber of fin and sei whales, *Mar. Pollut. Bull.,* 18(2), 70, 1987.

77. Peakall, D., *Animal Biomarkers as Pollution Indicators,* Chapman & Hall, London, 1992, 291.

78. Goksøyr, A., Solbakken, J. E., Tarlebø, J., and Klungsøyr, J., Initial characterization of the hepatic microsomal cytochrome P-450-system of the piked whale (minke) *Balaenoptera acutorostrata, Mar. Environ. Res.,* 19, 185, 1986.

79. Goksøyr, A., Andersson, T., Forlin, L., Stenersen, J., Snowberger, E. A., Woodin, B. R., and Stegeman, J. J., Xenobiotic and steroid metabolism in adult and foetal piked (minke) whale, *Balaenoptera acutorostrata, Mar. Environ. Res.,* 24, 9, 1988.

80. Watanabe, S., Shimada, T., Nakamura, S., Nishiyama, N., Yamashita, N., Tanabe, S., and Tatsukawa, R., Specific profile of liver microsomal cytochrome P-450 in dolphin and whales, *Mar. Environ. Res.,* 27, 51, 1989.

81. Fossi, M. C., Marsili, L., Leonzio, C., Notarbartolo di Sciara, G., Zanardelli, M., and Focardi, S., The use of non-destructive biomarker in Mediterranean cetaceans: preliminary data on MFO activity in skin biopsy, *Mar. Pollut. Bull.,* 24, 459, 1993.

82. Arnason, U., Bellamy, H., Eypórsson, P., Lutley, R., Sigurjónsson, J., and Widegren, B., Conventionally stained and C-banded karyotypes of a female blue whale, *Hereditas,* 102, 251, 1985.

83. Mathews, E. A., Multiple Use of Skin Biopsies Collected from Free-Ranging Gray Whales, *Eschrichtius robustus:* Sex Chromatin Analysis, Collection and Processing for Cell Culture, Microbiological Analysis of Associated Organisms, Behavioral Responses of Whales to Biopsying, and Future Prospects for Using Biopsies in Genetic and Biochemical Studies, Master's thesis, University of California, Santa Cruz, CA, 1986, p. 118.

84. Lambertsen, R. H., Baker, C. S., Duffield, D. A., and Chamberlin-Lea, J., Cytogenetic determination of sex among individually identified humpback whales *(Megaptera novaeangliae), Can. J. Zool.,* 66, 1243, 1988.

85. Brown, M. V., Helbig, R., Boag, P. T., Gaskin, D. E., and White, B. N., Sexing beluga whales *(Delphinapterus leucas)* by means of DNA markers, *Can. J. Zool.,* 69, 1971, 1991.

86. Baker, C. S., Lambertsen, R. H., Weinrich, M. T., Calambokidis, J., Early, G., and O'Brien, S. J., Molecular genetic identification of the sex of humpback whales *(Megaptera novaeangliae), Rep. Int. Whaling Comm. (Special Issue),* 13, 105, 1991.

87. Andersen, L., and Grandl, B., unpublished data.

88. Whitehead, H. and Gordon, J., Methods of obtaining data for assessing and modelling sperm whale populations which do not depend on catches, *Rep. Int. Whaling Comm. (Special Issue),* 8, 149, 1986.

89. Gordon, J. C. D., A simple photographic technique for measuring the length of whales from boats at sea, *Rep. Int. Whaling Comm.,* 40, 581, 1990.

90. Focardi, S., Notarbartolo di Sciara, G., Venturino, C., Zanardelli, M., and Marsili, L., Subcutaneous organochlorine levels in fin whales *(Balaenoptera physalus)* from the Ligurian Sea, in *European Research on Cetaceans,* Evans, P. G. H., Ed., 5, 93, 1991.

91. Focardi, S., Marsili, L., Leonzio, C., Zanardelli, M., and Notarbartolo di Sciara, G., Organochlorines and trace elements in skin biopsies of *Balaenoptera physalus* and *Stenella coeruleoalba,* in *European Research on Cetaceans,* Evans, P. G. H., Ed., 6, 230, 1992.

SECTION EIGHT

Remarks on
Nondestructive Biomarkers

CHAPTER 12

The Rational Basis for the Use of Biomarkers as Ecotoxicological Tools

Michael H. Depledge

TABLE OF CONTENTS

0-87371-648-5/94/$0.00+$.50
© 1994 by Lewis Publishers

I. INTRODUCTION

The key problem faced by ecotoxicological scientists is to recognize the damaging effects of chemical pollutants on natural biota. On many occasions (for example, after accidental chemical releases or at sites of continuous point source discharges) damage is clearly visible,[1-3] but insidious, long-term pollutant effects are more difficult to identify. During the last 20–30 years, ecotoxicologists have been compelled to attempt to predict whether a range of new chemicals that might potentially enter the natural environment is likely to affect ecosystems adversely. If such predictions suggest that certain chemicals pose real threats, managers and legislators must then be provided with information concerning the concentrations that will not elicit effects in the hope of setting safe limits and preventing environmental degradation. Current strategies of environmental management rely on:

1. Assessment of the potential toxicity of chemicals based on their physicochemical properties and, in particular, similarity to known toxicants — This is the so-called structure-activity relationship approach (SAR and QSARs).[4]
2. Assessment of actual toxicity under controlled laboratory conditions, with a limited number and variety of test organisms — Toxicity tests principally determine lethal concentrations of the chemical following uptake (usually from water in aquatic animals or from food in terrestrial animals) and no observable effect concentrations (NOECs), although increasing efforts are being made to assess effects on life cycles, fecundity, growth, etc.[5] as well as effects on several species exposed simultaneously (multispecies toxicity tests).[6]
3. Monitoring studies — There are evaluations in which concentrations of the chemical in water, substrate, and selected biota are determined to evaluate the extent of contamination and to assess bioaccumulation and biomagnification.[7]
4. Ecological assessments — These are evaluations in which various ecological parameters (for example, species diversity, the presence or absence of key indicator organisms, nutrient cycling rates, etc.) are used to evaluate the impact of pollutants on various ecosystems.[8]

While these procedures have certainly helped in making preliminary assessments of pollutant toxicity and in monitoring the environmental fate of chemicals, they have done little to help us understand how pollutants actually perturb communities of animals and plants. When ecosystem damage is not obvious following relatively short-term exposure to pollutant concentrations that are below legislative limits (derived from NOEC values found in laboratory investigations), it is frequently assumed that insidious degradation of populations and communities will not occur in the long term. This assumption has yet to be validated. Equally important, when ecosystem degradation is found, we have little knowledge about how to restore the biotic communities to their original condition, or how to measure the success of our attempts. Thus, it is clear that current environmental protection strategies are inadequate.

To remedy this situation, a number of ecotoxicologists have looked for better ways of evaluating the status of ecosystems and their components, and of detecting the early adverse effects of pollutants on organisms in situ. One of the most promising approaches in this regard is that involving the use of biomarkers. In the following account, I will discuss some of the important factors that should be considered when designing biomarker investigations and identify some of the unresolved questions that must be answered before we can confidently interpret biomarker data. In particular, the advantages of using nondestructive biomarkers approaches will be highlighted.

II. HISTORICAL PERSPECTIVES

Before describing contemporary biomarker approaches, it is worthwhile considering how their use arose.

The idea of measuring a biological parameter as an indicator of the well-being of an organism is very old indeed. In ancient China, physicians were able to assess the health of their patients by tasting urine, by smelling feces, and by examining their patient's hands and eyes. Modern medicine has, of course, refined this approach extensively. As well as carefully noting the physical appearance of the patient and recording signs such as breath odor and facial skin coloration, measurements of heart rate, breathing rate, body temperature, blood counts, hematocrit, etc. are now commonly used to monitor the patient's health. With the development of sophisticated analytical equipment over the last 30–40 years, it has become possible, using small tissue and body fluid samples taken by nondestructive biopsy, to measure the levels of electrolytes, enzyme activity, and metabolites to more accurately determine physiological state. Conventionally, it has been these latter biochemical measurements that have been viewed as constituting the biomarker approach. Biomarkers are now used extensively in all branches of medicine, but perhaps they have found most utility in the field of oncology where markers of tumor growth are routinely used to monitor the course of a disease and the effectiveness of treatment. For example, in choriocarcinoma, the tumor cells secrete human chorionic gonadotrophin (HCG) which can be detected in the serum and urine by radioimmunoassay. Levels of HCG just above the background value indicate a tumor burden of approximately 1 million cells.[9] Other important tumor biomarkers include α-Fetoprotein and carcinoembryonic antigen (CEA).[10]

Biomarkers have also been used extensively in the evaluation of occupational health risks.[11] Here, biochemical and physiological responses are used to indicate exposures to environmental pollutants. For example, acetylcholinesterase (AChE) inhibition has been used as a biomarker of pesticide exposure in humans.[12] The strategy in these risk assessments is to detect potentially toxic exposures long before adverse effects arise.

The possibility of using biomarkers in the context of ecotoxicology has been recognized many times over the last 30–40 years, but was not explicitly formulated into a coherent approach until attempts began to clearly define biomarkers. Recent definitions and progress in this area were reviewed by Leonzio and Fossi in Chapter 13. Wherever possible, attempts are made to establish a relationship between the level of exposure to the pollutant and the magnitude of the biomarker response. It should be clearly understood that biomarkers do not provide information regarding the significance of the pollutant exposure for the well-being of the exposed organisms. However, Sanders[13] and Depledge et al.[14] have pointed to the possibility of identifying biomarkers of declining health (so-called effect biomarkers) which have much greater ecological relevance. At present, this original approach has not been widely used in ecotoxicology, but it should be noted that this represents a potentially important extension of the current biomarker concept.

When planning the practical application of biomarker studies in ecotoxicology, at least one other important new concept should be considered. Usually, biomarker responses are measured in a sample of the population under investigation. The mean response of the individuals in the sample is then used to estimate the likely mean response of the whole population and to derive a suitable estimate of the variability in response. Depledge[15] pointed out, however, that the successes of natural populations and communities are determined by the range of attributes of their component individuals. Interindividual differences within populations arising due to the various influences of genetic, ontogenetic, and environmental factors are ultimately manifest in a range of phenotypes. With regard to the survival of a population exposed to pollutant stress, the key factors that will determine the continued survival of the population are:

1. the bioavailable concentration of the pollutant to each component individual
2. the phenotypic characteristics of each individual in relation to the pollutant load it acquires
3. the proportion of individuals in the population adversely affected by pollutant exposure
4. the proportion of individuals required to sustain the population at a particular locality
5. the phenotypic composition of the group of individuals required to maintain a reproductively viable population at a particular locality.

With regard to points 1, 2, and 3, it is important to recognize that the biomarker response of the individual will depend both on the pollutant load the organism acquires and the capability of the metabolic machinery of the organism to produce a response. In other words, when comparing two individuals, different biomarker responses may indicate differences in the level of pollutant exposure they have been subject to, or that the organisms have received identical exposure, but have different capabilities to mount biomarker responses.

With regard to points 4 and 5, mean biomarker responses clearly do not provide the information we require. Supposing we measure a statistically significant 10% increase in the mean concentration of a biomarker that signals severe adverse effects in a sample of the test population. If the standard deviation attached to this value represents a small percentage of the mean, there is a high probability that 68.3% of the organisms (mean ± 1 SD) has been affected. However, should this give rise to concern? There is still considerable doubt whether all the organisms in the remaining 31.7% of the population are significantly affected by the pollutant exposure. Indeed, it might be that a small number of "resistant" members of the population are able to maintain the species at the given locality despite a decline in overall population numbers.[15] This will also be dependent on whether the resistant individuals have an appropriate composition (i.e., an appropriate proportion of sexually viable males and females, sufficient genotypic diversity, etc.) and whether their offspring are able to grow and mature at the polluted site. In formulating new biomarker approaches, it is thus vital that we move beyond measuring only mean responses to exposure or the mean effects that ensue, if we aim to interpret the ecological relevance of our data. A cautious (and in my view appropriate) course of action for environmental managers to take would be to limit pollutant concentrations to levels that do not initiate biomarker responses in more than a very small proportion of the population monitored in situ. Nevertheless, we should not assume that because a high proportion of individuals exhibits biomarker responses and is adversely affected that the population's existence at a particular locality is in jeopardy. In summary, we must try to carefully relate biomarker responses to particular degrees of ecological change. How much ecological change is acceptable is a subjective issue.

Recently, there has been discussion of whether biomarkers might be used to assess higher level biological effects. For example, can biomarkers be identified which provide an integrated view of the well-being of whole populations, or indeed, whole communities? Such approaches offer exciting challenges for the future and are discussed in greater detail later.

III. DOSE-RESPONSE RELATIONSHIPS

The cornerstone of toxicological research is the dose-response relationship, and it is therefore important to see how the biomarker approach relates to this concept.

In determining dose-response relationships, experiments are performed in which groups of organisms (or in some cases, cultured cells or excised tissues) are exposed to one of several concentrations of a test chemical, either as a known injected dose or as an exposure concentration in food or the surrounding environment (air or water).[16] Most commonly, the extent of mortality within each group is assessed after a predetermined time (usually 48 or 96 hr) and these data are used to measure the dose (or exposure concentration) that kills 50% of the test organisms (LD_{50} or LC_{50} values). There are, of course, many variations on this basic theme. For example, lethal doses that kill 25 or 5% of the experimental animals might be measured or alternative sublethal endpoints may be adopted, such as a 10% decrease in oxygen consumption or a 25% decrease in growth rate, etc.

Toxicity tests of the kind described above have a number of major limitations when applied in ecotoxicology that are not so evident in medical toxicological studies:

1. In medical toxicology, the goal is to determine toxicity thresholds that protect the most vulnerable individuals of the species (i.e., man). In ecotoxicology, emphasis is placed on protecting some of the individuals in a very wide range of species and also the interrelations among the species and their environment. Obviously, it is extremely difficult to extrapolate with confidence from laboratory toxicity test results to achieve these latter goals.
2. As pointed out earlier, toxicity tests usually determine the mean response of a laboratory-maintained test population where every effort has been made to achieve a homogeneous response, instead of examining how a very broad range of individuals with different phenotypic characteristics respond. Thus, the ecological relevance of conventional toxicity tests is very low indeed.
3. Tests are performed over short time periods and therefore do not assess long-term effects of subtle changes in competitive abilities among the test organisms, or alterations in ecosystem processes.
4. They usually (but not always) involve testing one species at a time. Consequently, any deleterious effects on the relationships among groups of species associated with pollution exposure cannot be detected.
5. They are conducted in controlled, stable conditions so that the influence of fluctuations in physicochemical conditions (that are characteristic of natural environments) on the ability of organisms to adapt to or tolerate pollutant stress is not assessed.
6. Often they do not take account of variable tolerance of organisms associated with natural biological phenomena (e.g., diurnal rhythms, estrous cycles, life stages, etc.).

The question arises therefore: is the dose-response concept as currently applied optimal for recognizing and predicting both the direct and indirect effects of pollutants on individual organisms, populations, and communities; or is some other concept more appropriate?

IV. THE MULTIPLE RESPONSE CONCEPT

An alternative basis for assessing and predicting chemical toxicity to organisms is to identify early responses to pollutant exposure at the level of individuals; and by analyzing interindividual variability in responses and effects on Darwinian

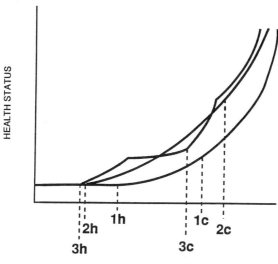

FIGURE 1. The relationship between health status and physiological condition. In the upper panel, hypothetical changes in the physiological condition of an organism are shown during progressive deterioration in health status. As the organism is exposed to increasing doses of a toxicant, compensatory physiological responses are mounted (compensation zone), which are not normally seen during homeostasis. Eventually, physiological compensation is insufficient to maintain bodily functions and pathological consequences ensue (noncompensation zone). In the lower panel, a variety of biomarker responses are illustrated which can be used to chart changes in physiological condition as the health status of the organism deteriorates. Thus, biomarkers can be used in the assessment of where an organism lies on the health status-physiological condition curve.

fitness, to establish mechanistic links with subsequent effects at the population and community level.

With regard to effects on individuals, Hatch,[17] working in the field of environmental medicine, formulated the hypothesis that a reproducible sequence of physiological changes occurs during occupational exposure to toxic agents before overt disease is detectable. This concept has been extended independently by others[18,19] and was reformulated most recently in the context of ecotoxicology by Depledge[20] and Depledge et al.[14] It is assumed that a healthy individual organism exposed to increasing pollutant loads will suffer a progressive deterioration in health which is eventually fatal (Figure 1). Early departures from health (measured on the health status scale — Figure 1) are not apparent as overt disease, but are associated with the initiation of biochemical and physiological compensatory responses (shown on the physiological condition scale). Within the compensatory zone, the survival potential of the organism may already have begun to decline because the ability of the organism to mount compensatory responses to new environmental challenges may have been compromised. If the organism has acquired a pollutant load that cannot be tolerated, detoxified, or excreted, then pathological processes will result in overt disease and finally death (Figure 1).

Nonetheless, within the noncompensatory zone, if conditions improve sufficiently, an organism may still be able to recover as repair mechanisms continue working to restore compensatory responses. In other words, an organism can return from a diseased to a healthy state.[20]

It is within the framework of this so-called multiple response concept that the use of biomarkers might be developed and extended in the future. Biomarkers raise the possibility of determining where an organism lies on the health status curve (Figure 1); and thus potentially offer an early warning of when pollutant exposure induces early, reversible departures from homeostasis or from the compensatory zone. This is illustrated in Figure 1 (lower panel) where a number of hypothetical biomarker responses are shown in relation to the health status curve (upper panel). Biomarker B1 might be a biochemical or physiological response (or even a behavioral response) that more or less follows the pattern of the health status curve. The greater the B1 response, the further the organism has progressed along the health status curve. To be useful, the biomarker response must be reproducible and quantifiable so that it can be related specifically to a certain degree of physiological disability. Biomarker B2 illustrates a transitory response. If such a response is detected, this signals that the organism is within the compensatory zone (see Figure 1). More precise determination of the position of an organism on the health status curve requires serial sampling; otherwise it is impossible to establish whether a particular value of B2 is on the rising or falling part of the biomarker response curve. A biomarker response as illustrated by B3 might be used to determine the location of an organism within the compensatory zone, and flattening of the curve indicates the loss of compensatory responses (Figure 1). Within the noncompensatory zone, however, B3 conveys little additional information. Biomarker B4 appears transiently and is a clear signal that the organism is close to the limit of compensation; however, in this case, the response is likely to be so transient that it may be detectable in only a small proportion of a group of organisms at any one time. Finally, biomarker B5 makes two appearances as physiological condition progressively deteriorates. The small transitory response within the early part of the compensatory zone is analogous to the response of B2. The reappearance of the B5 response in the noncompensatory zone signals something quite different, namely, that the organism has suffered irreparable damage. How then can these two responses be distinguished from one another? First, a distinction might be made concerning the magnitude of response. The early B5 response is never greater than 30% of the late B5 response. Thus, any response above the 30% level must signify that the organism is in the noncompensatory zone. Second, when the late B5 response occurs, it is evident that biomarker B3 will also be measurable; while in the early B5 response, a B3 response cannot be detected. This simultaneous use of more than one biomarker to help to determine where an organism lies on the health status curve illustrates an important principle, namely, that suites of bio markers offer greater possibilities for the detection of effects than do single biomarkers alone.

Returning to the dose-response concept for a moment, it is vital that biomarkers that have been shown to reflect the well-being of organisms in accordance with the multiple response concept, should also be related as accurately as possible to specific degrees of pollutant exposure. This serves to illustrate the complementary nature of the dose-response and multiple response concepts. Note that the dose-response concept usually focuses on the mean response of populations (of organisms or sometimes cells) whereas the multiple response concept tends to focus on responses at the individual level and below.

V. APPLICATION OF THE BIOMARKER APPROACH

The general concept outlined above offers a rational basis for the use of biomarkers, but it is emphasized that few experiments have yet been conducted to really substantiate its validity. The shape of the Health-Status curve is unknown. In any case, it is likely to vary from one organism to another, and more especially, among different species. The shape of the curve will reflect the extent to which biochemical/physiological compensatory mechanisms can protect an organism before overt signs of pathological damage and disease appear. In some species, these early responses may be very limited and difficult to detect, while in others a broad range of responses may be exhibited before the individuals approach the limit of compensation. Organisms of the latter type are most suitable for inclusion in biomarker studies.

With regard to the shape of the curve, it may be smooth for some organisms (as is shown in Figure 1), but could equally well be a stepwise curve in which an organism can compensate for pollutant disturbance up to a point and then slips into a new physiological status rather suddenly (Figure 2). Such events may be repeated several times during progression to the limit of compensation. Clearly, curves with other shapes can also be envisaged.

The rate at which an organism progresses along the curve is very important. This point is considered again later in a detailed discussion of the time scales of progression from early, sublethal responses to eventual mortality.

Peakall[21] finds it difficult to see how biomarker investigations might be applied to evaluate the adverse effects of mixtures of pollutants, although he does concede that the multiple response concept is scientifically sound and applicable for single toxicants. There is, however, no obvious reason why the "limit of compensation" should not be determined for any mixture of pollutants. Indeed, it is one of the strengths of the approach that both the "limit of homeostasis" and the "limit of compensation" are reached as a result of the integrated impact of pollutants, and other biotic and abiotic factors on the well-being of the organism. It seems reasonable to hypothesize that simultaneous responses to pollutants will impose added physiological and biochemical energetic costs on the responding organism. This may in some circumstances accelerate progression toward the limit of compensation. For example, an organism simultaneously exposed to toxic concentrations of a trace metal and an organic pollutant will have to mount a range of detoxification responses, such as synthesis of metal-binding proteins (metallothioneins) and mixed function oxidases to render the organic compound

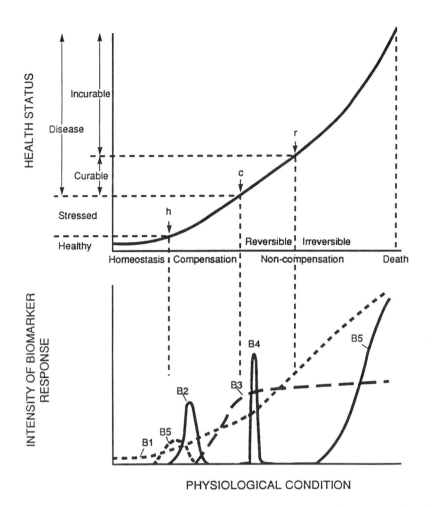

FIGURE 2. Examples of hypothetical health status-physiological condition curves. Experiments are required to ascertain the precise relationship between the onset of overt disease and underlying alterations in physiology and pathology. The curves illustrate interorganismal differences in the above relationship. In particular, differences in the limit to homeostasis (h) and the limit to compensation (c) are evident. The shape of the curve, whether smooth or irregular, has yet to be determined through experimentation.

harmless (see Reference 22). Other situations might be envisaged in which simultaneous exposure to pollutants may ameliorate the damage that would be expected from a single exposure. Thus, organochlorine insecticides have been shown to protect against the acute toxicity of several organophosphorus insecticides. Apparently, stimulation of liver microsomal enzymes by organochlorines results in enhanced biotransformation of organophosphorus compounds and increases the availability of noncatalytic binding sites.[23]

Peakall[21] also expresses the view that even if the limits of homeostasis and compensation can be established, they will be of little use in determining when to initiate remedial action to reduce pollutant levels in the natural environment. On the contrary, if biomarker responses can be monitored in a number of key components of ecosystems and confirm that the physiological condition (and perhaps Darwinian fitness) of a significant proportion of those organisms sampled is impaired due to exposure to specific pollutants (either singly or in combination), this should be viewed as a clear signal that remedial action is required. Of course, such issues might be obscured, as currently occurs, by environmental administrators who may be compelled to apply subjective criteria in environmental management. Thus, even though populations and communities of organisms can be shown to be degraded by pollutant exposure, socioeconomic and political factors may influence decision making so that chemical discharges are still permitted. The cost of these short-term expedient policies in terms of lasting environmental damage is still unassessable, but the use of biomarkers may provide a means of keeping track of the progress of ecological change.

Another attractive feature of the biomarker approach is that it may be useful for assessing the effects of repeated exposures to single pollutants or mixtures.[24] At present, this is seldom attempted in toxicity tests because they focus primarily on lethality. At least for many invertebrate species, more than 95% of the survivors of an exposure to an LC_{50} (96 hr) pollutant concentration die during the following 2 weeks when returned to clean conditions (e.g., for *Gammarus* and *Carcinus* exposed to chromium and mercury, respectively (Depledge, unpublished). Information regarding survivorship of organisms following repeated exposure to pollutants is scanty. Greater efforts are required to provide information in this area.

Finally, biomarkers may in the future serve a role in bioremediation — the active restoration of damaged ecosystems. The success of cleanup procedures and the effectiveness of recolonization by various organisms should be evident in the return of biomarker values to within predetermined normal ranges. The use of biomarkers in this way is analogous to their use in medicine to chart the effectiveness of treatments for various diseases and the return of the patient to health.

VI. BIOMARKERS, DARWINIAN FITNESS, AND ENERGETIC COSTS

Biomarkers can be identified that indicate organisms have been exposed to stress (general biomarkers), specific categories of stress (natural or related to pollutant exposure), and indeed specific types of stressors (metallothionein responses following metal exposure, acetylcholinesterase inhibition following exposure to certain organohalogen pesticides, etc.). It is apparent that while signaling that an exposure has taken place, these biomarkers contribute little to the prediction of consequences of the exposure for the organisms in question. For this to be true, we must be able to relate a biomarker response directly to a change in Darwinian fitness. In other words, a particular biomarker response should — if

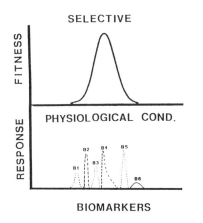
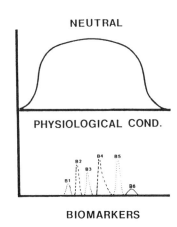

FIGURE 3. Changes in physiological condition in relation to Darwinian fitness. In the left panel, changes in physiological condition are associated with changes in Darwinian fitness. Biomarker responses signifying differences in physiological condition can therefore be used to chart changes in fitness. In the right panel, altered physiological condition and biomarker responses are not associated with significant changes in fitness. Under these circumstances, biomarkers tell us little about ecologically significant effects, but may still be useful for signaling exposure to pollutants. h = the point at which departure from the normal homeostatic response range is initiated. c = the limit at which compensatory responses can prevent development of overt disease. r = the limit beyond which pathological damage is irreversible by repair mechanisms.

possible — be related to a given degree of impairment of growth, or reproductive output or energy utilization which directly affects the survivorship of an organism. If the degree of biomarker response can be simply related to the extent of reduction in fitness, then so much the better. Examples of two extreme cases in which biomarker responses either indicate or fail to indicate changes in Darwinian fitness are shown in Figure 3.

Many of the biomarkers that are — or might be — measured are biochemicals that are intimately involved in the handling, biotransformation, and/or excretion of the pollutants that provoke the response. Metallothionein concentration can be used as a biomarker of exposure to particular trace metals because this protein is often induced by such exposure, and is involved in the binding and hence detoxification of free metal ions within cells (see Reference 25). Similar detoxification roles are attributable to mixed function oxidase (MFO) systems, stress protein responses, etc. that have also been used as exposure biomarkers.[13,26] I will refer to biomarkers that play a part in the detoxification or pollutant tolerance response as "functional biomarkers." Some have argued that the expenditure of energy on such responses inevitably deprives normal metabolic systems of some of their normal energy supply.[27] Consequently, biomarker responses which involve synthesis of biochemicals involved in detoxification may utilize energy that would otherwise be spent on growth and reproduction, and may directly indicate impairment of Darwinian fitness. There are, however, numerous unresolved issues arising from such a hypothesis. At present, there is little information available concerning the actual energetic costs involved in, for example, the synthesis of metallothionein or stress proteins, or in enhanced MFO activity. In humans, the costs of such protein synthesis is thought to be very small indeed compared with maintenance costs (Garby, personnal communication). Thus, it may be argued that such costs are trivial and can be met from savings in other homeostatic reactions, without significantly impairing growth or reproductive output or placing additional energy acquisition burdens on the organism. Impaired growth and reproduction in organisms exposed to pollutants might, in some cases, be a direct consequence of energy deprivation in normal metabolic pathways as energy is diverted into pollutant detoxification and excretion. However, it may also indicate that detoxification systems are not sufficiently effective to prevent direct interference with, or impairment of, normal metabolic pathways by pollutant molecules that have not been eliminated or rendered nontoxic (see Reference 28). The induction of some detoxification responses may even constrain other normal biochemical and physiological responses to environmental fluctuations with adverse consequences.

Related to this issue is the question of "trade-offs." Is it more advantageous for an organism to utilize energy on protecting itself against pollutant toxicity, or to channel energy into growth and reproduction in the hope of establishing the next generation? It is possible to envisage a vast range of scenarios in which one or the other alternative is most appropriate. Thus, with short-term pollutant exposure applied to slow growing, viviparous organisms that produce only a few offspring, it is clearly advantageous to expend energy on tolerating the pollutant stress until conditions improve. Production of a few offspring in such conditions is not likely to meet with great success because juvenile organisms are usually more susceptible to pollutants than are adults. Alternatively, organisms with rapid generation times that produce large numbers of offspring that are then widely dispersed might be better off channeling energy into growth and reproduction instead of pollutant detoxification systems. Production of offspring on time (in relation to seasonal cycles of food availability, etc.) may result in high mortality due to pollutant toxicity; however, if some offspring survive to encounter cleaner environments, then the population may be preserved. In summary, some of the key factors determining which strategy is most appropriate are the rates of growth, numbers of offspring, dispersal methods, costs of pollutant detoxification, importance of producing offspring at appropriate times in relation to particular seasonal cycles (e.g., in food availability, predatory pressure, etc.), and relationship between the duration of pollutant exposure and the generation time of the organism.

The above factors are also highly relevant to the design of biomarker monitoring strategies involving the measurement of biochemicals and physiological processes that are directly involved in the responses to pollutant exposure. This can be illustrated with an example. Fossi et al.[29] performed an interesting study on two populations of seabirds *(Larus ridibundus)*, one feeding in a relatively unpolluted coastal lagoon, and the other on a municipal rubbish dump in Tuscany, Italy. MFO activity was markedly raised in birds feeding at the rubbish dump, and this was subsequently related to pesticide and organic pollutant exposure. Bearing in mind this discussion, one might be tempted to conclude that because of the additional energetic costs of pollutant handling and detoxification, the "pollutant-exposed" birds would grow less rapidly and might have less reproductive success. In fact, this was not the case because the rubbish dumps provided a very rich food supply which apparently more than compensated for the additional cost of handling pollutants. This case illustrates how carefully one must select biomarkers and how prediction of adverse effects is intimately linked to assessment of Darwinian fitness.

VII. THE PROBLEM OF LINKING SUBLETHAL AND LETHAL RESPONSES

So far, a rationale has been described based on the assumption that varying degrees of sublethal biochemical and physiological changes precede pathological changes associated with toxicity or disease, and that these latter changes precede death. In other words, there is a particular sequence in which the deterioration in the condition of an organism proceeds, and there is also a reasonably consistent time schedule for the progression of the events. While there is considerable evidence to support the first assumption,[14] the second assumption is more difficult to support because the time scales separating sublethal and lethal events may be highly variable. For example, a biomarker that is indicative of chemically induced carcinogenesis may be detectable many years before a malignant disease ensues, and this may itself be a rare event of little ecological significance. At the opposite extreme, biochemical changes associated with pollutant exposure are sometimes followed almost immediately by physiological changes and the rapid onset of morbidity and mortality, perhaps within a matter of minutes or hours.[30] In these circumstances, measurements of biomarker responses are of little practical value. It is therefore desirable that biomarker responses indicative of sublethal effects should be detectable sufficiently in advance of pathological events that lead to death, but not so far in advance as to decouple the early response from the late effect. Furthermore, the biomarker response should be detectable at significantly lower doses (or exposure concentrations) than are associated with eventual, toxicity-related mortality. This latter point is also difficult to address because low-level pollutant exposures that are reflected in biomarker responses can often be toler-ated for the life time of the organism without significant adverse, population level effects. Indeed, they will be viewed as false alarms by some, or too sensitive indicators of potential problems to be practical by others. Unless these exposure/adverse effect time separations are achieved, the biomarker approach will have limited practical utility for recognizing or predicting ecologically significant events. It should, however, be noted that in an ecological context biomarkers of sublethal impairment may be just as relevant as those which predict death at higher pollutant exposure levels. Certain pollutants may be directly toxic to organisms at concentrations that are never likely to occur in situ. Nonetheless, they may cause a small degree of physiological impairment at the concentrations occurring at polluted localities. Thus, while there is little risk of mortality directly associated with pollutant exposure, the pollutant may well interfere with population dynamics and interspecies competition to such an extent that ecological change ensues. Moriarty[31] puts this case rather forcefully and concludes that "the fact that a pollutant kills, say, half of the individuals in a species population may be of little or no ecological significance, whereas a pollutant that kills no organisms, but retards development may have a considerable ecological impact." Biomarker monitoring strategies are required which can illuminate such differences.

VIII. PERSISTENCE OF BIOMARKER RESPONSES

Many biomarker responses are transient. An exposure to a pollutant may elicit a response that lasts for a matter of hours. However, some biomarker responses persist for weeks or months with continued exposure, and others may be detectable throughout the lifetime of the organism. It is therefore vital to establish in advance what duration of response can be expected. We must have confidence that the absence of a biomarker response among the individuals of a population indicates that the population is not exposed to or affected by pollutant exposure. If a biomarker response occurrs but then recedes despite continued exposure or adverse effects, we may be mislead into believing that the population is unaffected.

Another problem with regard to the timing of biomarker investigations concerns pollution-induced selection. When a population of organisms is first exposed to a pollutant, a range of biomarker responses will be exhibited depending on the level of exposure that each individual receives and the ability of each organism to respond. We might expect a rapid rise in the number of individuals responding, and the occurrence of a more or less "normal distribution" of responses. A few individuals will exhibit a very strong response and a few, a rather weak response. For the majority of individuals the response will be between these extremes. However, if continued pollutant exposure results in deletion of susceptible individuals from the population over time, we might expect to find a change in the distribution of biomarker responses. If strong responders have received the greatest exposure, then they may be lost first so that the mean biomarker response may appear to return to a lower level at a time when damage is most severe. Of course, many other scenarios might be envisaged. The point is that we need to consider such effects when attempting to interpret biomarker responses. Interindividual variability in biomarker responses is discussed further below.

IX. INTERINDIVIDUAL VARIABILITY IN RESPONSE

One of the most difficult factors to deal with in making ecological predictions based on biomarker analyses is to take account of interindividual differences in response within the population of indicator organisms. In many cases, among the individuals tested, some will exhibit a weak biomarker response while others will exhibit a very marked response. This may indicate that the test population has not been homogeneously exposed to the stressor as was noted above. In natural ecosystems this is very likely because changes in physicochemical and biological characteristics often vary over very small distances. Such heterogeneity will then be reflected in small-scale differences of the bioavailability of pollutants. An equally important factor may be the inherent differences in the morphology and biochemical/physiological status of exposed organisms. Size, age, genotype, ontogenetic stage, environmental conditioning, etc. are all important sources of interindividual differences in response.

X. ESTABLISHING NORMAL VALUES FOR BIOMARKERS

To be confident in assessments based on biomarker measurements it is absolutely vital that analytical procedures should be reliable and reproducible, and that the range of inherent variability in the biomarker measurements in healthy organisms is well known. Precise quantification of biochemical biomarkers in easily accessible body fluids and tissue samples or of physiological and behavioral biomarkers, and establishment of "normal" reference values are integral parts of the biomarker approach. This database can then be used in detection and monitoring of exposure to pollutants and in screening for toxicity and disease. Much diagnostic weight may rest on single determinations and on patterns of biomarker response as is the case in medical science.[10] While it is important to quantify biological variations among healthy organisms, it is also essential to define inherent variability in the laboratory method and to recognize the errors of sampling that influence the results of most determinations. In analyzing biomarker test results, the assumption that biological data conform to a Gaussian distribution may be inappropriate. Most biological data are not symmetrically distributed and require statistical tools that are based on other distributions or that are independent of distribution form (see for example, Reference 32). The "normal range" of biomarker values might best be defined as the mean ± 2 SD. This is, however, not the most common method applied when using biomarkers in a medical context. The values most ordinarily quoted are in fact ranges in a skew distribution calculated to include 95% of the values obtained from a healthy population. There has not been an adequate discussion of these matters in the context of ecotoxicology; but whatever criteria are used, the normal range is confounded by both intra- and interindividual, and irreducible analytical error. More elaborate statistical handling of biomarker data can be performed for individual tests, but is not necessarily required for concluding that pollutant exposure has occurred since "suites of biomarkers" can be used to signal abnormalities as is evident in altered relationships among the biomarkers rather than focusing on absolute values. By way of warning it should be noted that if the criterion for healthy organisms is that a biomarker test result falls within the 95% confidence limits for "normals," then when multiple biomarker tests are conducted on any individual, surprising findings may emerge. For example, Giles and Ross[10] point out that when 12 biomarker tests are conducted simultaneously, only 50% of the normal individuals will be found to be "healthy" (i.e., will have biomarker responses within the normal range in all 12 tests).

XI. THE SPECIAL CASE OF NONDESTRUCTIVE BIOMARKERS

The foregoing account addresses general questions related to the use of all types of biomarkers. However, there are special advantages in making measurements on tissue and body fluid samples that have been obtained nondestructively.

Biomarker test procedures are often limited by the size of the tissue sample available for analysis or — if physiological measurements are being considered — by the size of the organism to which transducers can be attached. With regard to invertebrates, their small size often renders measurements on individuals difficult and usually requires destruction of the whole organism (and perhaps even pooling of several organisms) when biochemical analyses are required. However, such organisms are usually readily obtained and are often plentiful so that destructive techniques, although not desirable, offer a pragmatic solution without affecting in situ population numbers significantly. The same is true of some vertebrate species. However, many of the larger vertebrates present the possibility of obtaining small tissue or body fluid samples without significantly damaging or affecting the well-being of the whole animal. The advantages of this nondestructive approach are:

1. Nondestructive sampling allows repeated measurements to be made on the same individual (providing that sampling frequency is not so high as to affect the biomarker response itself). Such repeated measurements may allow an individual to be used as its own control.
2. The persistence of biomarker responses in relation to pollutant exposure can be more accurately assessed.
3. Nondestructive sampling does not involve loss of animals from the population. This is especially important if a rare or threatened species is being investigated.
4. Long-term studies may be conducted on individuals as they pass through different ontogenetic stages.
5. Nondestructive studies facilitate the detection of interindividual variability in response to pollutant exposure.

Blood, urine, feces, semen, hair, feathers, scales, claws, tissue, and organ biopsies may be suitable for a broad range of biochemical biomarker measurements. Nondestructive physiological biomarkers include oxygen consumption, ventilation, cardiac activity (and the relationships among them), urine output, feces output, body temperature, growth, and reproductive output.

XII. BIOMARKER HIERARCHIES

Until relatively recently, most emphasis in the biomarker approach has been focused on biochemical biomarkers, that is, alterations in the chemical composition of tissues and body fluids. However, Depledge et al,[14] proposed that biomarkers of pollutant exposure and effects might be recognized at all levels of biological organization, ranging from the molecular level to the ecological community level. Previously, McCarthy and Shugart[33] made the point that biomarkers at the different levels have inherent strengths and weaknesses. Thus, the molecular biomarkers are valuable because pollutants initiate adverse effects by altering molecular components of the cell leading to adverse effects on cellular metabolism. The extent of damage can be related in a fairly precise manner to the "dose" of pollutant. The major drawback of biomarkers at this level is that it is difficult to determine the significance of varying degrees of pollutant damage in cells for the physiological function of the tissue in which the cells reside and for whole animal physiology. Taking this argument further, to predict consequences for individuals, populations, and community by extrapolating from cellular biomarker responses becomes an almost impossible proposition. Conversely, we might measure biomarkers of population well-being, such as composition (proportions of males and females, juveniles, mature and senescent adults, etc.), turnover, and generation times, etc. Unfortunately though, a host of natural abiotic and biotic factors affect these indices; and it is usually impossible to relate exposure to pollutants to specific effects or to establish dose-response relationships.

New strategies should be employed to overcome these difficulties. The first involves moving away from scientifically unsound extrapolations from effects at one level of biological organization to predict consequences at another level. Generally, such extrapolations rely on statistical correlations of specific levels of pollutant exposure with certain consequences for populations and communities. The old adage "correlation is not causation" should be remembered here since there are numerous examples of pollutant exposures being linked to alterations in ecosystems — only to find later that other, more important factors actually gave rise to the observed ecological change. For example, Schindler[34] discussed an incident in which a lake (in the Experimental Lakes Area of Canada) was artificially acidified and changes in selected ecological parameters were subsequently followed for several years. Alterations in lake transparency were initially attributed to acid effects, but later analysis revealed that much of the observed change was due to unusually low rainfall in the area over the experimental period (see References 34 and 35 for a fuller discussion of this case). To avoid similar misinterpretations, we must try to establish the mechanistic links by which pollutant effects at one level of organization give rise to higher level effects. Demonstrating that, for example, cellular stress protein responses in specific tissues are associated with specific physiological consequences that result in altered growth and/or reproduction of a known proportion of a pollutant exposed population, might enable prediction of population (and perhaps community) level consequences. In this context, it is interesting to note that it is implicit in our approaches that we can predict high level (population and community) responses from measured low level (cellular, physiological, and individual) responses, but not vice versa.

A second hierarchical approach that might be envisaged hinges primarily on the specificity of different biomarkers. As was mentioned earlier, a natural hierarchy of biomarkers might be constructed ranging from those which are nonspecific (for example, behavioral changes associated with detection of unfavorable environmental conditions whatever they may be) to the very specific (for example, a highly specific stress protein response at the cellular level following exposure to a particular xenobiotic). This should be considered during the design of biomarker monitoring strategies. In many cases, it may not be cost-effective or worthwhile to perform a very wide range of highly specific biomarker tests, if a sensitive biomarker of generalized stress or fitness impairment fails to signal that something is amiss. Nonetheless, when general problems are identified, their solution may depend on establishing which environmental contaminant (or combination of contaminants) is responsible for the adverse effects and what the concentration is. Otherwise, environmental management agencies will be unable to make rational decisions as to which chemical discharges to limit and to what extent.

In summary, hiercharchies can be contructed based on the level of biological organization which is being monitored (cellular, physiological, individual, population, community) or based on different degrees of specificity of response (ranging from signals of general stress to those relating exposure to a specific chemical to a particular degree of a certain biomarker response). In practice, it will probably be found that these two hierarchies in fact operate in parallel. Nonspecific indicators of stress will probably be found at the population level and above, while specific biomarkers are more likely to be recognized at the physiological level and below. Finding the exceptions to these general rules should prove to be most interesting.

XIII. BIOMARKERS OF STRESS OR ADAPTATION?

As was noted earlier, many biomarker responses are integral parts of the detoxification mechanisms dealing with pollutants that have been taken into the tissues. The question then arises: "when is a biochemical/physiological response to pollutant exposure to be regarded as an adaptation to or a reflection of stress?" Regrettably, this question is usually ignored.

With regard to biochemical/physiological adaptation, most organisms are capable of responding to changes in their environment to ensure that they continue functioning normally. Examples include behavioral and physiological responses to fluctuations in environmental temperature, food and water availability, respiratory gas composition, etc.[36] Depending on the species being investigated, it may be that the absence of a functional biomarker response (or a weak response) might signify minimal exposure to a pollutant; but it might also indicate that although the organism is threatened by a pollutant, various constraints (genetic and/or environmental) influence responses. For example, a number of studies have been performed in which organic pollutant exposure has been assessed in fish using the degree of induction of the MFO system as a biomarker.[26] Such an approach may be satisfactory for crudely monitoring the bioavailability of the pollutant. However, the question arises, does MFO induction signify that the fish is adversely affected (stressed) or that it has dealt with and adapted to the pollutant exposure? Can we be sure that among the individuals studied, those exhibiting the highest degree of MFO induction would be the most stressed? An alternative explanation might be that a high degree of MFO activity indicates that the organisms have very effectively dealt with the pollutant threat through biotransformation and excretion. A low level of induction may indicate that organisms are unable to mount an adequate detoxification response and are really threatened. The vital lesson here is that we must know something of the strategy of the species with regard to pollutant handling, the repertoires of responses of its component individuals, and the significance of the responses in relation to Darwinian fitness parameters before we can implement proper "effect biomarker" studies.

XIV. BIOMARKERS OF LATENT EFFECTS

So far, discussions have centered on "exposure" and "effect" biomarkers; however, there is one other area that deserves special attention, namely, that concerning changes in the capacity of organisms to adapt to future environmental challenges. Situations may be envisaged in which organisms have suffered a transient exposure to a pollutant, but then returned to a favorable environment where they apparently recover completely. Thus, although both exposure and effect biomarkers may have been induced by the pollutant, they return to normal values. The question is, are the organisms that received the transient exposure compromised in any way? Is their ability to respond to future environmental challenges reduced compared to before the pollutant exposure? This problem has been largely ignored although there is evidence from humans and lower vertebrate and invertebrate biomarker studies that it is important. For example, the World Health Organization[36] has identified the need to protect against decrements in homeostasis or functional capacity; in other words, chemical exposures that might not give rise to overt signs and symptoms, but over the course of time may result in lowering of the bodies defenses to infections, impaired athletic performance, slower recovery from injuries, etc. Grandjean[11] cites a case in which workers who were exposed to lead as a result of their occupation had normal hematological findings and appeared healthy. Nonetheless, their reserve capacity for replacing blood after donation was markedly reduced. Similarly, Hansen et al.[37] found that when trout, *Oncorhyncus mykiss,* were transiently exposed to copper in brackish seawater, recovery over the following weeks was apparently complete. No major abnormalities in biochemical or physiological indices could be detected. However, transfer of copper-exposed trout to dilute seawater was accompanied by leakage of sodium out of the animals, posing a threat to their continued survival. Trout that had not been exposed to copper and whose osmoregulatory systems remained intact were able to adapt completely following such transfer. This example demonstrates that while copper exposure resulted in changes in the gills of trout in brackish water, no significant adverse effects were seen. Nonetheless, the reserve capacity of trout to adapt to osmotic stress was severely compromised. Biomarkers which illuminate latent effects of the type described above would clearly be very valuable.

Among invertebrate organisms, the concepts of "scope for activity" and "scope for growth" have been widely used.[38] In some ways, these may also be regarded as "latent effect biomarkers" since organisms may survive perfectly well with a slightly reduced scope for activity or scope for growth as is evident from the wide range of scope values found in natural populations.[15] The question arises, are individuals with relatively low scope values (with reduced energy stores and presumably a reduced reserve capacity to adapt) more vulnerable to pollutant exposure than are individuals with a high reserve capacity? These issues require detailed investigation before further progress can be made in this area.

XV. A NEW DEFINITION AND CLASSIFICATION OF BIOMARKERS

With the additional uses of biomarkers proposed here it is necessary to extend the definition proposed by the National Academy of Sciences.[39] This may be inappropriate with regard to the use of biomarkers in a medical context and thus it is proposed that the term "ecotoxicological biomarkers" be adopted to convey this difference.

An ecotoxicological biomarker is a biochemical, cellular, physiological, or behavioral variation that can be measured in tissue or body fluid samples or at the level of whole organisms (either individuals or populations) that provides evidence of exposure to and/or effects of one or more chemical pollutants (and/or radiations).

It may also be helpful to develop a working terminology to facilitate biomarker discussions. Thus, ecotoxicological biomarkers may be of four classes:

Class 1. Exposure biomarkers — which signal exposure of an organism, a population, or a community to chemical pollutants — may range from generalized indicators of pollutant stress to specific indicators of exposure to a certain amount of one particular pollutant. (Examples are metallothionein induction and elevated 7-ethoxyresorufin-O-deethylase (EROD) activity following exposure to metals and organohalogen compounds, respectively.)

Class 2. Effect biomarkers — which signal that an organism, a population, or a community has been affected (usually adversely) by one or more pollutants — do not necessarily provide information concerning the nature of the pollutant stress to which the organism(s) was exposed. (An example is altered gill Na, K-ATPase activity as a measure of osmoregulatory dysfunction in fish.[40])

Class 3. Exposure/effect biomarkers not only indicate that an organism, population, or community has been exposed to one or more pollutants, but specifically link the exposure to an effect. A range of possibilities exists with regard to the degree of pollutant specificity and exposure level that can be determined with such biomarkers. (Examples are acetylcholinesterase inhibition in mammals and birds following exposure to carbamate and organophosphate insecticides, and eggshell thinning in birds following exposure to dichlorodiphenyl-trichlorethane (DDT) and its derivatives.[21])

Class 4. Latent effect biomarkers indicate that apparently normal organisms have been exposed to a pollutant which in other circumstances may limit the ability of the organisms to adapt or survive. (An example is altered Scope-for-Growth.[40])

Finally, the term functional biomarker has been introduced earlier in this chapter, to identify biomarkers that are integral components of the compensatory response or detoxification mechanism initiated by pollutant exposure. Examples include stress protein responses, metallothionein induction, MFO induction, etc. These contrast with nonfunctional biomarkers which are biochemical, physiological, or behavioral changes that signal pollutant exposure and/or effects, but do not contribute to the amelioration of potentially toxic effects. Examples include acetylcholinesterase inhibition; reductions in blood cell counts; appearance of protein in the urine; and other signs of pathological damage, impaired physiology, or altered behavior.

XVI. SUMMARY

Current ecotoxicological test procedures focus on ranking the relative toxicities of chemicals that might (or have) gained entry into the natural environment. No observable effect levels are established with a few test species in controlled laboratory conditions. By determining the fate of chemicals in ecosystems and their concentrations in water, air, soils, sediments, and biota, predictions are then made — based on extrapolations from toxicity tests — as to whether adverse effects are likely to occur. Legislation and management policy are then directed at ensuring that environmental levels of chemicals are maintained below critical thresholds that might give rise to toxicity. The validity of these procedures is constantly being questioned.

A more direct approach is to focus on the well-being of individual organisms, populations, and communities in situ. Nondestructive techniques are especially suitable in this regard. The biomarker approach aims to provide signals that organisms have been exposed to pollutants and also information regarding the adverse effects of pollutants on the organisms. Attempts are being made to refine biomarker tests so as to quantify degrees of exposure and severity of damage to organisms. By establishing the mechanistic links between exposure, effects, and biomarker responses, there is clearly the possibility of using biomarkers prospectively. Thus, not only will we be able to assess the well-being of organisms at any moment in time by nondestructive sampling in situ, but also we may be able to predict what will happen to them and over what time scale, if they suffer continued exposure to a pollutant. We may also be able to assess their chances of recovery if returned to clean conditions. This direct approach, placing the biota we are trying to protect at the center of our studies, offers great promise for future environmental management.

REFERENCES

1. Ramade, F., *Ecotoxicology,* John Wiley & Sons, Chichester, 1987, 262 p.
2. Howarth, R. W., Determining the effects of oil pollution in marine ecosystems, in *Ecotoxicology: Problems and Approaches,* Levin, S. A., Harwell, M. A., Kelly, J. R., and Kendall, K. D., Eds., Springer-Verlag, New York, 1989, p. 69.
3. Dingledine, J. V. and Jaber, M. J., A review of current field methods to assess the effects of pesticides on wildlife, in *Pesticide Effects on Terrestrial Wildlife,* Somerville, L. and Walker, C. H., Taylor and Francis, London, 1990, p. 241.
4. Kenaga, E. E., Correlation of bioconcentration factors of chemicals in aquatic and terrestrial organisms with their physical and chemical properties, *Environ. Sci. Technol.,* 14, 553, 1980.
5. Boudou, A. and Ribyere, F., *Aquatic Toxicology: Fundamental Concepts and Methodologies,* Vol. 1, CRC Press, Boca Raton, FL, 1989.
6. Cairns, J., Jr., Ed., *Multispecies Toxicity Testing,* SETAC Special Publication Series, Pergamon Press, Oxford, 1985, p. 261.
7. Samiullah, Y., *Biological Monitoring of Environmental Contaminants,* GEMS-Monitoring and Assessment Research Centre, Technical Report No. 37, Kings College, London, 1990, 767 p.
8. Ford, J., The effects of chemical stress on aquatic species composition and community structure, in *Ecotoxicology: Problems and Approaches,* Levin, S. A., Harwell, M. A., Kelly, J. R., and Kendall, K. D., Eds., Springer-Verlag, New York, 1989, p. 99.
9. Galton, D. A. G., Medical aspects of neoplasia, in *The Oxford Textbook of Medicine,* Weatherall, D. J., Ledingham, J. G. G., and Warrell, D. A., Eds., Oxford University Press, Oxford, 1984, chap. 4, p. 79.
10. Giles, A. M. and Ross, B. D., Normal or reference values for biochemical data, in *The Oxford Textbook of Medicine,* Weatherall, D. J., Ledingham, J. G. G., and Warrell, D. A., Eds., Oxford University Press, Oxford, 1984, chap. 27, p. 3.
11. Grandjean, P., Effects on reserve capacity: significance for exposure limits, *Sci. Total Environ.,* 101, 25, 1991.
12. Brock, A. and Brock, V., Plasma cholinesterase activity in a healthy population group with no occupational exposure to known cholinesterase inhibitors: relative influence of some factors related to normal inter- and intraindividual variations, *Scand. J. Clin. Lab. Invest.,* 50, 401, 1990.
13. Sanders, B., Stress proteins: potential as multitiered biomarkers, in *Biomarkers of Environmental Contamination,* McCarthy, J. F. and Shugart, L. R., Eds., Lewis Publishers, Boca Raton, FL, 1990, p. 165.
14. Depledge, M. H., Amaral-Medes, J. J., Daniel, B., Halbrook, R., Kloepper-Sams, P., Moore, M., and Peakall, D. B., The conceptual basis of the biomarker approach, in *Biomarkers: Research and Application in the Assessment of Environmental Health,* Peakall, D. D, and Shugart, L. R., Eds., NATO ASI Series H: Cell Biology, Vol. 68, Springer-Verlag, Berlin, p. 15.

15. Depledge, M. H., New approaches in ecotoxicology: can inter-individual physiological variability be used as a tool to investigate pollution effects? *Ambio*, 19, 251, 1990.

16. Klaasen, C. D. and Doull, J., Evaluation of safety: toxicologic evaluation, in *Toxicology, the Basic Science of Poisons*, Doull, J., Klaasen, C. D., and Amdur, M. O., Eds., Macmillan, New York, 1980, p. 11.

17. Hatch, T., Changing objectives in occupational health, *Ind. Hyg. J.*, Jan.-Feb., 1, 1962.

18. Jager, K. W., *Aldrin, Dieldrin, Endrin and Telodrin. An Epidemiological and Toxicological Study of Long-Term Occupational Exposure*, Elsevier, Amsterdam, 1970.

19. Wilson, K. W., The laboratory estimation of the biological effects of organic pollutants, *Proc. R. Soc. London, Ser. B*, 189, 459, 1975.

20. Depledge, M. H., The rational basis for detection of the early effects of marine pollution using physiological indicators, *Ambio*, 18, 301, 1989.

21. Peakall, D. B., *Animal Biomarkers as Pollution Indicators*, Ecotoxicology Series 1, Chapman & Hall, London, 1992, p. 291.

22. Lemaire-Gony, S. and Lemaire, P., Interactive effects of cadmium and benzo(a)pyrene on cellular structure and biotransformation enzymes in the liver of the European eel *Anguilla anguilla, Aquat. Toxicol.*, 22, 145, 1992.

23. DuBois, K. P., Potentiation of the toxicity of organophosphorus compounds, *Adv. Pest Control Res.*, 4, 117, 1961.

24. Depledge, M. H., Conceptual paradigms in marine ecotoxicology, in *Proc. 12th Baltic Mar. Biol. Symp.*, Bjørnstad, E., Hagerman, L., and Jensen, K., Eds., Olsen & Olsen, Fredensborg, Denmark, p. 47.

25. Depledge, M. H. and Rainbow, P. S., Models of regulation and accumulation of trace metals in marine invertebrates, *Comp. Biochem. Physiol.*, 97C, 1, 1990.

26. Addison, R. F., Hepatic mixed function oxidase (MFO) induction in fish as a possible biomonitoring system, in *Contaminant Effects on Fisheries*, Cairn, V. W., Hodson, P. V., and Nriagu, J. O., John Wiley & Sons, New York, 1984.

27. Calow, P., Physiological ecotoxicology: theory, practice and application, in *Proceedings of the First European Conference on Ecotoxicology*, Løkke, H., Tyle, H., and Bro-Rasmussen, F., Eds., Polyteknisk Forlag, Odense, Denmark, 1989, p. 23.

28. Forbes, V. E. and Depledge, M. H., Cadmium effects on the carbon and energy balance of mudsnails, *Mar. Biol.*, 113, 263.

29. Fossi, M. C., Leonzio, C., Focardi, S. and Renzoni, A., The Black-headed Gull's adaptation to polluted environments: the role of the mixed function oxidase system, *Environ. Conserv.*, 15, 221, 1988.

30. Doull, J., Klaasen, C. D., and Amdur, M. O., *Casarett and Doull's Toxicology: The Basic Science of Poisons.*, 2nd ed., Macmillan, New York, 1980, p. 778.

31. Moriarty, F., *Ecotoxicology: the Study of Pollutants in Ecosystems.*, Academic Press, London, 1983, p. 233.

32. Depledge, M. H. and Bjerregaard, P., Explaining phenotypic variation in trace metal concentrations in selected marine invertebrates: the importance of interactions between physiological state and environmental factors, in *Phenotypic Response and Individuality in Aquatic Ectotherms*, Aldrich, J. C., Ed., Japaga Publishers, Wicklow, Ireland, 1989, p. 121.

33. McCarthy, J. F. and Shugart, L. R., *Biomarkers of Environmental Contamination*, Lewis Publishers, Boca Raton, FL, 1990.

34. Schindler, D. W., Detecting ecosystem responses to anthropogenic stress, *J. Fish Aquat. Sci.*, 44, 6, 1987.

35. Levine, S. N., Theoretical and methodological reasons for variability in the response of aquatic ecosystem processes to chemical stresses, in *Ecotoxicology: Problems and Approaches,* Levin, S. A., Harwell, M. A., Kelly, J. R., and Kimball, K. D., Springer-Verlag, New York, 1989, p. 145.

36. World Health Organization, Recommended health-based limits in occupational exposure to heavy metals, Technical Report Series, 647, Geneva, 1980.

37. Hansen, H. J., Olsen, A. G., and Rosenkilde, P., The effects of Cu^{++} on osmoregulation in rainbow trout *(Oncorhyncus mykiss)* assayed by changes in plasma salinity and gill lipid metabolism, in preparation.

38. Warren, G. E. and Davis, G. E., Laboratory studies on the feeding, bioenergetics and growth of fish, in *The Biological Basis of Freshwater Fish Production,* Gerking, S. D., Ed., Blackwell Scientific Publictions, Oxford, 1967, p. 175.

39. NRC (Nuclear Regulatory Commission), *Biologic Markers in Reproductive Toxicology,* National Academy Press, Washington, DC, 1989.

40. Mayer, F. L., Versteeg, D. J., McKee, M. J., Folmar, L. C., Graney, R. L., McCume, D. C. and Ratner, B. A., Physiological and nonspecific biomarkers, in *Biomarkers; Biochemical, Physiological and Histological Markers of Anthropogenic Stress,* Huggert, R.J., Kimerle, R. A., Mehrle, P. M. and Bergman, H. L., Eds., Lewis Publishers, Boca Raton, FL, 1992, p. 5.

CHAPTER 13

NONDESTRUCTIVE BIOMARKER STRATEGY: PERSPECTIVES AND APPLICATIONS

Claudio Leonzio and M. Cristina Fossi

TABLE OF CONTENTS

0-87371-648-5/94/$0.00+$.50
© 1994 by Lewis Publishers

I. INTRODUCTION

The study of the distribution and effects of environmental contaminants began in the last few decades. In this short time, however, methods and analytical techniques have evolved at a constant pace, and the new field of ecotoxicology has acquired many new tools. Biomarkers are the most recent and rapidly developing of these. They promise to provide much information in monitoring programs for surveillance, hazard assessment, compliance with regulations, and remedial documentation.[1-4]

The contributors to this book propose further "fine tuning" of the biomarker methodology, with nondestructive (ND) applications. The advantages of nondestructive methods of assessment of chemical insult to communities and populations of vertebrates have been detailed in the first chapter of this book.

The book examines biomarker techniques applicable to nondestructive samples and sampling techniques specifically designed for conservative studies in vertebrate species.

On the basis of the suggestions of various contributors, this chapter summarizes the factors to be considered in the use of nondestructive biomarkers (NDB), reviewing the main methodological aspects associated with this technique. Particularly, the difficulties associated with the conservative sampling of biological materials and the biomarker procedures applicable to the different samples are discussed. In the final part of the chapter, a field study program in which NDB are applied is proposed. It is articulated in three phases: evaluation of the state of contamination of ecosystems or communities, identification of sensitive species exposed to contaminants, and evaluation of hazard to populations of vertebrates belonging to critical species.

II. THE CONCEPT OF NONDESTRUCTIVE BIOMARKER

Biomarkers *per se* have been defined many times in recent years. Conceptually a biomarker is the measure of a biological change that can be identified as the effect of chemical insult. A first definition of biomarker by the National Academy of Sciences is given in Chapter 1 of this book. A recent definition adds "ecotoxicological" to the term "biomarker" to distinguish it from medical and clinical contexts (see Chapter 12).

An NDB is a biochemical, cellular, physiological, or behavioral variation that can be measured in tissue or body fluid or in the whole organism or population, which provides evidence of exposure and/or effects of one or more chemical pollutants without causing damage or prolonged stress to the organism or population.

An unusual feature of NDBs is the sampling strategy used to measure biological changes. The study of natural ecosystems should involve harmless and nonstressful sampling techniques for populations and communities. Biomarkers detectable in the noninvasive sample should be sufficiently sensitive and possibly diversified. This can be achieved by using specific and nonspecific series of biomarkers at different levels of complexity (molecular, cellular, physiological) and residue analysis. In principle, the effect required for a positive verification of this strategy should tend to obtain maximum information with minimum disturbance of the population.

III. NONDESTRUCTIVE SAMPLING METHODS

Environmental investigations using NDB require sampling methods that are essentially different from conventional methods. They must by definition be nondestructive and are the primary aspect that qualifies the whole method. The choice of the species to be studied and biological materials to be used depends on a series of criteria governed by:

- the difficulty of capturing and/or sampling and the type of stress inflicted
- the information obtainable from ND samples by application of biomarker methods and residue assays
- integration of biomarker and residue data
- previous experience and the development of sampling methods

A. Capture Techniques and Sampling Stress

Chapter 1 gives a general picture of the biological materials usable as ND samples and a detailed description of possible ND sampling techniques. There are basically two strategies with different impacts on the study population:

- the use of biological material taken *in vivo* (blood, biopsy specimens, feathers, antlers, etc.) or fertile eggs, and the use of organisms for physiological and behavioral tests
- the use of discarded or ejected biological products (abandoned eggs, feathers, feces, fur) or carcasses of animals dying by accident (hunting, stranding)

Sampling involving the capture of animals is more likely to weaken populations than the use of excretory or reproductive products. However, more biomarkers can be measured and more information obtained from biological materials such as blood and biopsy specimens. These factors mean that the species, material, and test must be decided, balancing the quantity of information available with the degree of disturbance to the animal or population.

The choice of species is the first step in a monitoring program and is dictated by the type of investigation to be performed.

The study of an ecosystem or community is based on the use of indicator species chosen according to criteria in line with an NDB approach. On the other hand, studies of populations at risk require sampling techniques in line with the characteristics of the particular species. Sensitivity to capture and manipulation varies greatly from species to species. The main parameters to consider are morphology (size, structure), physiology, behavior, and psychology.

The criteria to which ND sampling in nature should conform can be summarized as follows:

- the possibility of sampling in different periods and in chosen areas with a statistically significant number of individuals
- the possibility of obtaining individuals of the same sex and size
- the possibility of having enough biological material or sufficiently long manipulation times in the case of physiological or behavioral tests

In endangered populations or those in a critical state, the sampling program should be planned with specialists of the particular species (wildlife biologists, ethologists) who could suggest the best methods of capture for population studies that can then be adapted for sampling for NDB studies.

B. Biomarkers for Nondestructive Samples

In this book many examples are given of biomarkers applicable to ND biological samples. A concise picture is given in Chapter I. The examples given in this volume refer mainly to blood. Experiences with skin biopsies in cetaceans and suggestions for diversified use of the eggs of nonmammalian vertebrates are also described.

From the analytical techniques reported in many chapters of the book, it emerges that much information can be obtained from blood. Biological and cytological changes occurring as a result of exposure to the main classes of contaminants can be determined in small quantities of this ND material. Apart from the many tests from clinical biochemistry (Fairbrother, Chapter 3), other proposed markers include esterases (Thompson and Walker, Chapter 2), porphyrins (De Matteis and Lim, Chapter 4), hemoglobin adducts (Shugart, Chapter 7), and cytological and morphometric analysis (Bickham, Chapter 6 and Moore et al., Chapter 8). Blood clearly occupies a prime position among ND samples, also because in all classes of vertebrates there are species capable of undergoing blood sampling with little stress. The main disadvantages are:

- lack of understanding of the relation between indications obtained from blood and those of the target organ (e.g., brain AChE vs serum BChE)
- lack of comparative experimentation in vertebrates
- possibility of blood data being extremely variable with respect to circadian cycles (e.g., circadian variation in esterases)

Aguilar and Borrell (Chapter 11) and Lambertsen et al. (Chapter 10) deal with the specific question of cetacean studies. In this case, the approach is based on the need to obtain information on one or more particular species. The authors propose a method based on the use of skin biopsies obtained with a dart shot from an air gun or crossbow. Biomarkers performed on this material have been the subject of very recent studies and concern the mixed function oxidases[5] and the determination of chlorinated hydrocarbon residues (Aguilar and Borrell, Chapter 11) and heavy metals.[6] Much experience has been gained taking biopsy samples in cetaceans; this technique seems to be sufficiently tested and is one of the best examples of ND sampling adapted for the study of species at risk.

Peakall (Chapter 9) gives original indications on the use of eggs, especially on the possibility of incubating fertile eggs and studying the embryos for biochemical and teratogenic effects. This technique could be very interesting if applied to amphibians and fish. The enormous numbers of eggs produced by each female makes the results highly significant. Even in reptiles in danger of extinction (e.g., marine turtles), a study based on eggs as NDB is a possibility for ecotoxicological evaluation.

Some biomarker techniques are at least potentially applicable to excretory products. In feces it is possible to measure certain metabolic products that can be related to the general state of the organism (e.g., fat content) and to exposure to different classes of contaminants (porphyrin profiles). Feces have also been used in residue-based monitoring studies for heavy metals and organochlorines.[7,8]

C. Residue Analysis in Nondestructive Samples

Residue evaluation is a different approach to that of biomarkers. Residues are indicators of exposure only for persistent and bioaccumulating chemicals. It is impossible to use residue analysis when contaminants are metabolized rather than accumulated (organophosphates, carbamates) (Figure 1).

Residue analysis can, however, be integrated or replace certain uses of ND biological materials. For example, discarded or external material (feces, fur, unhatched eggs, antlers, feathers, hair) can provide a limited number of biomarkers. Another use is to analyze temporal and historical trends using museum material.

D. Development of Sampling Methods

The nondestructive approach is relatively new in ecotoxicology. Future progress in the sector will depend on the development of new techniques and instruments, including sampling techniques.

As previously mentioned, current experience suggests that blood is the best ND material. Anatomical sampling sites for various classes of vertebrates are reported in Table 1. Skin biopsies are interesting samples that can be taken *in vivo* and can give data unavailable from blood samples. Existing sampling techniques, however, are limited to cetaceans. NDBs in skin and mucosa biopsies should be developed in amphibians, birds, and mammals.

Liver biopsies are a precious source of data because the liver is the primary target of most contaminants and the organ in which most biomarkers are evaluated. Under controlled conditions, a small quantity of liver tissue can be aspirated through a special needle without severe stress to the animal. Biomonitoring

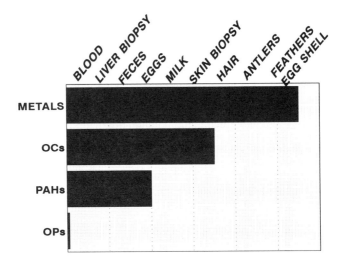

FIGURE 1. Main ND samples and relative possibility of analysis of residues.

Table 1. Main Anatomical Sites for ND
Blood Sampling in Vertebrates

Vertebrate class	Sampling point
Fish	Caudal vein
Amphibians	Ocular sinus, heart
Reptiles	Ocular sinus
Birds	Brachial vein
Mammals	Femoral vein, carotid

studies based on this type of sampling do not seem to have been performed despite their evident potential.

Some types of ND material (feathers, fur, feces, milk) have been used in biomonitoring.[9-12] Studies of historical trends of pollution based on museum specimens provide information on levels of persistent contaminants in feathers, fur, skin, bones, and teeth in exposed vertebrates.

IV. BIOMARKER TECHNIQUES

The methodological characteristics of the biomarkers discussed in this book are described in detail in the various chapters. Here we consider them in more general terms and discuss their suitability for monitoring programs having an ND approach.

The most important requisites of biomarkers are reported in many recent articles on the subject.[1-4] There are at least four basic criteria for the acceptance of NDBs in field studies:

- **Quality of response** — They must respond to the main classes of pollutants.
- **Sensitivity of response** — They must have good sensitivity of response.
- **Sensitivity of methods** — Sufficiently sensitive techniques must be available to measure the biological parameter in nondestructive samples.
- **Level of response** — The level at which the biological response occurs must be considered.
- **Ecotoxicological meaning of response** — It must be possible to correlate the biological response (biomarker) with harm to the species, population or ecosystem.

A. Quality of Response

The tests proposed in this book represent biological responses to the main classes of environmental contaminants (Tables 1, 2, and 3, Chapter 1). Chlorinated, aliphatic, and polycyclic hydrocarbons; heavy metals; and organophosphate and carbamate insecticides are the most common pollutants of the global ecosystem. Their effects may be, at least partly, monitored with NDB. The specificity of the response varies greatly from test to test. These differences in specificity enable the development of tests identifying responses to specific pollutants (ALAD, protoporphyrin) and to others of broader spectrum (immunotoxicology, stress proteins).

B. Sensitivity of Response

The sensitivity of biomarkers depends on the sensitivity of the biological response to environmental concentrations of the contaminant. In theory this is a modulated by dose-response function of all the variables present in nature. The biological response is generally chosen for its capacity to reveal exposure to environmentally realistic concentrations. Nearly all biomarkers as such must be sufficiently sensitive. This intrinsic sensitivity of response must be matched by a method of analysis and measure of the response itself which is just as sensitive (sensitivity of method). The power of a biomarker depends on these two characteristics. At least one of the two must hold if critical quantities of biological sample are to give high sensitivity of response or of method or of both.

C. Sensitivity of Methods

The sensitivity of methods may be a fundamental difference between destructive and ND biomarkers and a limiting factor from the practical and methodological points of view. In general, it can be said that ND samples are much smaller than destructive ones. This means that methods must be very sensitive, that is, able to detect very subtle biochemical and cell changes. The two needs arising from this are:

- to increase the sensitivity of existing methods in relation to small quantities of material, often together with low levels of response
- to seek new biomarkers that are easily determinable in ND samples

Many of the methods presented in this book are highly sensitive and suited to use on small quantities of material. Some of the most sensitive are methods for the determination of hemoglobin and deoxyribonucleic acid (DNA) adducts in blood (Shugart, Chapter 7). The detection of adducts in DNA measured by the postlabeling technique can be as little as 1 in 10^{10} normal nucleotides. Such methods have been used in studies reported by the same author. The methods proposed for blood esterases (Thompson and Walker, Chapter 2) allow the determination of activities in a few microliters of serum. This opens many possibilities for interesting natural investigations involving the evaluation of organophosphates and carbamates.

Clinical biochemistry methods make it possible to process extremely small specimens and obtain much information (Fairbrother, Chapter 3).

In the chapter of De Matteis and Lim (Chapter 4), the porphyrins are proposed as NDB. Exposure to many environmental contaminants causes changes in porphyrin metabolism that can be correlated with exposure to certain substances. One interesting possibility to develop regards the analysis of porphyrins in external biological samples (hair, feathers, skin, eggs) and in museum tissues.

D. Level of Response

In wild animals from contaminated habitats or organisms experimentally exposed to toxicants, the magnitude of the organism's response can be measured at four levels:

• molecular, biochemical
• cellular or histological
• physiological
• behavioral of an individual or population

Most of the biomarkers presented belong to the first category, some to the second (Bickham, Chapter 6 and Moore et al., Chapter 8), and a few to the third and fourth[13] (Peakall, Chapter 9). We have many early response systems and few of the higher response levels. Practically, the techniques have been developed with the aim of measuring biochemical variations. One of the reasons for this was that the first effect of a contaminant is observed at the biochemical-molecular level. The measure of this variation is a very early signal of interaction between the molecule and the organism (early warning system). The size of this response can often be correlated with the degree of exposure. The advantage of an early response is counterbalanced, however, by the difficulty of predicting, even in the short or medium term, the negative effects on individual and population. Biomarkers involving esterase inhibition are an exception. Brain esterases can be regarded as a reliable index of damage, giving three degrees of effect: no effect, reversible damage, irreversible damage[1] (Depledge, Chapter 12). However, blood esterase activities, the only ones that can be regarded as NDB, cannot be clearly correlated with brain esterase activities (Thompson and Walker, Chapter 2).

Studies at higher hierarchical levels are proposed in two chapters of this book. Moore et al. (Chapter 8) propose a histopathologic study in which the response to chemical insult is measured in terms of adaptive and pathological changes in the membranes. The method can be considered ND for intermediate levels of response due to the development of noninvasive liver biopsy techniques in fish. In Chapter 6, cytomorphic characteristics of blood cells are proposed as NDB.

The responses of the highest hierarchical level are slower to manifest since they are often the result of long exposure. Exposure to certain substances can be correlated with physiological or behavioral variations in an individual or population, but this can hardly be regarded as a causal relationship. For example, studies on the reproductive success of vertebrate populations exposed to contaminants are an interesting application.

Peakall makes some interesting suggestions in Chapter 9. He studied the embryos of populations of nonmammalian vertebrates for development capacity, for evaluating malformations and development anomalies, and for certain biomarkers in vivo (genotoxicity tests, MFO, esterases, porphyrins). Thus, the reproductive potential of populations can be evaluated by comparison with controls or by means of laboratory models.

E. Ecotoxicological Meaning of the Response

Correlation of early biological signals such as enzyme induction or inhibition with higher levels of response, up to physiological or population, is the goal and limit of studies based on biomarkers. Only very rare cases exist in which the measure of a signal represents a precise index of physiological impairment (see brain AChE in Thompson and Walker, Chapter 2 and ALAD in Fairbrother, Chapter 3).

Most responses supplied by destructive and ND biomarkers reflect exposure to a contaminant without the concrete possibility (at least for studies in nature) of transforming the latter into a function of health status. The variables of natural systems prevent evaluation based on a theoretical dose-response model. As suggested by Depledge (Chapter 12), it is better to consider the phenomenon as a family of dose-response relationships based on the combination of chemical and environmental stresses to which the organism is exposed. The main sources of variability in biomarker responses can be divided into three groups:

- **environmental factors** — bioavailability of chemicals, synergistic or antagonistic effects of chemical mixtures, route and time of exposure, food availability, photoperiod, temperature, solar radiation, etc.
- **interspecific factors** — different physiology and biochemistry of wild species exposed to toxicants (e.g., absence of A esterases in birds, lack of inducibility of aldrin epoxidase in fish, variability of enzyme activity in the same class of vertebrates)
- **intraspecific factors** — age, sex, hormone cycles, nutritional status, health status, migration and movement, capacity to adapt of groups or populations of species with a high ecological valency (biochemical acclimatization)

The extreme complexity of natural situations means that care must be taken in defining models for evaluating danger and risk. Nevertheless, for the application of the NDB, further research and suggestions for their rational use are required.

The methods suggested by Depledge are based on the use of a suite of biomarkers chosen for different levels of biological response (early, intermediate, late). This multiple response concept defines the position of the organism on the health status curve (homeostasis, compensation, noncompensation). The problem with this is the difficulty of defining the limits of homeostasis and compensation that in studies in nature are greatly influenced by all the above parameters.

For the practical applications of studies based on NDB to have meaning, research in four directions is required:

1. The development of laboratory models should define the limits of homeostasis and compensation. These would enable the effects of different chemicals to be tested in the different classes of vertebrates and in species of the same class, under controlled conditions, in order to obtain a response model based on three variables: toxicant-species-multiple response.
2. The definition of correlations between biomarkers in ND samples and biomarkers of target organs should clarify the reliability of the ND sample as a monitor of a response of other parts of the organism.
3. Laboratory models should be based on long exposure to environmentally meaningful doses in order to evaluate the effects and correlate molecular and cellular signals with higher levels of response.
4. Verification in nature of models obtained in the laboratory should be made through populations living in contaminated and control areas; the status of populations should be evaluated using compensation limits and verifying the model by evaluating the higher levels of response (reproductive success, population dynamics).

V. NONDESTRUCTIVE-BASED FIELD STUDIES

The main ecotoxicological application of the NDB approach is the hazard assessment of endangered or threatened species of vertebrates. Increasingly frequent incidents involving drastic reductions in populations linked to the presence of contaminants of anthropic origin in populations of higher vertebrates (e.g., marine mammals, endangered bird species, etc.) make it indispensable to develop techniques of biomonitoring and hazard assessment based on ND methods. As suggested in the overview section of the Workshop on Nondestructive Biomarkers in Vertebrates,[17] NDB strategies should be addressed to three aspects related to conditions and methodology: the study of critical communities and ecosystems (phase 1), the definition of critical species (phase 2), and the status assessment of critical populations (phase 3) (Figure 2).

In practice, this involves the progressive refinement of impact evaluation in ecosystems subject to environmental pollution. The three phases of investigation constitute a sequence involving the acquisition of increasingly precise data up to specific targets and community sensitivity.

The study of species at risk can be performed outside an overall context of ecosystem study. This is true for species endangered or near to extinction for reasons that cannot always be correlated with pollution alone, and is probably the most appropriate application of studies based on NDBs.

A. Phase 1: Study of Critical Ecosystems (Table 2)

Ecotoxicological studies of ecosystems are performed when pollution by natural and synthetic microcontaminants is suspected. The aim is to determine whether human activity is affecting the integrity of the ecosystem by the emission or use of chemical substances. However, the complex interactions between polluting molecules and the ecosystem should first be investigated in an interdisciplinary study using ecotoxicological tools such as database evaluation of regional features, analysis of diffusion models, chemical analysis, and wildlife studies. This exploratory phase should be coupled with investigations based on a biomarker approach.

The aim of this first phase of study is to evaluate the status of the biological community and if possible assess hazard, in other words, establish a measure of present and future damage due to pollutants.

The research should be organized in three main steps:

1. Step 1:
Acquisition of Data on the Characteristics of the Area and the Main Chemical Compounds; Predictive Diffusion Models and Residues Screening

The preliminary, exploratory part of the study supplies the first indications of the potential levels of danger in the study area, on the basis of what it is possible to predict with diffusion models and chemical analysis in the various environmental compartments. A priori knowledge of these molecules and the models makes it possible to direct the first phases of the investigation to the risk compartments of the ecosystem. For example, molecules with an affinity for water such as certain herbicides (atrazine) will determine a potential risk in aquatic environments and their communities.[14] Fat soluble substances are readily accumulated in organisms and biomagnified in terminal consumers that are therefore the preferred subjects of the study.[15] Metals such as lead tend to be associated with the soil and sediments, accumulating in soil- and sediment-dependent organisms.[16]

2. Step 2:
Use of Biomarkers in Invertebrates and Abundant Species of Vertebrates; the Suite of Biomarkers Used to be Aspecific for the First Screening and More Specific in Later Screenings

Once the compartment has been identified, a series of responses of the biological community to chemical insult will be measured using a suite of biomarkers in a large range of species; in this way we have an *in vivo* verification of the effects of the substances or the combination of substances present. The biomarkers used will also give information on unknown or unpredictable chemicals in the study area. The evaluation criteria should be based on comparison of different biomarkers with physiological limits of compensation and/or responses of populations in

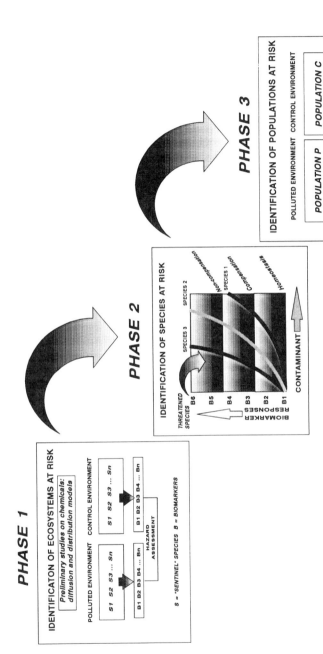

FIGURE 2. Three-phase approach in an NDB-based field study.

Table 2. Assessment of Critical Ecosystems

Scenario
Community-ecosystem suspected of being in a critical state
Objective
Knowledge of degree of disturbance to community by xenobiotic compounds
Methodology
Exploratory screening: data bank on use of chemicals in the area, diffusion models, residue analysis in ecosystem compartments
Step 1: Assessment of community using destructive biomarkers in invertebrates and any abundant class of vertebrates
Step 2: extension of study to vertebrate species suspected of being vulnerable on the basis of data obtained in step 1 and occupying critical trophic positions
Step 3: integration of results not only on the basis of presence-absence of the chemicals but also comparative analysis of biological responses taking into account the specific sensitivity of the various phyla
Expected results
Assessment of damage to community

control areas. These data will show how far the biological response is from physiological limits, giving an idea of the position on the health status curve (homeostasis, compensation, noncompensation). The samples used for the determination of biomarkers can also be used for genetic population research and residue analysis. Population studies should, however, be performed to evaluate danger on the basis of different levels of response.

3. Step 3:
Extension of the Study to Species Suspected of Being Sensitive to the Substances Identified in Phase 1, By Virtue of Their Trophic Niche

At this stage, nondestructive biomarkers should be used. The nondestructive biomarker results integrated with population studies should enable the effect of pollution on species at risk to be determined. This step is closely linked to phase 2.

B. Phase 2: Identification of Species at Risk

Species of vertebrates with a critical trophic position (e.g., predators, fish-eating birds, and marine mammals) are next investigated, the rationale being that within classes of vertebrates interspecific differences exist in the susceptibility to contaminants. For example, it is well known that different species of wild birds have different levels of tolerance to liposoluble xenobiotics. In some cases this may be due to differences in the capacity of the mixed function oxidase system, and may be expressed as a different adaptability or otherwise to survive in a highly polluted environment. If our goal is to identify the species at risk in a particular polluted environment, the use of noninvasive techniques (mainly biomarkers of the effects) should be directed to the evaluation of interspecific differences in response to the sum, known or unknown, of the polluting agents. A suite of

Table 3. Phase 2: Identification of Species at Risk

Scenario
  Environment contaminated by a specific substance or substances
  Several species of vertebrates with a critical trophic position or with habitat in
    target environmental compartment
Objective
  Identification of the species at risk in vertebrate classes in relation to trophic
    position and susceptibility to contaminants
  Estimate of ecological importance of the species at risk
Methodology
  Suite of NDB of effect to estimate how much biomarker values of different
    species differ in relation to the same level of environmental contamination
  Suite of biomarkers of high-level responses
  Residue analysis of suspected contaminants
Expected results
  Identification of the species at risk
  Determination of ecological importance of species at risk
  Identification of molecules responsible for toxic phenomena and their sources

nondestructive biomarkers may be tested in several species of the same class suspected of being at risk (e.g., fish-eating birds exposed to biomagnification in the marine food chain). An estimate of how much the biomarker values differ between species for the same level of environmental contamination enables the identification of species at risk and threatened species (Figure 2).

C. Phase 3: Identification of Populations at Risk

The last phase of this hierarchical approach concerns population risk evaluation in the target species. The main question to be answered is whether the population of this critical species is really at risk in this particular environmental situation.

In the assessment of a population of an endangered species suspected to be exposed to toxic substances, nondestructive biomarkers at biochemical, cell, and physiological levels may be applied together with population studies, as follows. A series of biomarkers may be tested in the population and compared with data from a control population (Figure 2). An estimate of how much the biomarker values of the endangered population differ from control population values gives a measure of the status of the population studied. In this case, the goal of biomarker research is to identify how biomarker responses correspond to different levels of departure from normal homeostasis, so as to evaluate the real risk to which the study population is exposed. Ecological methods of research based on the evaluation of the general state of the population (birth rate, mortality, fertility index, relationship between ages) are indispensable for interpreting links between biochemical and cell changes (biomarkers) and negative effects at an ecological level.

One of the main aims of this type of investigation is to define the real ecological consequences associated with the potential disappearance of species at risk. In

Table 4. Phase 3: Identification of Population at Risk

Scenario
Identification of one or more species at risk for their trophic position or a habitat in the target environmental compartment
Limited number of specimens available per species or protected species

Objective
Evaluation of risk or health status of target population(s) without use of destructive system of investigation

Methodology
Suite of NDB in the target population; comparison with a control population
Identification of biomarker responses with different levels of departure from normal homeostasis

Expected results
Assessment of risk of target population
Identification of molecules responsible for toxic effects and their sources
Report ecological changes to competent authorities (Ministry for Environment, Environment Control Unit) so that species can be safeguarded

fact, there are two types of ecological damage associated with the disappearance of a species from an ecosystem:

- The species can be regarded as a "key species," and its disappearance can upset the homeostatic equilibrium of the community.
- The species does not have a determinant ecological role because it overlaps with other species as far as trophic niche and habitat are concerned, but its disappearance causes a decrease in the genetic variability of the community to the detriment of the total genetic load.

It is important to combine biomonitoring programs based on NDBs with ecological studies (expressed in conventional terms) for the final risk evaluation of an ecosystem or community subject to a drastic decrease in numbers or the disappearance of one or more endangered species.

The final aspect to deal with is the potential role, in terms of prevention or remediation, of information obtained in such studies. It is the basic task of the ecotoxicologist to report to the competent authorities the species at risk in given environmental situations and the causes of these ecological changes (identification of contaminants and their sources), so that measures to safeguard them can be undertaken.

REFERENCES

1. Peakall, D. B., *Animal Biomarkers as Pollution Indicators*, Ecotoxicological Series 1, Chapman & Hall, London, 1992.
2. McCarthy, J. F. and Shugart, L. R., *Biomarkers of Environmental Contamination*, Lewis Publishers, Boca Raton, FL, 1990.

3. Hugget, R. J., Kimerle R. A., Mehrle, P. M., and Bergman, H. L., *Biomarkers: Biochemical Physiological, and Hystological Markers of Anthropogenic Stress,* Lewis Publishers, Chelsea, MI, 1992.

4. Peakall, D. B. and Shugart, L. R., *Biomarker-Research and Application in the Assessment of Environmental Health,* NATO ASI Series, Ser. H, Cell Biology, Vol. 68, Springer-Verlag, Berlin Heidelberg, 1993.

5. Fossi, M. C., Marsili, L., Leonzio, C., Notarbartolo di Sciara, G., Zanardelli, M., and Focardi, S., The use of non-destructive biomarker in Mediterranean cetaceans: preliminary data on MFO activity in skin biopsy, *Mar. Pollut. Bull.,* 24(9), 459, 1992.

6. Marsili, L., Focardi, S., Cuna, D., Leonzio, C., Casini, L., Bortolotto, A., and Stanzani, L., Chlorinated hydrocarbons and heavy metals in tissues of *Stenella coeruleoalba* stranded along the Apulian and Sicilian coasts (Summer 1991), *Proc. 6th Annu. Conf. Eur. Cetacean Soc.,* San Remo, February 20–22, 1992, 234.

7. Clark, D. R., Jr. and Prouty, R. M., Organochlorine residues in three bat species from four localities in Maryland and West Virginia, 1973, *Pestic. Monit. J.,* 10(2), 44, 1976.

8. Leonzio, C. and Massi, A., Biomonitoring metals in bird eggs: a critical experiment, *Bull. Environ. Contam. Toxicol.,* 43, 402, 1989.

9. Fitzner, R. E., Rickard, W. H., and Hinds, W. T., Excrement from heron colonies for environmental assessment of toxic elements, *Environ. Monit. Assess.,* 1, 383, 1982.

10. Thompson, D. R., Furness, R. W., and Barrett, R. T., Mercury concentrations in seabirds from colonies of the northeast Atlantic, *Arch. Environ. Contam. Toxicol.,* 23, 283, 1992.

11. Renzoni, A. and Norstrom, R. J., Polar bears and mercury, *Polar Rec.,* 26, 326, 1990.

12. Focardi, S., Leonzio, C., and Fossi, C., Variations of polychlorinated biphenyl congener composition in egg of Mediterranean waterbirds in relation to position in the food chain, *Environ. Pollut. Ser. A,* 52, 243, 1988.

13. Depledge, M., The rational basis for detection of early effects of marine pollutants using physiological indicators, *Ambio,* 18, 301, 1989.

14. Bacci, E., Renzoni, A., Gaggi, C., Calamari, D., Franchi, A., Vighi, M., and Severi, A., Models, field studies, laboratory experiments: an integrated approach to evaluate the environmental fate of atrazine (s-Triazine herbicide), *Agric., Ecosys. Environ.,* 27, 513, 1989.

15. Cornell, D. W., *Bioaccumulation of Xenobiotic Compounds,* CRC Press, Boca Raton, FL, 1990, chap. 7.

16. Forstner, U. and Wittann, G. T. W., *Metal Pollution in the Acquatic Environment,* Springer-Verlag, Berlin, 1983, p. 323.

CHAPTER 14

The Future of Nondestructive Biomarkers

John F. McCarthy

TABLE OF CONTENTS

I. INTRODUCTION

One of the objectives of the international Workshop on Nondestructive Biomarkers in Vertebrates was to develop a consensus of the participants on the future directions and key research needs in the area of nondestructive biomarkers in vertebrates. Working groups focused on three areas of discussion: scientific and regulatory applications of nondestructive biomarkers; application of nondestructive biomarkers in studies of endangered and protected species; and advanced methodologies and innovative technologies to improve nondestructive biomonitoring. The reports of the working groups were discussed by all participants, and the general consensus of the workshop in these areas is presented below.

II. SCIENTIFIC AND REGULATORY APPLICATIONS

A. Biological Criteria in Regulatory Applications

The goal of environmental regulation is protection of biota from the toxic effects of chemicals released into the environment. Given this goal, then, the use of biological criteria for regulatory applications seems appropriate, even though regulations have traditionally been based on chemical rather than on biochemical measurements. Environmental regulations in many countries appear to be moving in the direction of accepting biological criteria in addition to, or instead of, more traditional chemical criteria. For example, a recent report by the National Academy of Science in the United States has made the following observations: "The U.S. government may be wasting billions of dollars in its zeal to clean up the environment simply because it has failed to conduct studies that would provide a sound scientific underpinning to judge which of the many potential toxins constitute a genuine and serious threat to human health...Biomarkers have not been adequately adopted for assessment of broad environmental exposures to hazardous waste."[1]

Regulatory goals and enforcement of compliance to environmental guidelines are traditionally based on measuring levels of toxic chemicals in soil, sediments, water, or biological tissues. While these criteria have advantages of being objective, analytically tractable, and enforceable, the chemical approach has a number of disadvantages:

1. It is very difficult to relate chemical concentrations in environmental media to biological effects.
2. Zero tolerance cannot be achieved using chemical criteria due to the exquisite sensitivity of modern analytical technology and the global dissemination of many pollutants.
3. Chemical analyses are expensive, especially if an attempt is made to be complete.

4. Chemically based compliance criteria do not address the problem of unknown chemicals and complex mixtures of known chemicals. Toxicological profiles have been established for only a few of the tens of thousands of chemicals released into the environment, and almost nothing is known about the toxic effects of mixtures of those contaminants.

In the context of regulatory applications, biological criteria and especially the measurement of biomarkers in exposed organisms have a number of advantages:

1. Biomarkers can demonstrate whether chemicals in the environment are biologically available in sufficient concentration to cause a biological change.
2. A regulatory goal of zero tolerance based on biological criteria is possible because a chemical may not be bioavailable, or may not cause a change to occur. Therefore, biomarkers can generate a case that no regulatory or compliance action is needed. If changes in biomarkers are seen, then a case for additional studies (including both chemical and biological) can be justified.
3. Biomarker criteria are more biologically relevant, since they are based on demonstrating exposure at molecular and biochemical targets, as well as indicating significance of effects of such exposures. In addition, monitoring changes in markers over time (as opposed to monitoring only chemical concentrations) provide fundamental information on toxicological mechanisms of response to (and recovery from) long-term exposure to toxicants.

B. Nondestructive Biomarkers in Scientific and Regulatory Applications

There are several important reasons why nondestructive biomonitoring should be adopted as a key tool for the assessment of exposure of wildlife to environmental pollutants and should, in the view of the workshop, largely replace destructive methods of biomonitoring. First, a nondestructive approach is very much more likely to receive public acceptance. Second, it would allow repeated sampling in the same individual, an enormous advantage in following the time course of exposure and efficacy of remedial operations. Also, this approach offers a direct analogy to human diagnostic medicine and can therefore draw from the vast experience available in human clinical chemistry. Finally, it offers the potential for an early diagnosis of a disordered state so that an appropriate therapy may be instituted, as already reported in some endangered species poisoned by lead.

Therefore, more research is urgently needed to devise new markers suitable for nondestructive monitoring of exposure, or to improve already established markers; however, priorities should be established based on recognition of the following general aims.

1. Mechanistic/Process-Oriented Understanding

1. Nondestructive biomarkers offer potential for improved mechanistic understanding since the temporal expression of the response, extent, and time course or recovery from insults for individual organisms can be determined.
2. There is a critical need to demonstrate and validate linkages between more traditional (and well-understood) destructive biomarkers and newer nondestructive biomarkers.
3. The increased genetic heterogeneity of field populations compared to that observed in the laboratory must be considered.
4. The use of nondestructive biomarkers can contribute to understanding of the genetic basis for variability in individual responses.

2. New and Improved Nondestructive Biomarkers

1. Equal value is place on both general indicators of environmental problems, as well as specific indicators of exposure and effects of specific contaminants.
2. Improvements are required to make nondestructive approaches as informative and sensitive as destructive biomarkers.

3. Ecological and Environmental Relevance

1. The use of nondestructive biomarkers can overcome statistical limitations in evaluating the impacts of environmental pollution since an adequate number of organisms can be sampled even if the population is small.
2. The effect of seasonal changes in nutrition, competition, and other ecological interactions can be examined since nondestructive biomarkers permit repeated sampling even at small sites (e.g., waste sites where population is limited by the size of the site).
3. Nondestructive biomarkers may be particularly useful in laboratory and in controlled field studies; however, it may be unreasonable to design experiments for repeated sampling in large systems of wild animals.

III. NONDESTRUCTIVE BIOMARKERS IN STUDIES OF PROTECTED OR THREATENED SPECIES

Much of the initial motivation for development and applications of nondestructive biomarkers arises from efforts to assess the health status of protected or threatened species that cannot be examined by destructive means. In fact, one of the first applications of nondestructive biomarkers was the use of acetylcholinesterase inhibition to assess pesticide toxicity in endangered species of birds (e.g., Thompson and Walker, Chapter 2). Recently, considerable public attention throughout the world has been focused on the mass mortalities of marine mammals, such as the striped dolphin epizootic in the Mediterranean, the seal die-off in the North Sea, and the dolphin strandings along the Atlantic and Gulf coasts of the United States. Although toxic pollution has never been identified as a principal cause of any mass mortality, questions have been raised about the possible role that pollution may have played in making the affected populations more vulnerable to stresses that may have been the proximal cause of the mortalities. For example, the striped dolphin epizootic in the Mediterranean in 1991 was the result of *Morbilli* virus, a disease agent related to measles and canine distemper viruses. Nevertheless, levels of pollutants such as polychlorinated biphenyls in the dolphins were quite high (Aguilar and Borrell, Chapter 11). As an exercise to focus discussion, the workshop considered the approaches that might be taken to use nondestructive biomarkers to assess the exposure of marine mammals such as the Mediterranean striped dolphin to environmental toxins.

Discussion considered three topics:

1. What constitutes an appropriate reference population as a basis of comparison to evaluate biomarker responses?
2. What type of biological material can be used as noninvasive samples?
3. What kinds of biomarkers can be performed on biological materials collected by noninvasive techniques?

A. Identifying a Suitable Reference Population

In most natural populations, little information is available about the normal levels of many of the biochemical parameters used as biomarkers. This difficulty is usually addressed by comparing the biomarker measurements of animals in the contaminated site with the responses of the same species of organisms in an uncontaminated site; ideally, that uncontaminated reference site is similar to the contaminated site in all ways other than the presence of toxic chemicals. This idealized comparison controls for variables such as season, nutrition, etc. so that differences in biomarker responses can be attributed to the effects of pollutant exposure. Thus, it is not necessary to know, a priori, the normal level of serum cholinesterase activity in a particular population; the significance of exposure is assessed in terms of how much the enzyme activity is depressed relative to that observed in the population of the same species at the reference site.

While this idealized matching of contaminated and reference sites is rarely achieved in nature, it is usually possible to identify a suitable reference population. Because of the importance of the reference population in comparisons between sites, multiple reference sites must be examined to demonstrate that the responses are not an artifact of some unrecognized factor at a single site. The problem is more complex in species that have a large spatial range (such as many marine mammals) or for assessing pollution that is geographically widespread (as in the Mediterranean or the North Sea). Questions were raised in discussion about whether the reference population must be from the Mediterranean? Although there would be differences in populations from outside the Mediterranean, even if they were the same species, equally valid questions could be raised in comparisons of populations within the Mediterranean. For example, do different populations of the same species exploit different food sources in different areas, and would this have an impact on biomarker responses? In general, it was felt that for many biomarkers, such as biomarkers of genetic damage, these concerns would be of secondary importance. In many cases, it would be possible to collect reference populations from the Atlantic Ocean or near Japan. Issues of possible genetic differences between such widely spaced populations could be addressed by analysis of mitochondrial DNA. The general topic of the significance of population differences to biomarker measurements, and more specifically, decisions about what constitutes a valid reference population stirred considerable discussion. No firm consensus was achieved, largely because so little information is available about the biomarker responses of marine mammals or differences in the response of different populations.

B. Noninvasive Sampling

The issue of marine mammal mortality has achieved such high visibility because many of the sick animals are stranded and die on the beaches or dead animals are washed ashore. These organisms represent an obvious source of biological material for biomarker analysis without consideration of nondestructive or noninvasive techniques. However, dead animals, or even stranded live animals, are unlikely to be useful for biomarker analyses. Statistical variability in responses are likely to be very high because the sample material from stranded animals is likely to include animals that may have died from different causes or in different stages of disease progression or tissue autolysis following death. Some of the dead animals may simply have died in a net and been washed ashore, and their tissues may be representative of healthy animals in the population. In mass mortalities, animals found on the shores must be considered unrepresentative, unless the exact cause of the mortality is known, along with the consequences of that disease state on the biomarker response.

The amount of trouble and expense that should be invested in stranded animals — and the danger of generating unreliable and possibly misleading data — depends on the condition of the specimen. Vitreous and aqueous humor or the eye may be useful materials since they do not degenerate as rapidly as does serum. Likewise, biomarkers that measure structural changes (such as the presence of DNA adducts of mutagenic contaminants or porphyrin profiles) may be useful even in stranded animals, while biomarkers that measure functional changes (such as enzyme activities) may be considerably less reliable.

In spite of concerns about the use of stranded animals, biological material from these organisms can be useful in validating fundamental relationships, especially if the material is collected shortly after the death of the stranded animals (within 30 min). Under such circumstances a variety of biomarkers can be analyzed, such as acetylcholinesterase activity in brain tissue; mixed function oxidase activity; porphyrins; DNA damage in the liver; and porphyrins, esterases, and retinols in the blood. These data can be used, for example, to compare the level of mixed function oxidase activity measured in a skin sample (that might be collected in a biopsy) and in liver as a useful comparison between information provided by destructive and nondestructive biomarkers. Correlations between organochlorine contaminants in skin and deep blubber, or between methyl mercury levels in the brain vs skin may all be useful data that will aid in validating noninvasive sampling techniques.

Clearly, sampling live animals is preferable to relying on data from dead or stranded animals. Biopsies may represent the best samples. By-catches (animals accidently caught) during fishing operations offer a potentially useful source of material, but fishermen do not readily offer this proscribed material to the biologist. Also, the by-catches may be of variable freshness and not all species may be represented. Capture of smaller marine mammalian species expressly for biological sampling is possible, but some species react better than others to capture. If the animal struggles violently, some measurement of some biomarkers may be compromised. Some methods of capture, such as tail grabs, are more traumatic than others, such as slipping a net under a dolphin.

C. Biomarker Analyses in Biopsy Samples

A biopsy of the skin of marine mammals can be recovered with minimal trauma to the organism (Lambertsen et al., Chapter 10). The biopsy sample, which contains skin cells and blubber, may be analyzed for a variety of biomarkers, including the following:

Skin: Mixed function oxidase system
          (enzyme activities or measurement of protein concentrations using
          monoclonal antibodies)
          Porphyrins
          Genotoxic assays
          (including measurement of DNA adducts, DNA strand breaks, or
          other techniques)
          Histopathology
          Determination of the sex of the organism
          Population genetics data
          (allele frequency, DNA probes, or other techniques to determine
          the interrelatedness of populations or the genetic diversity of popu-
          lations)
          Body burdens of contaminants
          (including polycyclic aromatic hydrocarbons, organochlorines, and
          heavy metals)
Blubber: Lipid content and composition
          Body burdens of organic contaminants

D. Relating Biomarker Responses to Higher Level Effects

The goal of biomarker research on endangered and protected species is to aid in evaluation of the health status of the population. Specifically, biomarkers can help provide data that might establish linkages between impaired health status and exposure to toxic chemicals in the environment. However, interpretation of the biomarkers must be part of a larger investigation that includes consideration of the higher order effects manifested at the population level. Such higher level effects include growth impairment, frequency of disease (immunocompetence), morphological abnormalities, and reproductive impairment. Such population level studies have been attempted with certain species, but results are not always clear-cut. For example, seal populations have been affected by polychlorobiphenyl (PCB) levels, but the role of other pollutants (such as dioxins) has not been examined. In dolphins, parasite and virus infections have been observed in stranded animals, but little is known about the occurrence of the diseases in the population as a whole. One of the difficulties in population studies is the need for large sample sizes; yet even when large numbers of animals have been studied, interpretation of the results is not unambiguous. For example, Lambertsen's[2] investigations of 220 fin whales found a nematode which resulted in serious localized lesions; however, the potential role of environmental pollution remained unresolved. Skeletal abnormalities have been observed in the skull and spine of seals; while such abnormalities could reflect the effects of certain pollutants on calcium metabolism, clear linkages cannot be drawn. Reproductive rates are critical to the viability of populations, but they can be affected by a variety of ecological and behavioral factors, in addition to any direct effect of pollutants on reproductive physiology. Perhaps one of the most sensitive measures in populations may be observation of their behavior. A possible measure of relevant behavior may be simply the fraction of time that is spent in play, since play represents an abundance of available energy. These behavioral observations may represent an untapped resource for evaluating health status of at least some marine mammal populations. In many coastal areas, whale and dolphin watching is a popular activity, and no great expertise is required.

The higher level manifestations of toxic effects may be the first to be observed. In birds, this is sometimes possible because changes in population size and age structure can be directly observed. For cetaceans, the problem is much more difficult because these organisms have such a large geographic range, and thus little historical data is available. Reduced number of organisms may indicate that the animals have failed to reproduce, or may mean only that they have chosen not to return to a particular area. Comparing reproductive rates between populations in different areas can provide an answer, but this information is difficult to estimate and changes in reproduction may or may not be related to pollution.

The role of biomarker research is to address this difficult issue of establishing specific linkages between exposure to environmental pollutants and higher level manifestations of population impairment. Biomarkers measured nondestructively in natural populations of endangered and protected species can provide specific evidence of exposure to toxicants. Body burdens of contaminants are another key indicator that toxic chemicals may be an underlying cause of a higher order impairment. However, levels of hydrophobic organic contaminants in blubber are difficult to evaluate since it is not clear to what extent these toxicants become mobilized to reach sites of toxic action. By directly probing the molecular and cellular targets, biomarkers provide information that is more toxicologically significant. Suites of biomarkers should be employed to probe low and intermediate levels of response. Ideally, investigations should begin by measuring simple and aspecific biomarkers (such as DNA strand breaks or stress protein responses), before moving on to more sophisticated techniques that can identify specific chemicals or classes of contaminants. At present, capabilities to measure and interpret reliable and sensitive nondestructive biomarkers are limited, but development of new and more sophisticated biomarkers was another subject of the workshop discussion.

IV. ADVANCED METHODOLOGIES AND INNOVATIVE BIOMARKER TECHNIQUES

Workshop participants agreed that advanced methodologies and innovative techniques should focus on both assessment of the current extent of pollution and prediction of the likely effects of that pollution. Several new approaches to improve biomarker technology, and biomonitoring in general, were proposed and discussed. Those approaches are outlined here, arranged roughly from improvement of exposure estimates to predictions of higher order effects.

Biomarkers of exposure — It was a strong consensus of the workshop that additional research was needed to identify and develop biomarkers of different types of pollutant exposure. These should signal exposure to specific pollutants or to mixtures of pollutants. Furthermore, biomarkers which can distinguish between single acute exposures, repeated exposures, and continuous, chronic, low-level exposure should be sought.

Sampling procedures — There are many cases in which samples can be obtained without affecting the overall well-being of the organism. For example, hair, feathers, claws, and in some cases small biopsies of parts of the body provide useful material for the analysis of biomarkers. Likewise, body fluids, such as urine or saliva, can provide sample material as can feces. This approach generally requires specialized analyses because only limited amounts of tissue are available. Research in this area should receive more attention in the future.

Molecular biological techniques — There is no doubt that recent advances in molecular biology will offer tremendous opportunities for measuring changes in proteins, DNA, etc. in very small tissue samples (which can be obtained without significant damage to test organisms). Particularly promising examples are stress protein responses and identification of activated multi-drug-resistant genes.

Labels — Labeling potentially hazardous chemicals (or surrogate compounds expected to have similar transport behavior in the environment) should be explored as a means of enhancing our ability to detect and track pollutant movement and effects in biological systems. Compounds could perhaps be labeled with unusual isotopes, dyes, etc.

Telemetry — Telemetry and satellite tracking offer tremendous possibilities in the future for the continuous collection of ecological data on a global scale. Through the use of suitable transducers that monitor sessile or caged organisms at several localities, or transducers and telemetry transmitters on larger, free-moving organisms, a picture might be assembled of changes in biomarker responses over extended periods, reflecting fluctuations in the quality of the environment and the well-being of selected populations of organisms.

Challenge tests — Challenge tests were proposed for enhancing the sensitivity of the biomarker approach. Abnormalities associated with pollutant toxicity may not be evident in organisms maintained in standard laboratory conditions. However, when additional stresses (challenges) are imposed, damage caused by earlier pollutant exposure may become evident. This approach offers a practical supplement to conventional toxicity testing.

Transplant experiments and genetically-engineered organisms — The introduction of genetically engineered organisms into localities contaminated with different concentrations of chemicals may offer a way of investigating the selective pressures exerted by pollutants over successive generations. Genetic biomarkers (i.e., specific genes, altered gene frequencies, etc.) may be used to study the evolution of tolerance or the narrowing of the genetic diversity of a population. On a more conventional scale, transplant of organisms from clean to polluted sites and visa versa may also provide a means of studying adaptation to polluted environments. Biomarkers are required to identify different genotypes (using ribonucleic acid [RNA] probes, alloenzyme analysis, etc.).

Noninvasive physiological and behavioral monitoring —Biomarkers have been developed principally at the molecular/biochemical level. However, recent technological advances may now permit changes in whole organism physiology and behavior to be related to particular degrees of pollutant exposure. Noninvasive techniques involving computer acquisition of data are now available. Physiological and behavioral biomarkers offer the advantage that they may be more ecologically relevant than molecular/biochemical biomarkers, as they may reflect altered Darwinian fitness. However, such responses are more difficult to relate to specific pollutants at particular concentrations than are biochemical biomarker responses. Combined approaches involving biochemical, physiological, and behavioral biomarkers appear promising for future development.

Ecosystem structure and function — There is an ongoing debate as to whether changes in ecosystem structure or function provide the best means of detecting adverse effects. Since changes in these parameters reflect integrated effects of pollutants and environmental stressors, they should be explored further. Biomarkers which signal such changes should be identified. Special attention will have to be given to relating particular degrees of pollutant exposure to responses in ecosystems.

Plant biomarkers — Although the focus of the workshop was on higher vertebrates, the working group noted that biomarker responses in plants have barely begun to be addressed and should be the focus of extensive studies in the future.

Simple tests for use in developing countries — Finally, it was noted that with predicted population growth in developing countries and associated increases in resource use, it is highly likely that ecotoxicological problems in such regions will escalate alarmingly. Furthermore, many of the most useful and sensitive biomarker techniques will not be available in such countries for many years. The working group concluded that efforts to identify simple biomarkers that are inexpensive to measure and that are easy to interpret are urgently required. Obviously, there will have to be a trade-off between the simplicity of the techniques and the quality of data they provide.

V. SUMMARY AND CONCLUSIONS

The final discussion of the workshop focused on developing a group consensus on the current status and future directions of biomarker research. Was the field of biomarker-based biomonitoring on the proper course, and what directions seemed to be the most productive?

In general, the participants at the workshop felt comfortable with our current approaches to, and interpretation of, several biomarkers of exposure. These included the well-documented responses of acetylcholinesterase activity ([AchE] a measure of exposure to neurotoxic pesticides), DNA adducts (chemical modification of DNA by a pollutant), aminolevulinic acid dehydrase [ALAD] a measure of exposure to lead), and ethoxyresorufin-O-deethylase (a measure of mixed function oxidase activity). In addition, the group felt that a small number of biomarkers, such as AchE, ALAD, and eggshell thinning, were well-developed and reliable indicators of the effects of exposure.

One of the obstacles to acceptance of biomarkers is a perception of them as an exotic and specialized subject of research, instead of a standard tool of environmental assessment. Acceptance of biomarker measurements might be increased if they began to be incorporated into screening bioassays as an adjunct to currently accepted toxicity tests. For example, DNA strand breaks could be measured at the end of a standard fish larval growth and survival bioassay. However, there was some disagreement over this point because many of the bioassay tests are of quite short duration and may not be the best exposure conditions for biomarker endpoints.

In general, there was agreement that the current trend in biomarker research is appropriate and that progress is slow but acceptable. Continued emphasis on (1) integrating several research approaches and levels of measurement, (2) use of organisms in situ (as opposed to limiting research to laboratory studies), and (3) development of ecophysiological teams of interdisciplinary researchers was encouraged. There was a strong encouragement for hypothesis testing vs descriptive or observational approaches. Development of specific and testable hypotheses about the direction and magnitude of biomarker responses should be the focus for interactive laboratory-field studies. A key concern was the need to promote development of databases of "normal" responses from reference sites as a criteria against which to compare responses from experimental or contaminated sites. The workshop participants were united in advocating a research strategy that combines laboratory and field studies, and relates biomarker responses to more conventional measures of pollutant effects. It was noted that there are several studies going on now that reflect this philosophy. Examples include studies that are relating traditional and accepted endpoints of effect (e.g., scope for growth in mollusks as affected by Cd) with biochemical markers (e.g., stress proteins). These studies provide a better mechanistic understanding of the biomarkers and are recommended for the future. While recognizing current limitations in our ability to predict effects of pollutants, the group felt that we should not restrict ourselves to studies focused solely on exposure in individuals, but also try to seek the relationships between biochemical responses and population level effects through integrated laboratory-field interactions.

Finally, the workshop strongly encouraged the education of a broader cadre of researchers. This would include both education of researchers, industry, waste-site managers, etc. in the methods and interpretation of the biomarker approach, as well as organization of other workshops focusing on QA/QC and methodology of measuring specific biomarkers.

ACKNOWLEDGMENTS

The leaders of the various working groups contributed notes and summaries of the discussions within their groups, and their contribution to this summary is gratefully acknowledged. Those contributing in this way include M. Depledge, A. Fairbrothers, C. Fossi, C. Leonzio, and N. Mattei. The Oak Ridge National Laboratory (ORNL) is managed by Martin Marietta Energy Systems, Inc. under Contract No. DE-AC05-84OR21400 with the U.S. DOE, publication No. 4023 of the Environmental Sciences Division of ORNL.

REFERENCES

1. NRC, *Environmental Epidemiology*, National Research Council, Washington, DC, 1991.
2. Lambertsen, R. H., Disease biomarkers in large whale populations of the North Atlantic and other oceans, in *Biomarkers of Environmental Contamination*, McCarthy, J. F. and Shugart, L. R., Eds., Lewis Publishers, Chelsea, MI, 1990, p. 395.

INDEX

Index

327

right, 251
sei, 221, 260
sperm, see Sperm whales
stranded, 238
Whale watching, 320
White blood cells, 74, 149
White-faced ibis, 14
White-footed mice, 166
Wilcoxon sign rank tests, 229, 234, 236
Window technique, 204–205, 213
Wings, 11, 19, 27
Woodchucks, 166

Xenobiotics, 289
  blood esterases and, 44
  in cetacean biopsies, 256, 258
  in fish, 172, 175–177, 179, 181–184,
    189–194

genotoxic responses to, 140–141, 148
in hazard assessments, 9
hemoglobin adducts and, 160–162
in immunotoxicology, 79
species at risk and, 309
Xenopus, 209
Xenopus laevis, 203, 209
X-rays, 152

Y chromosomes, in cetaceans, 261
Yolk, 205–206, 208
Yolk replacement, 205

Zebra fish, 209
Zero tolerance, 315
Zinc, 19, 107, 117
Zinc porphyrin, 102–103, 108, 114, 117
Zoo animals, 74, 76